DATE DUE

APR 3 - 1996			
APR 9 - 1997			
FEB 1 9 2010			

POLYMER SCIENCE AND TECHNOLOGY SERIES

Synthetic fibre materials

POLYMER SCIENCE AND TECHNOLOGY SERIES

SYNTHETIC FIBRE MATERIALS

H BRODY

Longman Scientific and Technical,
Longman Group UK Ltd,
Longman House, Burnt Mill, Harlow,
Essex, CM20 2JE, England
and Associated Companies throughout the world.

Copublished in the United States with
John Wiley & Sons, Inc., 605 Third Avenue, New York, NY 10158

Trademarks
Throughout this book trademarked names are used. Rather than put a trademark symbol
in every occurrence of a trademarked name, we state that we are using the names only
in an editorial fashion and to the benefit of the trademark owner with no intention of
infringement of the trademark.

First published 1994

ISBN 0 582 06267 5

British Library Cataloguing in Publication Data
A catalogue record for this book is available from the British Library

Library of Congress Cataloging-in-Publication Data
A catalog entry for this title is available from the Library of Congress

ISBN 0-470-23427-X   (USA only)

Set by 16JJ in 10/12½pt Times

Printed and bound by Bookcraft (Bath) Ltd

Contents

List of contributors

Dr H Brody, IRC in Polymer Science and Technology, University of Leeds, Leeds LS2 9JT, UK

Professor D W Lloyd, Department of Industrial Technology, University of Bradford, Bradford BD7 1DP, UK

Professor D M Lewis, Department of Colour Chemistry and Dyeing, University of Leeds, Leeds LS2 9JT, UK

Dr S M Burkinshaw, Department of Colour Chemistry and Dyeing, University of Leeds, Leeds LS2 9JT, UK

Dr I Holme, Textile Industries, University of Leeds, Leeds LS2 9JT, UK

Dr E D Williams, 820 Summerset Drive, Hockessin, Delaware, 19707, USA

Dr P A Smith, Textile Industries, University of Leeds, Leeds LS2 9JT, UK

Dr S K Chawla, New Materials Processes Department 401G, The Goodyear Tire and Rubber Company, 142 Goodyear Boulevard, Akron, Ohio 44305, USA

Dr V Gabara, E I Du Pont de Nemours, Spruance Plant, Richmond, Virginia 23261, USA

Dr D C Prevorsek, Allied Signal Inc, Corporate Technology, PO Box 1021, Morristown, New Jersey 07962-1021, USA

Dr G Desitter, Kermel, BP 3106, 69398 Lyon Cédex 03, France

Mr R Cassat, Kermel, BP 3106, 69398 Lyon Cédex 03, France

Dr D J Hannant, Civil Engineering Department, University of Surrey, Guildford, Surrey GU2 5XH, UK

Mr B M McIntosh, ICI C and P Ltd, Zyex Group, Brockworth, Gloucester GL5 4HP, UK

Preface

There have been many books containing reviews of the materials made from synthetic fibres. Although they are helpful in listing the fibres involved in many end-uses, very little attempt has been made to address the question why particular synthetic fibres are involved in certain applications. This book is aimed at explaining the science behind the use. The intention is not just to produce an encyclopaedia of which fibres are used for different purposes; the authors of each contribution have been asked to explain why certain fibres are used. They have attempted to describe the mechanism governing the use and to give postulated theories and models of why they work—the physics and chemistry of performance—wherever possible. These are the underlying science–application relationships. There is, of course, another very important property governing use, and that is the price. However, the price the manufacturer or consumer is prepared to pay becomes more understandable when the technical factors governing the use are known.

Each chapter is devoted to a unique field of application and is written by a world expert in that field. The book attempts to cover the whole spectrum of synthetic fibre applications. For the purpose of this book as part of a series on synthetic polymers, the synthetic fibres exclude regenerated cellulosics such as rayon, cellulose esters, etc. Natural fibres such as wool and cotton, or alternatives such as carbon fibre and steel, are mentioned where comparisons are useful.

The book is essentially divided into three sections, background science in Chapters 1 to 4, textiles in Chapters 5 to 7 and industrial uses in Chapters 8 to 11, although there may be a blurring of distinction between the two fields of application, as in nonwovens or thermostable fibres, for example. Chapters 12 to 15 cover some important specialized end-uses.

The first two chapters trace the formation of structures from the molecular level controlling the physics of fibres in Chapter 1 to the science of fibre assemblies in Chapter 2. Chapters 3 and 4 deal with the science of dyeability of polyamides, polyacrylonitrile and polyesters, which determines many end-uses. Textiles cover clothing, fabrics and carpets, with nonwovens bridging the gap between textile and industrial uses. The chapter on rubber composites is primarily concerned with tyre cords, the most important area of application. Additionally in this chapter the physics of such composites is treated in a similar fashion to much of the work on fibre-reinforced resins. High performance fibres have been divided into two distinct sections, rigid aromatic polymers, primarily aramids, and high performance polyethylene, which is a successful demonstration of the use of a flexible polymer. The chemical basis underlying the behaviour of thermostable and fire-resistant fibres is discussed in the next chapter. Finally, four major specialized end-uses have been selected, covering engineering and medical applications, cement reinforcement and monofils.

The book is intended for all workers in the field of synthetic fibres, from students to people in industry. The layout is intended to lead into understanding of the mechanisms of the end-uses via the groundwork of the initial basic science, and so should be useful to nonspecialists as well as specialists.

CHAPTER 1

Thermal and mechanical properties

H BRODY

1.1 Introduction

The purpose of this chapter is twofold. Firstly, although there is much published on the definition and measurement of thermal and mechanical properties, there is little discussion about which properties are the most important in determining the practical application of fibre materials. It is planned to present such properties with this in mind, with emphasis on those that could have the most important ramifications for end-use. Included are complex properties such as those of fatigue, which could well be more important than straightforward properties such as extensional strength.

Secondly, various views on the molecular and structural origin of such properties will be presented where possible at the same time to provide a complementary and linked review of physical polymer structure, instead of treating this as a separate topic.

1.2 Thermal properties

1.2.1 Melting point

A key property that determines end-use function is the maximum working temperature at which the desirable properties gradually or suddenly fall. The most common attitude is to view this as the melting point, but in fact this is not usually the case. Physical properties decline steadily as the temperature is raised, and there is a large change in properties such as stiffness at the so-called glass transition temperature, T_g. However, the melting point does undoubtedly govern the level of performance, since other thermal properties are related, and it certainly constitutes an upper bound for integrity. For this reason it is valuable to consider the factors that determine it.

Synthetic polymers are a mixture of crystalline and noncrystalline or amorphous regions. Crystallinity depends on regularity of molecular packing, and statistically high crystallinity is difficult to achieve from melt or solution for most flexible polymers, but can be high where there is little steric hindrance and bond rotation is easy, e.g. in polyethylene and polypropylene. For nonflexible polymers of the 'liquid crystal' type, such as aramids, crystallization incipiently occurs in the spinning solution, and crystallinity is very high. For flexible polymers there is a range of crystal sizes and therefore a melting range, which can be as much as 10°C. The melting point is usually taken to be the peak in the melting endotherm in a thermal melting apparatus.

The major factor controlling the melting point is chemical structure. Other factors are molecular mass and degree of branching. Low molecular mass gives many chain ends and these defects lower the melting point. True melting is a first-order thermodynamic transition that depends on the crystal properties. The thermodynamic factors can be summarized in the following equation:

$$\Delta F = \Delta H - T \, \Delta S \qquad [1.1]$$

where F is the free energy; H is the internal energy or enthalpy; S is the entropy and T is the temperature. Melting occurs when ΔF is zero, i.e. when the free energy is the same in the crystal and the melt so that melting and crystallization processes are balanced. Whence the equilibrium melting temperature T_m is given by

$$T_m = \frac{\Delta H}{\Delta S} \qquad [1.2]$$

T_m is difficult to measure and is usually lower than theoretically predicted. Many factors affect it, such as crystallite size, rate of heating, etc.

ΔH is determined by chemical structure, since it is a function of interchain forces. In general the entropy, ΔS, dominates T_m, and this in turn is a function of the stiffness and flexibility of the polymer molecule in the melt. The stiffer the molecule, the smaller the ΔS in the transition from solid to melt, and consequently the higher the T_m. For example, cellulose has a very stiff molecule and consequently the theoretical melting point is about 400°C higher than for PET (poly(ethylene terephthalate)), even though the ΔH values are of similar magnitude. Cellulose therefore decomposes on heating.

Stiffness is largely determined by ease of rotation about the chemical bonds. The series –O–, –O–O–, –CO–O, has increasing flexibility. Stiffness is increased by *p*-phenyl groups and *trans* double bonds, and decreased by irregularity such as *m*-phenylene substitution and *cis* orientation. Side-groups reduce chain flexibility, making polypropylene less flexible than polyethylene.

The first-order transition produces a discontinuity of enthalpy or volume with temperature, as shown in Fig. 1.1. The melting point can

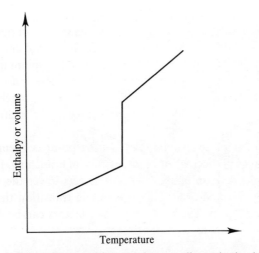

Figure 1.1 A first-order transition produces a discontinuity between enthalpy, or volume, with temperature.

thus be determined, but evaluation of the maximum in the endothermic peak by differential scanning calorimetry (DSC) is the most common method. This is relatively easy for solid polymers, but for fibres special precautions have to be taken, such as wrapping them around formers, to prevent shrinkage.

1.2.2 Glass transition

Noncrystalline or amorphous regions are characterized by a degree of disorder where molecular register does not occur. Some polymeric materials are totally amorphous. They can be viewed as supercooled glasses, since glass itself is a supercooled liquid. Typical of these are polystyrene and poly(methyl methacrylate). These materials exhibit a second-order transition, e.g. a change in slope, or first derivative, of volume or enthalpy vs temperature, as shown in Fig. 1.2. This is a softening point rather than a true melting point, and by analogy with glass is called the glass transition point, T_g.

Similarly to the melting point, it can be measured by following the change of volume in a dilatometer. However, it is not a true thermodynamic parameter as it depends on the rate of heating or cooling, as shown in Fig. 1.2. The lower the cooling rate the lower the T_g. Typically the T_g is measured by the change in elastic properties with temperature, usually the dynamic modulus or loss tangent. At the transition, the molecules in the glassy region are freed and can absorb energy from an imposed oscillation because intermolecular mobility can occur where work is

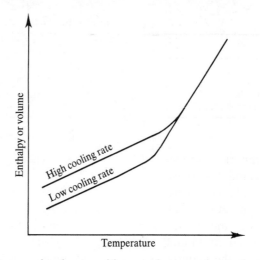

Figure 1.2 A second-order transition produces a change in slope enthalpy, or volume, with temperature.

performed. In fact there are usually different transitions with increasing
temperature, depending on the type of molecular motion that is freed.
For instance for PET there are three transitions: γ at about $-165°$C due
to hindered rotation of methylene groups; β at about $-60°$C due to
complex movements involving the carboxyl group; and the principal
transition α at about $70°$C due to large scale segmental motion in the
amorphous regions.[1] It is the α which is the largest and most important,
and this is the true 'T_g' for practical purposes.

The effect of chemical structure on T_g is similar to that on T_m, and the
most important is again chain flexibility. For this reason there is a rough
correlation between T_g and T_m, where the ratio of T_m to T_g varies from
about 1.3 to 2.0—the more symmetric the molecule the higher the value.
This depends on the method of T_g measurement. Values in kelvins for
some typical fibres are given in Table 1.1.

At temperatures greater than T_g, molecular motion is facilitated and so
the work loss is reduced. Since such motion is restricted in the glassy
state, the T_g occurs at a peak of work loss absorption when measured by
dynamic methods. From a fibre point of view, because of the maximum
loss of energy here, fibre recovery from small strains is a minimum. In
fact, as a corollary of this, because of the changed molecular state and
increased motion, there are many ways of measuring T_g by physical
measurements as well as thermal, though they may not all agree owing
to other factors such as rate of oscillation, rate of heating, etc. The
molecules cannot respond as rapidly to higher rates of oscillation, and so
the T_g will be raised.

Table 1.1. Glass transition temperatures

Fibre	T_m/K	T_g/K	$T_g : T_m$
PET	538	343	0.64
Nylon 6.6	538	333	0.62
Modacrylic	498	423	0.85
Isotactic polypropylene	435	268	0.62

A most important property change from a practical point of view is reduction of modulus. An example is shown in Fig. 1.3 for PET of different levels of crystallinity,[1] where the shear modulus was measured at 1 cycle/s. The drop in modulus is due to the change from a 'glassy' to a more 'rubbery' state, and it can be quite appreciable as can be seen. These results show that crystallinity stiffens the rubbery phase, preventing flow, which occurs if the material is totally amorphous. This then is another good method of measurement and is very useful because it relates to practical situations where the modulus change can be so severe that it restricts the use, for example in the case of tyre cords. It is obvious that in many applications T_g is more important than the melting point in restricting the range of operation. For instance, the work loss is manifested as heat, an important practical consideration in dynamic situations, especially for tyre cords.

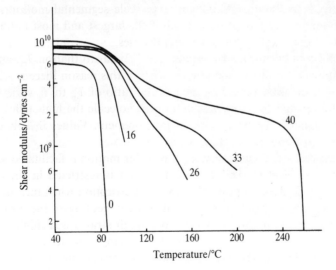

Figure 1.3 Change of modulus of PET with temperature and effect of crystallinity (shown by percentage levels on each curve). (Redrawn after K H Illers and H Breur, *J. Coll. Sci.*, **18**, 1 (1963) by courtesy of Academic Press)

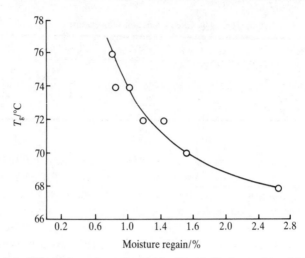

Figure 1.4 Effect of water on the T_g of nylon 6. (Redrawn from N V Bhat and D S Kelkar, *J. Macro. Sci. Phys.*, **B28**, 375 (1989) by courtesy of Marcel Dekker Inc.)

Addition of plasticizer would be expected to reduce the intermolecular attraction, and thus lower the T_g. One of the most important plasticizers practically, and the simplest, is water. This lowers the T_g of textile fibres, which is an important factor in dyeability, affecting the diffusion rate at which dye is transported. Small amounts of water are highly effective, e.g. 1.5 per cent water reduces the T_g of PET from 80°C to 60°C. Organic solvents have a similar effect: 4.5 and 10.1 per cent of dioxane reduce the T_g of PET to 40°C and 20°C, respectively; 3.2 per cent benzene gives 40°C while 19 per cent nitrobenzene gives the low value of -10°C.[1]

Polyamides are similarly affected. Nylon 6 has a T_g at 70–80°C which is also attributed to segmental motion of the chains in the amorphous regions, similarly to PET. This has an activation energy, E, which is necessary to create free volume for the segmental movement. It has been found that water considerably reduces E, the value being almost halved by 10 per cent water.[2] The effect of water on the T_g of nylon 6 is shown in Fig. 1.4, where the T_g of the nylon 6 was determined from sonic modulus measurements.[3] The determination of the sonic modulus is a method used extensively for fibres using proprietary instruments available. Actually water bonds to polyamides in various ways, but at high regains of greater than 4 per cent it pushes the molecules apart.[4]

Orientation in the amorphous regions would be expected to impede segmental motion, thus raising the T_g. It is here that the structural origins of T_g can be very important to fibres, since orientation is such an important variable. Measurements on PET fibres spun at increasing

Table 1.2. Effect of wind-up speed on the T_g of PET

Wind-up speed	Crystal fraction			
yards min^{-1}	%	$\tan \delta_{max}$	f_a	$\dfrac{T_g}{°C}$
3000	6.3	1.35	0.13	96
3800	6.6	0.52	0.35	97
4400	10.1	0.22	0.50	121
5000	21.4	0.15	0.56	124
5600	31.9	0.13	0.62	125
6200	37.1	0.13	0.66	126

(Reprinted from R W Miller and T Murayama, *J. Appl. Polymer Sci.*, **29**, 933 (1984) by courtesy of John Wiley & Sons, Inc.)

wind-up speeds have demonstrated this effect.[5] The T_g was obtained from the maximum in $\tan \delta$ (the work loss). The amorphous orientation function, f_a, was determined from a combination of wide angle X-ray diffraction, density and sonic modulus measurements. Table 1.2 shows the results obtained. The position of T_g is increased from 96°C to 126°C by an increase in f_a from 0.13 to 0.66.

In practice there occurs the effect of both water and orientation. This has been studied by Acierno *et al.* on nylon 6 fibres.[6] To produce completely unoriented fibres the as-spun fibres were shrunk in an oven at 170°C. For the most oriented they were drawn to a draw ratio of 4. The T_g was measured by sonic velocity. The effect of orientation on the three samples, unoriented, as-spun and drawn, was to give T_g values of 64°C, 71°C and 82°C, respectively. The effect of water content was studied on the unoriented sample. Three conditions were used: dry, 3 min at ambient and 72 h at ambient. The T_g values were 60°C, 50°C and 30°C respectively.

1.3 Mechanical properties

1.3.1 Failure

In any end-use a key objective is to prevent fibre failure. In practice this is usually by a complex combination of tension, fatigue, abrasion, torsion, creep, etc. The contribution of each of these factors has therefore to be considered individually. They are usually measured under standard conditions, but should really be determined in the actual end-use environment, taking into account the temperature, rate of loading, presence of water, etc.

1.3.2 Tensile properties

For synthetic fibres there is no unique strength in practice. A range of
properties can be obtained at processing, depending on the end-use
requirements. To increase strength, melt spun fibres are subsequently
drawn, wet and dry spun fibres may be stretched, while aramids and other
liquid-crystal types may be heat treated with stretch. It is the maximum
strength obtainable that determines the relative merits for industrial
applications. However, for many uses properties have to be optimized.
Increasing the strength, e.g. by drawing, usually reduces the extension-to-
break, and hence the toughness. This may be acceptable if the working
strains are kept low, e.g. in composites. For most industrial applications
a compromise between toughness and strength is usually necessary. On
the other hand, the tensile properties required by textile fibres are not so
stringent. A major requirement is to survive the processing stages of
weaving, knitting, etc., and here toughness is important for survival of
load surges. In use, very high extensions are not expected and in fact
would produce permanent deformation owing to the drop in recovery
that occurs above strains of about 5 per cent.

The degree of orientation, e.g. after drawing, and the consequent
extensibility of textile fibres form a fine balance. High orientation reduces
molecular mobility and thus dyeability; or the consequent high strength
could lead to 'pilling', small entanglements of fibres that are so strong
they cannot break off. Low orientation could lead to an unstable material
that stretched in use, and poor wash-fastness of dyes, which could easily
diffuse out. As an example of the range of drawn properties of PET that
can be obtained, the two properties tenacity/extension can range from
about 40 cN tex^{-1}/35% to 80 cN tex^{-1}/10%.

As pointed out previously, the thermal properties may limit the
maximum working temperature; but also elevation of temperature reduces
the desirable tensile properties. For example, the tenacity of nylon 6 falls
fairly linearly to about half its value at 100°C; PET is similar, and the
tenacity falls linearly to about a third of its value at 200°C.

The usual way to express tensile properties is as a load–strain curve,
instrumentally obtained by increasing the strain at a fixed rate and
recording the load. This could be converted into the more conventionally
accepted engineering approach of a stress–strain curve by taking into
account the large reduction in cross-sectional area that occurs because of
the large strains that are usually involved, but for practical purposes it is
usually used as obtained. Examples of these curves for melt spun fibres
are given in Fig. 1.5. For fibres spun at relatively low wind-up speeds, e.g.
below 1000 m min^{-1}, there is an initial steep portion followed by a clear
yield point and then a nearly horizontal portion, more so if load is the

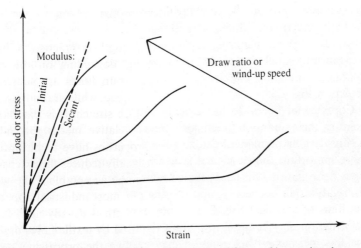

Figure 1.5 Load–strain or stress–strain curves of fibres of increasing draw ratio or wind-up speed.

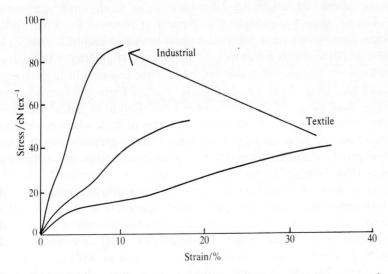

Figure 1.6 Stress–strain curves of drawn PET fibres showing range of behaviour from textile to industrial fibres.

vertical axis, and then a further upturn as some form of strain hardening occurs. As the draw ratio or wind-up speed is increased the curves steepen and the horizontal portion gradually disappears. Typical stress–strain curves for commercial PET fibres covering the range from textile to industrial fibres are shown in Fig. 1.6.

At the higher draw ratio of commercial fibres, the initial portions of the curves are not usually perfectly linear, though they may approach this for high performance fibres, and a tangent to the curve defines the initial modulus. This parameter is very important in practice since it defines the resistance to small strains. Alternatively, a secant modulus may be more useful in defining the conditions for an operating strain. These moduli are shown in Fig. 1.5.

Commercial fibres do not usually show such a clearly defined yield point as their undrawn precursors. However, the strain past which there is considerable permanent deformation is a critical threshold that can be more important than the final strength in many applications. Recovery from deformation will be discussed later. There is an analogy here with the relationship between T_g and melting point, where T_g usually indicates the limit of practical use. For textile purposes the initial resistance to deformation may be more important than the strength. If the modulus is too low, permanent deformation may easily occur. On the other hand, for industrial fibres strength is a specified requirement of use and a major selling point. Also the higher associated modulus, as shown by Fig. 1.6, is very important in limiting deformation.

The custom of expressing the strength of fibres as load per unit of linear density (the mass per unit length) is used. SI units, cN tex^{-1}, are now preferred, being 8.83 times the older units of grams per denier.

1.3.3 Fundamental theories of strength

The actual strengths of fibres are very much lower than theoretically predicted, probably only 5–10 per cent. The prediction is based on the strength of a single C–C bond and the calculated rupture of all the bonds in a single layer. Materials break because of the presence of flaws that amplify the applied load, so that there is localized failure leading to total catastrophic failure. Theories of strength are therefore usually based on the existence of flaws at certain levels of organization, which range all the way from the molecular level to considering the fibre as an imperfect solid body. It is not possible to consider all the work in this field, but some key examples will demonstrate the range of ideas. The process of this discussion is particularly useful in fulfilling one of the purposes of this chapter by giving a picture of the structural organization of fibres.

At the most basic level is a consideration of the oscillations of atoms by Zhurkov.[7] He showed that the lifetime of a polymer under constant load could be described by an Arrhenius type equation:

$$t_f = t_0 \exp(U - v\sigma)/RT \qquad [1.3]$$

where t_f is the time to failure; t_0 is related to the reciprocal of the molecular oscillation frequency; U is the molar activation energy; v is the activation

Figure 1.7 Fibrillar model of fibrous structure. (Redrawn from A Peterlin, Plastic deformation of polymers with fibrous structure, *Colloid Polymer Sci.*, **253**, 809 (1975) by courtesy of Steinkopff Verlag.)

volume and σ is the applied stress. The values of t_0 are 10^{-13} to 10^{-12} seconds, close to the period of thermal oscillations of atoms in solids. U has a value close to that necessary for chain rupture of covalent bonds by thermal degradation. There should be a linear relationship between σ and $\log t_f$, which is found for many polymers. Therefore it is concluded that the lifetime determination process is single chain rupture, where chains are broken gradually at stresses much less than the brittle strength of the material as atoms are displaced permanently to an activated state. In support of this it is pointed out that stretched rubber gives hydrogen gas long before fracture,[8] while polystyrene, poly(methyl methacrylate) and polypropylene give volatile products in mechanical breakdown that are similar to those from thermal degradation.[9] This theory thus involves gradual stressed breakdown, assisted by thermal fluctuations.

A fibre is considered to have fibrillar structure, as shown in Fig. 1.7 taken from reference 10, where the fibrils themselves are composed of bundles of microfibrils. Microfibrils explain the fibrous, longitudinal splitting of drawn fibres. They are postulated to have a width of 10–20 nm and a length of tens of micrometres, and to be easily separable.[10] They are also postulated to consist of alternating crystal, or 'ordered', and amorphous regions, the latter being considered as 'tie molecules' between the ordered regions, as shown diagrammatically in Fig. 1.8. For melt spun fibres like polyamides and PET of medium crystallinity, the tie molecules are clearly very important. This is still the case for highly crystalline fibres like polyethylene, since they interrupt ordered sequences. They would not

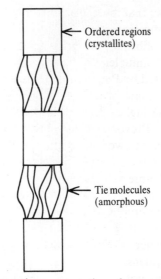

Ordered regions
(crystallites)

Tie molecules
(amorphous)

Figure 1.8   Diagrammatic representation of sequential ordered and amorphous regions along a microfibril, where the amorphous regions are considered as 'tie molecules'.

be considered for high performance fibres like aramids. To continue with the molecular approach, it is these tie molecules that will now be considered. They must have a distribution of lengths, as shown in Fig. 1.8, giving rise to a stress distribution biased towards the shorter lengths, which get straightened on extension. In fact it has been calculated that up to 30 per cent are initially taut.[11]

The straightening of these tie molecules has interested many authors. Peterlin points out[12] that measurements by electron spin resonance (ESR) show the number of radicals produced from loading polystyrene, poly-(methyl methacrylate), PET, nylon and silk is much greater, by many magnitudes, than if all chains were cut by a single fracture plane. The value is actually equivalent to a chain cut up to 1000 times per centimetre of fibre length. Therefore chain rupture proceeds throughout the fibre as the most stressed tie molecules are broken, and the distribution of damage is irregularly distributed. If there is more chain rupture at one point, there is a higher stress concentration on the remaining molecules. Microcracks would then start here, leading to fibre rupture. A more biased distribution of tie molecules from PET fibres spun at high wind-up speeds compared with low wind-ups speeds has been postulated as the reason for lower ultimate strength.[13]

Another view of these tie molecules is to consider them capable of a glass–rubber transition,[13] the glass phase being brittle. Nuclear magnetic resonance (NMR) shows that molecules in highly drawn fibres lose their mobility owing to 'vitrification'.[14]

Considering the microfibrillar level of organization, if the microfibrils are postulated to have a finite length then one source of flaws will be the ends. It has been suggested by Peterlin that there are disclike microcracks at these sites which initiate failure.[15] High strength fibres may not show this effect because higher draw ratios could eliminate such point vacancies.[10] In contrast, Prevorsek believes that microfibrils do not have abrupt ends but form an interwoven structure, with branching and fusion of ends.[16] In this picture the amorphous domains of microfibrils are the weakest element of the fibre structure, and the precursor of the microcrack has a diameter the same as that of the microfibril. This was supported by photomicrographs of stained nylon 6.

There may well be a rather profound difference between the highly crystalline addition polymers like polyethylene and polypropylene, studied by Peterlin, and the condensation polymers, nylon and PET, studied by Prevorsek, so that it may not be possible to generalize from polyethylene and polypropylene only. In the olefinic addition polymers there is a predominantly microfibrillar structure. In the less crystalline nylon and PET, where it is thought that microfibrils are embedded in an amorphous continuum, it can be seen why interfibrillar connections could govern the strength. In this case the increase of fibre strength on drawing is considered to be due to the transformation of weak microfibrils into strong, highly ordered interfibrillar extended chains.[17] This is considered the most important phase of the fibre because it controls the key fibre properties of strength, modulus, shrinkage, etc. The relatively weak microfibrils in this picture are only responsible for maintaining the dimensional stability of the fibre.

The possible formation of interfibrillar chains means that the picture shown in Fig. 1.7 is simplistic. Small-angle X-ray measurements of the width of crystals show that microfibrils in nylon 6 do not stretch the same as the fibre on drawing (draw ratios 3.5–5.4 times) but slip past each other, shearing the interfibrillar phase and leading to a three-phase model,[17] as shown in Fig. 1.9 where the interfibrillar phase is the most important, as discussed above. The microfibrils may not stretch as much as the fibre but are assumed to thin down during drawing, leading to the so-called 'Swiss cheese model' for a transverse section of the fibre, as shown in Fig. 1.10. From results for PET, the plateau region in Fig. 1.5 for undrawn fibres has been ascribed to the sliding of microfibrils past each other, terminating when the interfibrillar molecules have become taut.[13] Annealing of undrawn PET spun at different wind-up speeds leads to a reduced plateau region, which is assumed to be due to crystallization and consequent immobilization of the extended chains.[13] However, in

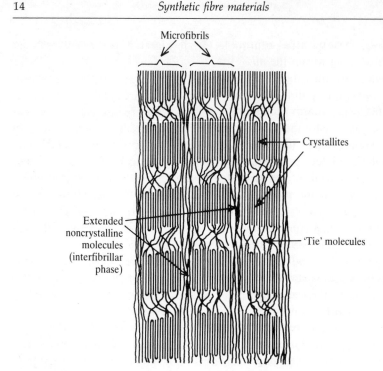

Figure 1.9 Three-phase model of microfibrillar structure. (Redrawn from D C Prevorsek *et al.*, *J. Macro. Sci. Phys.*, **B8**, 127 (1973) by courtesy of Marcel Dekker Inc.)

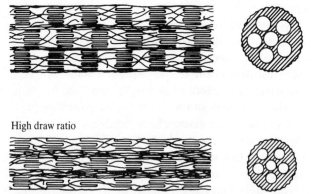

Figure 1.10 'Swiss cheese model' of fibre showing longitudinal and cross-sectional views of low draw ratio and high draw ratio fibres. (Redrawn from D C Prevorsek *et al.*, *J. Macro. Sci. Phys.*, **B8**, 127 (1973) by courtesy of Marcel Dekker Inc.)

the latter work eventual rupture is still ascribed to the intrafibrillar tie molecules lying within the microfibrils rather than extended interfibrillar molecules. It must be said that at our present level of molecular understanding it is difficult to assess the relative contributions of intra-molecular and intermolecular tie molecules, but such molecules would seem to have vast importance in controlling strength.

For completeness it is worth mentioning here the ideas of Hearle, particularly related to polyamides, who envisages a crystalline gel struc-ture.[18] The concept of the gel state observed in other physical systems is applied to polymeric molecular order. This is something like the early fringed micelle model, where molecules had local order in a micelle but they wandered from one micelle to another. It is also similar to Prevorsek's three-phase model, but with less emphasis on interfibrillar molecules. According to this model, at low to medium strains the disordered regions act as rubbers, where the cross-linking in nylons is due to hydrogen-bonding, while at large strains there is crystal yielding and fracture owing to breakage of tie molecules.

The fibre can be treated at a higher level of organization in terms of imperfect crystallinity. In this case the picture is that of a fairly uniform crystal structure containing crystal defects that are the 'amorphous' regions and the source of flaws.[19,20] It is interesting to note that the breaking strain of a fully stretched aliphatic chain has been calculated by Boudreaux to be about 33 per cent.[21] This comes from opening of the valence angle from 109° to 180° and stretching of the covalent bonds. This means that the localized breaking strain must usually be much greater than the bulk strain for C–C bond breakage. Peterlin therefore argues against this model by pointing out that nowhere is the strain high enough in such an imperfect crystal to actually produce bond breakage.[22] He also puts forward the same view for the 'Swiss cheese model'. He feels that only in the microfibrillar model is it possible for microcracks at the ends of the microfibrils to grow large enough.

Another treatment of the totally crystalline fibre takes a very different view, considering the effect of the stress itself on the structure.[23] It is axiomatic that molecules cannot be long enough to span the grips holding a test piece. It is therefore assumed that it is the applied stress that disrupts the crystal, depressing the melting point. When the load is sufficient to lower this to ambient temperature, the sample fails by 'melting' at this 'breaking load'. It is thus contended that high melting point, heat of fusion and modulus should give high strength, and molecular mass should not matter. This is a rather sweeping generalization to which exceptions can be found, though the relative degree of variation may be acceptable within the theory. For example, the clear effect of molecular mass on the strength of polyethylene films has been demonstrated.[24] This work takes the view

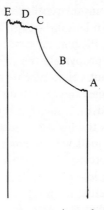

Figure 1.11   Diagrammatic representation of side view of fibre fracture surface. (Redrawn from J W S Hearle and P M Cross, *J. Mater. Sci.*, **5**, 507 (1970) by courtesy of Chapman & Hall)

that the number of molecules supporting the load also governs the modulus, and that therefore the strength is proportional to the modulus. Results for solution spun polyethylene are given and extrapolated to a theoretical maximum modulus of 250–300 GPa and a theoretical minimum strength of about 8 GPa. Again breakage of molecular chains is assumed.

The problem with molecular and supermolecular theories of strength is that they ignore the fact that fibres are not perfect cylindrical bodies but are imperfect solids with relatively large flaws, usually arising from the method of production. The situation here is one of fracture. Fibre breakage is actually a complex multistage process, as described in work where SEM (scanning electron microscopy) was used to study the broken ends.[25] Figure 1.11 shows a schematic side view of all the regions that might be observed. It is assumed that as the specimen is stressed, internal cracks or voids grow until they coalesce. This usually first happens near the surface, giving rise to region A, where crack initiation begins. Crack initiation is followed by a smooth region B, where the fibre spreads into a diamond-shaped notch.[26] At C there are sharp concentric steps, less smooth at D and finally very rough at E. The smooth–rough transition is a classical situation. The smooth region B dominates fibres broken slowly, and is due to relatively slow plastic yielding. The energy input is eventually too great for slow yielding and catastrophic failure occurs in the rough, transverse region, increasing in severity from C to E. This rough region resembles the breakage of a fibre-reinforced composite, probably owing to unstable internal voids or microfibrillar splitting.

Different fibres show all or some of these features, and many examples are pictorially illustrated in reference 27 by SEM micrographs. For example, undrawn nylon 6,6 bristles show all these features, while drawn nylon 6,6 has no region C. PET is the same as nylon 6,6, but region E is more pronounced. For polypropylene region D has a 'cobbled' appearance, probably because it does not void, and E is converted into an overhang.

Lamb has shown by measuring the relative sizes of the smooth and rough regions of nylon and PET that the breaking load is based on the cross-section that supports the fibre at rupture.[28] It is known that there is an increase of breaking load with strain rate. This is because the size of the smooth region decreases with strain rate, and thus a greater cross-section supports the load at failure. By this model the apparent reduction in strength of wetted nylon is due to the faster growth rate of the smooth area, a more ductile failure. The ultimate breaking stress is unchanged but the breaking load is reduced because of the smaller rough area. Analysis of these relative areas can obviously lead to greater understanding of breakage.

The results of Lamb were found to fit the Griffith crack theory, developed for brittle materials like glass fibres,[29] $S = (2E\gamma/\pi c)^{0.5}$ where S is the failure stress, E is Young's modulus, γ is the specific surface energy of the solid and c is the flaw size.

This is really the ultimate demonstration that the breakage of fibres is macro-flaw controlled. Smith found the strength of polyethylene was proportional to $E^{0.8}$ rather than $E^{0.5}$, which he ascribed to the slip of microfibrils that would reduce the stress concentration at the flaw and thus increase the strength.[24] This is an interesting interaction of micro and macro mechanics.

In fact, the last item above is an important straw in the wind. It is tempting to postulate a way of connecting all the various theories discussed above, which at first seem very independent. In this model the crack initiation would be macro-flaw controlled, almost corresponding to a Griffith approach, but with the stress concentration modified to some extent by particular polymeric fibre properties such as flow. This type of crack initiation introduces the flaw factors arising from the method of production. Subsequent behaviour then depends increasingly on fibre microstructure, where the final strength is dominated by the breakage of molecules within and around the microfibrils by the various molecular mechanisms discussed above. In this way it can be seen how efficiency of preparation can be very important, but how the eventual properties could indeed be controlled by molecular orientation and structure, and hence the high strength of aramids and gel-spun polyethylene. Alternatively it could be argued that these high performance fibres have less pronounced macro-flaws by virtue of their improved microfibrillar organization.

1.3.4 Loop strength

The preceding treatment of strength has dealt with the extension of straight fibres. This is the case in certain industrial applications, such as a radial belt in tyre cords and other composites. Departure from linearity reduces the strength as a result of the angle between the axial stress and the fibre axis. This is the thinking behind the development of polyethylene sheath-enclosed parallel arrays of high performance fibres such as aramids for high strength cables. However, as the curvature of the fibre increases, more than vector forces are involved because the fibre becomes damaged.

The extreme situation is where fibres are joined via loops, as is the case in certain industrial applications, e.g. in some belting, and especially in knitted fabrics. Another feature is the repetitive sliding nature of such joins. Reference 30 deals with this in some detail, and has many SEM photomicrographs showing fibre structure changes. The outside of a loop is in extension, the inside in compression. Initiation of deformation was observed on both these surfaces in this work. On the outside, adjacent cracks coalesce into jagged 'teeth'. On the compression side, there are many changes, such as flattening, the generation of diamond shapes owing to simultaneous compression and shear, and 'shingling'. As before, at failure a ductile crack eventually becomes unstable and gives a granular fracture surface. A new feature of repetitive sliding is longitudinal cracks, typical of fatigue, which will be dealt with later.

On the inside curvature there is simultaneous axial and lateral compression. Different fibres respond with their own peculiarities. For instance, PET has complex shear bands, aramids give platelet buckling, while surface shearing is particularly marked in nylon, giving overlapping shingles. Gross filament distortion eventually gives semiflattening.

Longitudinal splitting arises at the outside surface if a surface crack opens and the longitudinal shear stress exceeds the local shear strength. The surface crack migrates inwards and generates more longitudinal cracks. This is especially true of aramids, which may show extensive fibrillation at a loop. It is clear that transverse properties are very important in determining the propensity for longitudinal cracks to propagate. They are easily propagated if the local shear strength is low, i.e. if the attractive forces between the microfibrillar constituents are low and fibrillation is easy. Unfortunately this seems usually to be true for high performance fibres such as aramids, which have a low axial compressive yield stress.

In practice a loop does not maintain a fixed position. A sliding loop test is therefore recommended for evaluation purposes where surfaces alternate between severe bending and unbending, lateral compression and shear. The concept of a dynamic test for real behaviour leads to the consideration of fatigue, which is dealt with later.

1.3.5 Modulus

An extended linear polymeric chain, oriented in the fibre direction, represents the maximum contribution of the basic polymer unit. The moduli of polyethylene, nylon 6,6 and PET have been calculated from basic principles, using the force constants for bond stretching and valence angle deformation.[31] An experimental approach is to measure the extension of crystallites by wide-angle X-ray diffraction in a fibre under stress.[32] The assumption here is that the measured stress is transmitted to the crystals. These calculated and measured values are in quite good agreement.

Unfortunately, for other than some high performance fibres such as aramids, there is no overall crystal continuity of this nature along the fibre, although it may occur partially since crystal bridges between ordered regions have been postulated for polyethylene, or alternatively taut tie molecules that also maintain continuity. The general view, for flexible polymers, whatever the basic model, is of discontinuous crystalline and amorphous regions.

If the crystal regions lie parallel to the amorphous regions, as in Fig. 1.12a, then the overall modulus is simply diluted by the presence of the amorphous regions and mathematically is defined by

$$E = V_c E_c + V_a E_a \qquad [1.4]$$

where E is the overall modulus; V_c is the crystal fraction; V_a is the amorphous fraction; E_c is the crystal modulus; and E_a is the amorphous

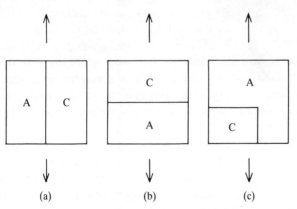

Figure 1.12 (a) Parallel, (b) series and (c) Takayanagi models of amorphous (A) and crystalline (C) regions.

modulus. Usually $E_c \gg E_a$ and so $E \simeq V_c E_c$. However, it is clear that such crystal continuity is not likely in practice.

If the crystallinity is interrupted by amorphous regions in series, as in Fig. 1.12b, the modulus is greatly reduced because the rubbery amorphous regions can extend considerably in comparison with the crystal regions. This is more the normal state of affairs. Mathematically this is defined by

$$\frac{1}{E} = \frac{1}{V_c E_c} + \frac{1}{V_a E_a} \qquad [1.5]$$

Such a reciprocal relationship is greatly dominated by a low value of E_a.

In practice the arrangement of crystalline and amorphous regions is a complex combination of these two simplified extreme conditions, although the series situation is usually predominant. The complex case has been treated mathematically by Takayanagi, according to the representation in Fig. 1.12c, where there are varying amounts of the parallel and series contributions.[33]

As with strength, the actual maximum fibre modulus obtainable is considerably lower than the theoretical maximum, usually of the order of a third. It is not depressed as much as for strength since flaws have a much more devastating effect.

The discussion so far has considered the case for full alignment of crystallites only, but the modulus of a fibre will depend considerably on the degree of molecular orientation achieved after drawing or heat setting. Other factors are the degree of crystallinity, the respective moduli of the crystalline and amorphous regions, and their orientation. This becomes quite complex, and in order to illustrate the factors involved the following expression is given:[34]

$$\frac{3}{2}\left(\frac{1}{E_u} - \frac{1}{E}\right) = \frac{\beta f_c}{E_{t,c}} + \frac{(1 - \beta)f_a}{E_{t,a}} \qquad [1.6]$$

where E_u is the isotropic modulus; $E_{t,c}$ and $E_{t,a}$ are the transverse moduli of the crystalline and amorphous regions, respectively; f_c and f_a are the orientation functions of the crystalline and amorphous regions, respectively; and β is the crystal fraction. Transverse functions are involved because of the tensor nature of the forces.

Also, in practice, apart from some high performance fibres there is usually a fall in modulus with extension, which has to be taken into account when considering end-use. Structurally this must arise from a progressive yield or slippage of components. Certain highly bonded structures like glass cannot slip in this way and exhibit a perfectly linear stress–strain behaviour.

1.3.6 Fatigue

Fibres rarely endure a fixed stress, and if they do then creep is a factor, as will be discussed later. Most usually the fibre is repetitively stressed. An important feature is the mode of deformation. Again it is rare that it will be simply tensile, although this may be the case for the belts in tyre cords. If loops are formed and unformed, then the fibre alternates between tension and compression. The effect of cyclic torsion is in the opposite sense. The most complex mode is produced by rotation of a bent, loaded fibre over a pin or edge, as might occur during a conventional flex fatigue test. The same part of the fibre is exposed to different amounts of compressive shear, tension and torsion. Since fibres are very likely to fail by fatigue rather than simple breakage, such failure will be complex because of the complex modes possible. Complex experimental approaches that involve all these factors are used in evaluation, but it is easier to consider simple modes initially.

The simplest mode is tension. Tyre cords represent a good example of fibres subject to highly oscillatory tensile loads, of maximum frequency about 50 Hz. This is therefore a useful frequency to use to study tensile fatigue.[35] The load is chosen to be some fraction of the simple tensile strength, and this load is maintained during cycling. A fixed strain is not used because the load would fall owing to increase in length by what can be viewed as interrupted creep. Experimentally the fibre is held between two jaws. One is attached to the vibrator and the other to a load cell, which maintains the load constant by length adjustment at the jaws.

At high loads there is eventually simple fracture of the type already discussed above. Successive loading enlarges the plastic deformation zone until there is catastrophic failure. In contrast, at low loadings a very different failure geometry occurs. Such failure by what can be considered real fatigue processes usually occurs when the maximum cyclic load is much less than that for tensile or creep failure. It is obvious, therefore, that this is very important in practice, since failure can occur under conditions of use that do not appear very severe.

The type of failure that can be generated is shown in Fig. 1.13, and it can be seen that this is remarkably different from that shown in Fig. 1.11. An initial notch at which failure becomes initiated is soon deflected after penetrating about 1 μm and a crack runs along the fibre at a slight angle to the axis. The deflection of the crack in this way is due to changes of material properties ahead of the crack, and the relief of shear stress. Eventually simple tensile failure occurs because the load-bearing section becomes reduced owing to the angle of the crack. One broken end thus has a long 'tongue' of fibre.

The lifetime decreases with increase of load. The final breaking load is

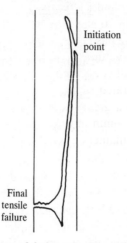

Initiation
point

Final
tensile
failure

Figure 1.13   General type of fatigue fracture. (Redrawn with permission
from A R Bunsell and J W S Hearle, *J. Appl. Polymer Sci.*,
18, 267 (1974) courtesy of John Wiley & Sons, Inc.)

less than if the fibre is stretched without fatigue. If a fibre is subject to
fatigue, the maximum safe tensile loading is probably about two-thirds
of the normal strength. The order of increasing severity of fatigue failure
in terms of the type of load experienced at the extremes of the cycle is
(1) tension–tension, (2) tension–no load, (3) no load–compression and
(4) tension–compression.

As would be expected, different fibres show different forms of behaviour.
Nylon 6,6, PET and Nomex all give the long 'tongue'.[35] A critical factor
seems to be the angle the 'tongue' makes with the axis. For PET there is
a long crack growth because the angle is low. The final failure region is just
like simple fracture, but occurs some distance from the fatigue crack, showing that creep has occurred.[36] This does not usually occur with nylon 6,6.
For most acrylics there is no slow crack growth and fracture usually
occurs across a radial plane. Some acrylics fibrillate, behaving like a bunch
of fibrils with poor cohesion.

The flaw that produces the initial notch is due to such factors as surface
damage, microvoids, stress corrosion, etc. The initial crack growth is
widespread in a fatigued fibre.[37] There are many cracks in various stages
of growth. The minimum load above which fatigue occurs is very constant,
but it is not known why.

As an aid to understanding the changes occurring during this type of
fatigue, the molecular structure has been studied by various techniques
for highly drawn industrial PET filaments.[36] Infrared spectroscopy and
wide-angle X-ray diffraction show there is a fall in crystallinity in the
crack regions. Transmission electron micrographs and electron diffraction

show that an amorphous band is created. This is probably due to break up of crystals rather than to melting, since there is not enough heat generated. The temperature has been calculated to rise by only about 6°C, not enough to affect the structure.[35] The fatigue crack is propagated along the amorphous band. Small voids are created just ahead of the crack, and the crack is actually generated by the joining up of these voids.

It is clear that there is a great need to design tests that simulate the end-use conditions. One common type of fatigue test is the single-fibre flex test, where the flexing filament is wound around a small diameter wire, e.g., 0.076 mm, at a standard applied load, e.g., 5.3 cN tex^{-1} (reference 38). As the fibre is flexed, the inside curvature undergoes compressive fatigue and the outside undergoes extension fatigue. The flex life depends exponentially on the applied load.[39]

For tests on fibres to be used in tyres, cords are embedded in rubber blocks and mounted on the periphery of two rubber discs, which are rotated together. The blocks are alternately extended and compressed at the same frequency as experienced in tyre operation.[39] The mechanism of failure is complex. Compression can give filament buckling. Abrasion is also an important contributor to failure.

Considering torsional fatigue, a major area occurs in the abrasion of fabrics, where the abrading medium rotates the filament in the stitch around the pinning points at each end. The looser the stitch, for example as in knitted fabrics, the more this is likely. The torsional fatigue of PET fibres has been studied at a frequency of 1.7 Hz under constant load.[40] The effect of draw ratio or orientation on the changes that occur is well illustrated. The governing factor is the propensity to fibrillate. In general, the higher the draw ratio and crystallinity, the greater the degree of fibrillation, as would be expected, and the greater the angle of twist between the fibrils.

At very low draw ratios little internal structure is generated after fatigue since there is no precursor microfibrillar orientation. At a draw ratio of 1.65 times, the morphological changes that occur are transverse cracking followed by longitudinal cracks, leading to a rope-like structure. At a draw ratio of 3.64 times, there is heterogeneity of microfibril formation, with outer layers more oriented. At 5.08 times, there is no transverse cracking, presumably because such cracks are prevented by the oriented microfibrils. Longitudinal cracking occurs by slippage between fibrils.

There has been an attempt to link torsional fatigue with the formation of 'pills' in fabrics,[41] small, unsightly entangled knots of fibres that form on all fabrics, especially those made from staple fibres, but are retained on stronger synthetic fibres since they cannot break off easily. When a high torsional strain amplitude at 5 Hz was imposed on commercial PET fibres, there was helical crack development with some fibrillation. In contrast, at a relatively low strain amplitude, the behaviour was totally different. A highly flaked surface was produced, where cracks were only skin deep. These were similar to those observed on the surfaces of PET fibres in pills, where the cracks were interlocked. It was thus assumed that pills were formed by a slow and gradual cyclic torsional fatigue mechanism. The cracks were assumed to be formed by the tensile component of pure shear on the surface.

Reality would be approached closely if a combination of different modes of fatigue deformation were used. One such study has included all the types of deformation discussed above.[42] Nylon 6,6 monofilaments of 16.7 tex were used. As a reference guide simple tensile, loop fracture, torsion and simple fatigue tests were also performed, and some points of interest are worth mentioning here in the general context of this chapter. Tensile fracture gave the type of fracture surface described above in section 1.3.3, but high speed extension, using the retraction of a stretched rubber band, gave a completely smooth fracture surface. In contrast, tensile fracture in liquid nitrogen gave a fibrillated break, showing weak lateral links between microfibrils. The loop strength was about 65 per cent of the tensile strength, and there were many circumferential cracks. Simple torsion to failure gave cracks that followed the lines of twist, and a typical V-notch. Tensile fatigue gave the 'tongue' at break.

Complex fatigue was performed by rotation of the bent fibre over a pin. This involves shear, bending, torsion and tension. Surprisingly, the crack development was similar to simple tensile fatigue, involving an initial V-notch and multiple splitting. However, there were also kink bands due to compressive stress. It was considered that all the failure mechanisms fitted the picture of nylon structure being composed of weakly linked pseudofibrils.

In another study of cyclic tension and compression, a loaded monofil was rotated over a pin, the rotating motion being horizontal.[43] High tenacity nylon 6,6, PET and regular polypropylene were examined. All developed helical cracks, due to compressive yield, which penetrated inwards before the monofil finally broke. The fatigue life of polypropylene was much greater than for nylon 6,6 or PET, which was considered to be due in some way to its crystalline nature.

1.3.7 Recovery

When fibres are strained in use, either (a) the applied stress is held constant, in which case creep occurs, which will be discussed later, or (b) the stress is removed, when recovery of length occurs. The amount of recovery depends on the degree of strain and the time deformed. The deformation may be purely extensional or may be more complex,

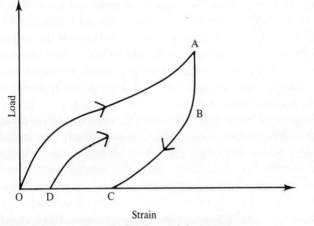

Figure 1.14 Load–strain cycle for determining recovery.

typically, bending, where a combination of fibre extension and compression occurs. One of the most important areas for such behaviour is fabrics, where deformation occurs in use, e.g. wrinkling, and it is desirable for the fabric to regain its original shape for appearance purposes.

In situations where extensional recovery occurs, the loading is temporary, and therefore tests involve applying a load or strain for a fixed time and then removing the load. These tests are usually performed on an Instron, and the time is of the order of minutes. A typical cycle is shown in Fig. 1.14. The strain is increased to point A and held constant, say for 2 minutes, during which time stress relaxation occurs and the stress falls to B. When the jaws are reversed the stress follows the path BC to point C at zero stress. At this point an increase in length or set remains, equivalent in amount to OC. Depending on the initial strain and the time held at this point of zero stress, this residual strain may totally disappear on recycling or may only partially disappear. In the latter case a second cycle of extension will commence from some point D. The strain DC is here called 'temporary set', while OD is called 'permanent set'. Clearly the magnitude of these qualities depends on the conditions used. In practice permanent set would usually be undesirable after allowing for a reasonable time to recovery.

One analysis of the molecular mechanism governing recovery assumes that it is due to portions of the molecular chains between ordered regions behaving like rubber molecules.[44] Recovery curves BC in Fig. 1.14 do in fact resemble the stress–strain curves of elastomers, and manipulation of these portions of Instron traces (which are usually discarded) was used to estimate the 'length' and 'modulus' of such elastomeric portions. Such rubbery behaviour might be considered a paradox according to the previous discussion since the main T_g of most synthetic fibres is usually above ambient temperature. This was resolved by postulating that 'free volume' is created on extension, allowing for segmental motion.[45]

This concept was used to explain the mechanism of stress relaxation (portion AB in Fig. 1.14), which is so important in governing the degree of set. It was postulated that localized ordering of appropriately placed segments occurs, maintained there by the imposed stress. It is well known that stress in an extended elastomer can be reduced by crystallization, and the two behaviours are seen as being similar. The difference is that the 'order' is only temporary and reversion occurs on the removal of the stress.

When the strain OA was gradually increased, it was found that for PET, poly(butylene terephthalate), nylon 6,6 and poly(vinyl chloride) the permanent set begins when the temporary set approaches a constant value. It was assumed this occurred when the greatest extension of the initially kinked rubbery chains was achieved, after which more severe chain slippage took place. Once it was initiated the rate of increase of the value of permanent set with increasing strain OA was found to be almost the same as the rate of increase of the level of strain.

Turning now to bending recovery, a number of methods have been used to measure this property. In the simplest approach, two ends of a fibre can be squeezed together between two plates and then released.[46] A major disadvantage here is that friction against the plates that keep the fibre in place can impede the recovery. A method that enables the fibre to recover without such impedance involves creasing it between metal plates and then allowing it to hang freely.[47] Typical values of the angles remaining after being bent fully to 180°C for a period of 5 minutes are 102° for nylon 6,6 and 121° for PET, i.e. 57 per cent and 67 per cent recovery, respectively.

A rather elegant method involves wrapping the fibre around a rod of small diameter with a large number of turns.[48] This imposes a high degree of bending curvature close to actual practice, without the extreme flattening of the previous method. A small weight is used to keep the fibre tensioned. It is then cut and the number of turns is counted after rebound.

The recovery is calculated as follows:

$$\text{Initial axial length} = n_1 \pi (D + d) \qquad [1.7]$$

where n_1 is the number of turns D is the rod diameter; and d is the filament diameter.

$$\text{Recovered loose coil length} = 2n_2 \pi p \qquad [1.8]$$

where n_2 is the final number of turns; and p is the recovered radius of curvature. Hence,

$$p = \frac{n_1}{n_2}\left(\frac{D + d}{2}\right)$$ [1.9]

This parameter can then be used in a relative way to rank various fibres, since such experimental values always have to be compared with practice in the actual end-use applications, e.g. carpet deformation and recovery, where bending curvature in the test can be made comparable.

It would be reasonable to suppose that bending would involve compression below the normal neutral axis and extension above, the reverse happening on recovery. However, it has been shown that the neutral axis is shifted because of the orientation in synthetic fibres.[49] It is moved towards the convex side of the filament, which unbalances the tensile and compressive stresses. The extent of shift was calculated by equating tensile and compressive forces across the section. It was concluded that the recovery behaviour of a single filament is independent of tensile modulus, but depends on the position of this neutral axis coupled with the yield strain.

1.3.8 Creep

Polymeric solids usually exhibit an immediate extension when loaded, and then a time-dependent increase in length, or creep. Synthetic fibres are particularly prone to this behaviour owing to the nonstable nature of the structure, as is evident from the models that have been discussed above. Various elements of structure can move relatively to each other, e.g. microfibrils and crystal regions can slide, molecular assemblies can be unfolded and disentangled, and so on. A good way to visualize these events is via the phenomenological representation of a polymeric solid as springs and dashpots shown in Fig. 1.15, the so-called standard linear solid.[50]

The immediate extension is mainly governed by the elastic spring of modulus E_1. In parallel is a slowly slipping dashpot containing a Newtonian fluid (for which the rate of shear is proportional to the shear stress) of viscosity η coupled to an elastic spring of modulus E_2. The strain ε at time t for a stress σ is given by

$$\varepsilon = \frac{\sigma}{E_1}\left(1 - \frac{E_2}{E_1 + E_2}\exp[-E_1E_2t/\eta(E_1 + E_2)]\right)$$ [1.10]

From this expression it can be seen that the instantaneous strain at time $t = 0$ is $\sigma/(E_1 + E_2)$. The final strain at time $t = \infty$ is σ/E_1. The dashpot can be equated to the slipping elements of structure discussed above.

It is useful here to introduce a concept that is used frequently, that of

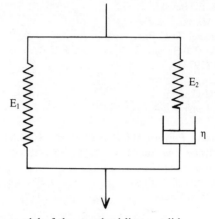

Figure 1.15   The model of the standard linear solid.

linear behaviour, incorporated in the expression given above for strain since the strain is proportional to the load, $\varepsilon \propto \sigma$. However, the practical significance of this point is that at significant strains polymers are not linear, and a much more complex nonlinear model is needed to describe creep behaviour adequately. For instance, one commonly used approach is to make the liquid in the dashpot non-Newtonian and use a power law to represent its behaviour.[51] This is necessary to better describe mathematically the complex structural changes that occur during the 'slippage' that is the essence of creep.

References

1. K H Illers, H Breur, *J. Colloid Sci.*, **18**, 1 (1963).
2. D C Prevorsek, R H Butler, H K Reimschuessel, *J. Polymer Sci.*, **A-2**(9), 867 (1971).
3. N V Bhat, D S Kelkar, *J. Macro. Sci. Phys.*, **B28**, 375 (1989).
4. B L Deopura, A K Sengupta, A Verma, *Polymer Commun.*, **24**, 287 (1983).
5. R W Miller, T Murayama, *J. Appl. Polymer Sci.*, **29**, 933 (1984).
6. D Acierno, F P La Mantia, G Titomanlio, A Ciferri, *J. Polymer Sci. Phys.*, **18**, 739 (1980).
7. S N Zhurkov, E E Tomashevsky, in Physical Basis of Yield and Fracture, Conf. Proc., Institute of Physics, London, 1966, p. 200.
8. E H Andrews, P E Reed, *J. Polymer Sci.*, **B5**, 317 (1967).
9. V R Regel, T M Muinov, O F Pozdynakov, in Physical Basis of Yield and Fracture, Conf. Proc., Institute of Physics, London, 1966, p. 164.
10. A Peterlin, *Colloid Polymer Sci.*, **253**, 809 (1975).
11. G Meinel, A Peterlin, *J. Polymer Sci.*, **B5**, 197 (1967).
12. A Peterlin, *J. Polymer Sci.*, **A2**(7), 1151 (1969).
13. H Brody, *J. Macro. Sci. Phys.*, **B22**, 19 (1983).

14. L I Slutsker, L E Utevskii, *J. Polymer Sci. Polymer Phys.*, **22**, 805 (1984).
15. A Peterlin, *J. Macro. Sci. Phys.*, **B7**, 705 (1973).
16. A C Reimschuessel, D C Prevorsek, *J. Polymer Sci. Phys.*, **14**, 485 (1976).
17. D C Prevorsek, P J Harget, R K Sharma, A C Reimschuessel, *J. Macro. Sci. Phys.*, **B8**, 127 (1973).
18. J W S Hearle, *Appl. Polymer Symp.*, **31**, 137 (1977).
19. E W Fischer, H Goddar, *J. Polymer Sci.*, **C16**, 4405 (1969).
20. E S Clark, *Polymer Prep.*, *Natl Am. Chem. Soc. Meeting, Chicago, Illinois*, August, 1977.
21. D S Boudreaux, *J. Polymer Sci. Polymer Phys.*, **11**, 1285 (1973).
22. A Peterlin, *Polymer Eng. Sci.*, **19**, 118 (1979).
23. K J Smith, *Polymer Eng. Sci.*, **30**, 437 (1990).
24. P Smith, P J Lemstra, *J. Polymer Sci. Polymer Phys.*, **19**, 1007 (1981).
25. J W S Hearle, P M Cross, *J. Mater. Sci.*, **5**, 507 (1970).
26. G E Chadwick, S C Simmens, *J. Textile Inst.*, **52**, 40 (1961).
27. J W S Hearle, *Textile Mfr.*, **99** (1972), Jan. p. 14; Mar., p. 12; May, p. 20; Aug., p. 40; Sep., p. 16; Oct., p. 40; Nov., p. 12; Dec., p. 36; **100** (1973), Jan., p. 24; Mar., p. 24; Apr., p. 34; May, p. 54; June, p. 44.
28. G E R Lamb, *J. Polymer Sci. Polymer Phys.*, **20**, 297 (1982).
29. A A Griffith, *Phil. Trans. R. Soc., London*, **221**, 163 (1920).
30. S C Smith, E K Lav, S Backer, *Textile Res. J.*, **48**, 104 (1978).
31. L R G Treloar, *Polymer*, **1**, 95 (1960); *ibid.*, 279 (1960).
32. W J Dulmage, L E Contois, *J. Polymer Sci.*, **28**, 275 (1958); I Sakurada, T Ito, K Nakamae, *J. Polymer Sci.*, **C15**, 75 (1966).
33. M Takayanagi, K Imada, T Kajiyama, *J. Polymer Sci.*, **C15**, 263 (1966).
34. I M Ward, *Mechanical Properties of Solid Polymers*, Wiley, Chichester, 1983, p. 291.
35. A R Bunsell, J W S Hearle, *J. Appl. Polymer Sci.*, **18**, 267 (1974).
36. C Oudet, A R Bunsell, R Hagege, M. Sotton, *J. Appl. Polymer Sci.*, **29**, 4363 (1984).
37. A R Bunsell, J W S Hearle, *J. Mater. Sci.*, **6**, 1303 (1971).
38. R E Wilfong, J Zimmerman, *Appl. Polymer Symp.*, **31**, 1 (1977).
39. J Zimmerman, *Textile Mfr.*, **101**, 19 (1974).
40. L Fu-min, B C Goswami, J E Spruiell, K E Duckett, *J. Appl. Polymer Sci.*, **30**, 1859 (1985).
41. B C Goswami, K E Duckett, T L Vigo, *Textile Res. J.*, **50**, 481 (1980).
42. B C Goswami, J W S Hearle, *Textile Res. J.*, **46**, 55 (1976).
43. J W S Hearle, B S Wong, *J. Textile Inst.*, **68**, 89 (1977).
44. H Brody, *J. Appl. Polymer Sci.*, **22**, 1631 (1978).
45. G M Bryant, *Textile Res. J.*, **31**, 399 (1961).
46. J Skelton, *J. Textile Inst.*, **56**, 454 (1965).
47. P Neelakantan, N C Patel, *Textile Res. J.*, **45**, 264 (1975).
48. W D Freeston, M M Platt, *Textile Res. J.*, **34**, 308 (1964).
49. G E R Lamb, R H Butler, D C Prevorsek, *Textile Res. J.*, **45**, 267 (1975).
50. J C Jaeger, *Elasticity Fracture and Flow*, Methuen, London, 1962, p. 103.
51. V B Gupta, S Kumar, *Textile Res. J.*, **47**, 647 (1977).

CHAPTER 2

Textile properties

D W LLOYD

2.1 Introduction

One of the characteristic features of textiles in general is that a hierarchy of products exists; fibres are used to make yarns, yarns are used to make fabrics and fabrics are made up into garments or other products. It is often convenient to ignore the actual physical structure of a yarn or fabric and to view it instead as a continuum for practical purposes. Then the properties of the 'continuum' are controlled by its actual physical structure and the 'continuum' properties of its structural elements. Thus yarn properties depend on the yarn structure and fibre properties, fabric properties on fabric structure and yarn properties, and so on.

The means used to obtain and control fibre properties are considered elsewhere, so attention here will be confined to the properties of yarns and fabrics, and to some of the properties of garments.

2.2 Yarn properties

2.2.1 Mechanical properties

It is appropriate to begin a consideration of yarn properties with a description of the mechanical properties of yarns in general. A yarn can be regarded as essentially one-dimensional in overall form, but with anisotropic material properties. It is usually assumed that one of the material principal axes is aligned with the yarn axis; then the constitutive equations for the tensile and bending modes of yarn deformation are uncoupled:

$$\begin{Bmatrix} F_w \\ M_u \\ M_v \\ M_w \end{Bmatrix} = \begin{bmatrix} E_w & 0 & 0 & 0 \\ & B_u & 0 & 0 \\ & \text{sym} & B_v & 0 \\ & & & B_w \end{bmatrix} \begin{Bmatrix} \varepsilon_w \\ k_u \\ k_v \\ \tau_w \end{Bmatrix} \qquad [2.1]$$

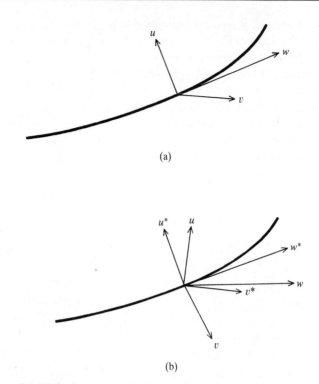

(a)

(b)

Figure 2.1   Principal axes of the mechanical properties of a yarn: (a) aligned with the yarn axis; (b) aligned at an angle to the yarn axis.

where u, v and w are the principal material directions and are defined in Fig. 2.1. F is the tension applied to the yarn, M_u and M_v are the bending moments applied about the u and v axes, and M_w is the twisting moment about the yarn axis.

If, however, none of the principal material directions is aligned with the yarn axis (Fig. 2.1) the same form of constitutive equations applies with respect to the principal material axes, but when transformed to a set of axes (u^*, v^*, w^*) aligned with the yarn axis the constitutive equations become more complicated:

$$\begin{Bmatrix} F_w \\ M_u \\ M_v \\ M_w \end{Bmatrix} = \begin{bmatrix} E_w & C_{wu} & C_{wv} & C_{ww} \\ & B_u & C_{uv} & C_{uw} \\ & \text{sym} & B_v & C_{vw} \\ & & & B_w \end{bmatrix} \begin{Bmatrix} \varepsilon_w \\ k_u \\ k_v \\ \tau_w \end{Bmatrix} \qquad [2.2]$$

where the C_{ij} are coupling terms that link the different modes of deformation. (Tensile strains normal to the yarn have no meaning in the current context and are ignored.)

In Eqns 2.1 and 2.2 it has been assumed that the matrix of properties is symmetric; this implies that it may be necessary to have a different set of material properties for loading and unloading. In general, these terms may be complicated functions of time and the deformation measures, if the viscoelastic or inelastic nature of real yarns is to be represented.

The relevance of these two descriptions becomes clear when real yarns are considered. In the context of synthetic fibres, it is necessary to distinguish between two classes of yarns: continuous filament and staple yarns. Continuous filament yarns also form two groups: monofilaments and multifilament yarns.

2.2.2 *Continuous filament yarns*

Monofilaments are produced in a wide range of linear densities and cross-sectional shapes, including coarse tape yarns. These yarns are normally twist free, and so correspond to Eqn 2.1. Yarns with circular or near-circular cross-sections will have values of the bending rigidities B_i of the same order of magnitude, whereas yarns, such as tape yarns, with exaggerated cross-sections will have bending rigidities with greatly different values, causing one direction of bending to be preferred. Such yarns will behave 'classically', with bending rigidities dependent on the tensile modulus of the material and the second moment of the cross-section.

Multifilament yarns consist of a 'bundle' of (normally fine) continuous filaments, either twist free or twisted (ignoring crimp or texturing for the moment). When it is twist free, the yarn properties can again be represented by Eqn 2.1, but the bending rigidities are much lower. This is a consequence of the yarn structure; the individual filaments are largely free to move to the neutral axis of bending by moving sideways. As a result, the bending rigidities essentially become the sum of the bending rigidities of the individual filaments. As twist is introduced into a multifilament yarn, the freedom of filaments to move is reduced as the yarn is given more cohesion. The bending stiffness of the yarn also increases, as filaments away from the neutral axis are forced to stretch.

Twist also has a significant effect on the tensile strength of the yarn and on its initial modulus. Twist causes the outer filaments to press on inner filaments, so that the filaments exert a pressure on each other. Tensile strength is increased initially by increasing twist, because the interfilament pressure allows the filaments to share load, so that if an individual filament breaks, friction with neighbouring filaments enables the broken filament to carry load a short distance away from the break. The ability of broken filaments to carry load is related to the interfilament pressure and hence to the level of twist. However, twist requires the filaments to follow a helical path, at an angle to the axis of the yarn. This angle increases with increasing twist, so that the filaments are increasingly less able to resist the yarn tension. Also, some fibre types, such as the aramids, are sensitive to transverse compression and hence to increasing levels of twist. These effects work to reduce tensile strength with increasing twist; at some twist level the increase in tensile strength with twist ceases and further increases in twist result in a decrease in yarn strength. The level of twist that corresponds to maximum tensile strength depends on fibre type; in the case of aramid yarns, this twist level is very low.

In twist-free yarns, the initial modulus depends directly on the filament modulus. In a twisted multifilament yarn, initial extension results in the helical filaments reducing their angle to the yarn axis by twisting and bending, and compressing the inner filaments of the yarn. Thus, the initial yarn modulus depends on the transverse compressibility of the yarn and on the bending and twisting moduli of the filaments. The net effect is that the initial yarn modulus decreases with increasing twist.

It is worth noting that a twisted yarn has its principal material axes aligned at an angle to the yarn axis; consequently, Eqn 2.2 holds. Such yarns attempt to untwist when extended, although this is usually neglected when their tensile properties are measured experimentally.

Further complications are introduced when the yarn is crimped or textured. Both treatments increase the linear density of the yarn and alter its mechanical properties. Planar crimp causes the yarn to possess a J-shaped load–extension curve; in the early stages of yarn extension the predominant mechanism is that of yarn bending, as the curved portions of yarn are straightened. Since this depends on the bending modulus, the yarn tensile modulus is low at this stage. As the yarn becomes almost straight, the mechanism of extension switches to that of filament extension. This depends on filament tensile modulus, so the yarn tensile modulus increases to a high value. Such crimp is characteristic of yarns in woven fabrics, and will be discussed again in that context.

In textured yarns, the filaments are tangled into complicated three-dimensional loops, or are set into separate helices, to give bulk to the yarn. This again results in J-shaped load–extension behaviour, as the early stages of extension involve the bending and twisting of individual filaments. As the filaments are separated, each is free to respond largely on its own, so the initial tensile modulus of the yarn is very low. The three-dimensional shape of the loops means that more yarn is contained in the loops than in planar crimp, so that greater extension is needed before the mechanism switches to filament extension and the yarn tensile modulus increases. In the case of helically set filaments, such as occur in false-twist textured yarns, individual filaments may have portions of opposite-handed helix joined by reversal points. Extension of the yarn will cause such reversal points to rotate around the yarn axis.

2.2.3 Staple yarns

Staple yarns are produced from fibres of finite length by means of twisting them together. The purpose of twisting is to give the yarn coherence; individual fibres follow roughly helical paths from the yarn surface to the yarn centre and back. As with twisted continuous filament yarns, fibres in the outer layers exert a pressure on the inner layers. This allows load-sharing between fibres through interfibre friction. The level of twist has similar effects on the mechanical properties of staple yarns as on continuous filament yarns, except that at low levels of twist staple yarns respond to extension by a different mechanism. At low levels of twist, as yarn tension rises it is able to overcome the friction between some fibres and their neighbours. These fibres then slide, or 'draw', past one another. This type of extension cannot be recovered and can quickly result in yarn failure.

Staple yarns can also be divided into two classes: those made from long fibres and those made from short fibres, respectively. Long fibres give rise to less hairy yarns, in general, than short staple yarns. Some yarns spun from long fibres, using the worsted system, have a structure composed of essentially parallel fibres. This imparts higher modulus and strength to these yarns. Some yarns made from long fibres are given extra bulk by a process of causing some of the fibres in the yarn to contract in length after spinning; this forces the remaining fibres to buckle into three-dimensional loops. Acrylic fibres are normally used for this purpose. Although this results in a lower initial modulus for the bulked yarn, the effect is not as great as for textured continuous filament yarns, as the fibres that have contracted to induce bulk are already essentially straight.

Short staple yarns exhibit hairiness as a consequence of the number of fibre ends in the yarn, a portion of which protrude through the yarn surface. These hairs can cause problems in subsequent processing when they can be broken off; the resulting fibre fragments can cause health problems if inhaled and can accumulate in the form of fluff or lint to cause processing faults. Short staple yarns generally require higher levels of twist to ensure yarn coherence.

Staple yarns can suffer from variations in their linear density along their length. Such variations affect other properties: twist levels are higher in thin places than thick, for instance. Two or more 'single' yarns may be twisted together to produce a multi-ply or folded yarn that is stronger and more regular than a single yarn of equivalent linear density. Folded yarns are normally produced using the opposite direction of twist to the component singles, to give a twist-balanced structure.

2.3 Fabric properties

Fabric properties depend on the fabric structure and the properties of the constituent yarns. The mechanical properties of fabrics control the ways they are used, so these will be considered first, followed by the related properties that affect the way fabrics make up into garments. Finally, properties will be considered that relate to end-use performance.

2.3.1 Mechanical properties

It is convenient to regard fabrics as two-dimensional sheets whose transverse properties can largely be ignored. The material properties will have two principal directions in the plane of the fabric in the general case, with three associated in-plane deformations (two tensile extensions and shear) and three deformations normal to the plane of the fabric (two bending deformations and twist). In general, this would require 36 mechanical properties, but if symmetry is assumed, this reduces to 18:

$$
\begin{Bmatrix} T_1 \\ T_2 \\ T_{12} \\ M_1 \\ M_2 \\ M_{12} \end{Bmatrix} =
\begin{bmatrix}
A_{11} & A_{12} & A_{13} & B_{14} & B_{15} & B_{16} \\
 & A_{22} & A_{23} & B_{24} & B_{25} & B_{26} \\
 & & A_{33} & B_{34} & B_{35} & B_{36} \\
 & \text{sym} & & D_{44} & D_{45} & D_{46} \\
 & & & & D_{55} & D_{56} \\
 & & & & & D_{66}
\end{bmatrix}
\begin{Bmatrix} \varepsilon_1 \\ \varepsilon_2 \\ \varepsilon_{12} \\ K_1 \\ K_2 \\ K_{12} \end{Bmatrix}
\quad [2.3]
$$

where T_1 and T_2 are the in-plane tensile stresses, T_{12} is the shear stress, M_1 and M_2 are bending moments, and M_{12} is the twisting moment, all referred to the principal material directions 1 and 2, as shown in Fig. 2.2; ε_i, ε_{12}, K_i and K_{12} are the corresponding strains and curvatures.

The terms A_{ij} are the in-plane moduli; the D_{ij} are the bending moduli; and the B_{ij} are coupling moduli that connect the in-plane and bending modes. As before, if symmetry is assumed, different moduli may be required for loading and unloading if the real viscoelastic or inelastic behaviour of fabrics is to be represented.

Equations 2.3 are too complicated for practical purposes, so simplifying assumptions are normally made. The coupling terms B_{ij} would be difficult (if not impossible) to measure, so it is normally assumed that $B_{ij} = 0$ for all i, j. This leaves the in-plane and bending modes uncoupled.

The next simplifying assumption is that the fabric is orthotropic; then

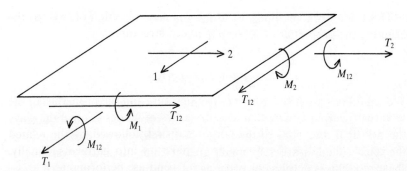

Figure 2.2 Stress resultants acting on an element of a fabric sheet, with reference to the principal axes of the fabric mechanical properties.

$A_{13} = A_{23} = D_{46} = D_{56} = 0$. This leaves five in-plane moduli, which are usually expressed as

$$
\begin{Bmatrix} T_1 \\ T_2 \\ T_{12} \end{Bmatrix} = \begin{bmatrix} \dfrac{E_1}{1 - v_1 v_2} & \dfrac{v_2 E_1}{1 - v_1 v_2} & 0 \\ \dfrac{v_1 E_2}{1 - v_1 v_2} & \dfrac{E_2}{1 - v_1 v_2} & 0 \\ 0 & 0 & G \end{bmatrix} \begin{Bmatrix} \varepsilon_1 \\ \varepsilon_2 \\ \varepsilon_{12} \end{Bmatrix} \qquad [2.4]
$$

where E_1 and E_2 are tensile moduli, v_1 and v_2 are Poisson ratios, and G is the shear modulus. The E_i are normally measured by measuring the equivalent compliance, by carrying out a tensile test in which the sample is free to contract in width, i.e. T_1, ε_1 known, $T_2 = T_{12} = 0$. The tensile moduli and Poisson ratios are connected by $v_1 E_2 = v_2 E_1$.

The bending behaviour of a fabric also requires five material properties on this model; this is usually expressed as

$$
\begin{Bmatrix} M_1 \\ M_2 \\ M_{12} \end{Bmatrix} = \begin{bmatrix} F_1 & \sigma_2 F_1 & 0 \\ \sigma_1 F_2 & F_2 & 0 \\ 0 & 0 & \tau \end{bmatrix} \begin{Bmatrix} K_1 \\ K_2 \\ K_{12} \end{Bmatrix} \qquad [2.5]
$$

where F_1 and F_2 are bending rigidities, and σ_1 and σ_2 are analogous to Poisson ratios and give rise to anticlastic curvature. Symmetry again requires that $\sigma_1 F_2 = \sigma_2 F_1$.

Anticlastic curvature, bending perpendicular to the direction of imposed bending, is observed in some fabrics, though usually indirectly as a result of stress relaxation in rolled coated fabric giving rise to synclastic curvature on unrolling. Normally, anticlastic curvature is insignificant, so $\sigma_1 = \sigma_2 = 0$ is assumed. The twist modulus, τ, is also normally assumed to be zero for practical purposes.

The framework provided by Eqns 2.4 and 2.5 will now be used to examine the properties of particular types of fabric.

2.3.2 Woven fabrics

Woven fabrics are formed by interlacing two mutually perpendicular sets of yarns—warp and weft. These natural directions are normally taken as the principal directions of the fabric. With reference to these directions, the tensile moduli are very much greater than the shear modulus; the shear modulus is of the same order as the bending rigidities, which is very much greater than the twist modulus, which will be assumed to be zero. Much of the behaviour of woven fabrics is a direct consequence of this mix of properties.

It is worth addressing the values of the Poisson ratios at this point. The Poisson effect in woven fabrics arises through the mechanism of crimp interchange, i.e. as one set of yarns is pulled straight the interlacing set of yarns is forced to become more crimped, with an equivalent decrease in the corresponding fabric dimension. The values of the Poisson ratios will therefore depend on the values of yarn crimp, and a very wide range of values is possible, including very large and very small values. Clearly, there is also an interaction with the yarn properties, as easily extended or compressed yarns reduce the need for crimp interchange. Poisson ratios are important in applications where large uniaxial or biaxial tensions are involved. The Poisson effect is most clearly observed in the 'waisting' that occurs during the strip tensile test, but it is also used to induce stabilizing biaxial stress in some frame-supported fabric buildings and to give shape to tensioned membrane structures.

The dominant in-plane property is shear, because the value of the modulus is small in comparison with the tensile moduli. Thus shear is the preferred deformation because it requires least energy. Shear is important for another reason. Woven fabrics are frequently used in situations that require them to adopt double curvatures smoothly, i.e. to drape well. It can be shown (from the differential geometry of surfaces) that for a fabric to take up double curvatures it must reduce its area. A woven fabric can do this most easily by in-plane shear.

The other properties involved in drape are the bending rigidities, which interact with shear to give the particular form of the draped fabric, and the fabric mass per unit area, as fabric weight often provides the force that results in drape.

It is instructive to compare woven fabric drape with that of paper and rubber sheet. Paper is approximately isotropic with high shear modulus; this does not provide any mechanism for area reduction on a large scale, so paper can only adopt a series of sharp peaks, involving local material failure, and areas of single curvature when forced into double curvatures. Rubber sheet can reduce its area through the Poisson effect of increasing its thickness, but this involves more strain energy than fabric shear, giving a different form of drape.

Shear modulus depends on the structure of the weave and the yarn bending properties. Shear involves a rotation of one set of yarns relative to the other. At first, friction at interlacing prevents rotation there, so the yarns are forced to bend in the plane of the fabric. As the bending moment rises, the friction is overcome and the yarns rotate at the interlacings. If the direction of shear is reversed, friction again locks the interlacings until the yarns bend sufficiently far in the opposite direction to overcome the friction at interlacings so that the yarns are able to rotate in the opposite direction. This results in the shear curves characteristic of woven fabrics (Fig. 2.3).

Fabric bending depends on the bending rigidities of its constituent yarns, modified to some extent by crimp. The fibres or filaments in the yarn need to overcome internal friction in the yarn and friction between the interlacing yarns before they can move to a position where bending can take place with minimum energy, so cyclic bending exhibits similar hysteresis to shear (Fig. 2.4). It follows that shear modulus and bending rigidity increase with the yarn linear density, yarn twist, the number of yarns per centimetre in warp and weft, and with the number of interlacings (i.e. the particular woven structure).

Tensile properties are important to apparel and industrial applications. If the yarns in a woven fabric are crimped, they impart a J-shaped load–extension curve to the fabric, but crimp interchange means that the initial modulus is higher than would be obtained from one set of yarns alone. The low initial modulus enables some stress redistribution to occur around stress concentrations at low stresses, and contributes to the comfort of garments. This initial low modulus region can be extended by the use of extensible yarns; if the yarns are elastic, so that this extension is fully recoverable, permanent deformations can be avoided. Some modern fabrics are made from yarns that include elastomeric fibres or filaments for just this purpose.

The tensile strength of the fabric depends on the strength of the yarns, but also on the fabric structure. If the fabric structure does not distribute loads equally between yarns, but allows stress concentrations to develop, the fabric strength will fall short of the product of the number of yarns and the yarn strength. Weaker yarns will fail first, followed by rapid failure across the fabric as load is shed from the failed yarns to the remaining yarns. If the fabric structure permits load sharing, so that yarns continue to carry some load after initial failure, then the strength of the yarns will be fully realized in the fabric strength.

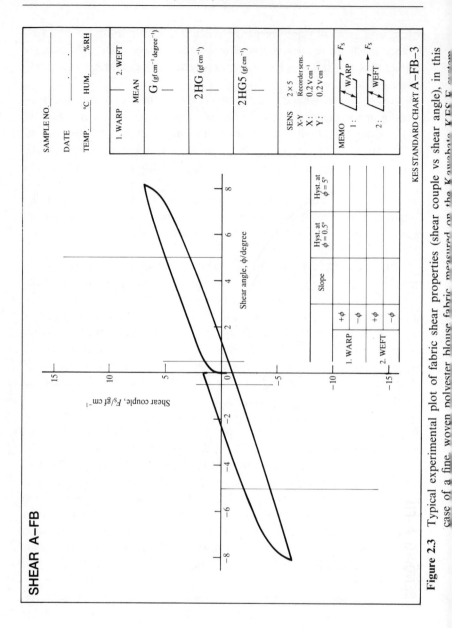

Figure 2.3 Typical experimental plot of fabric shear properties (shear couple vs shear angle), in this case of a fine, woven polyester blouse fabric measured on the Kawabata KES-F system.

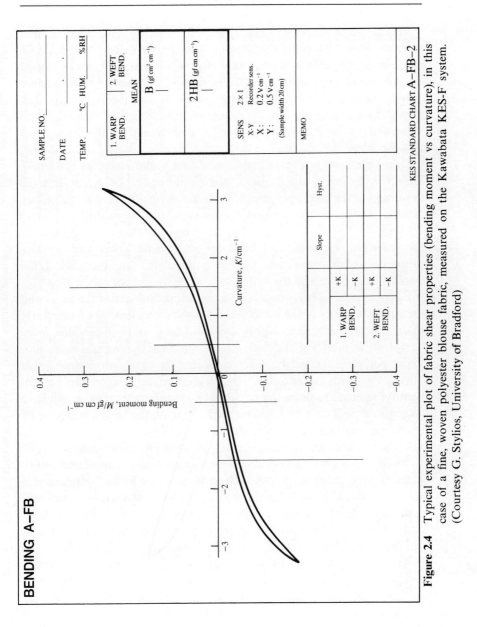

Figure 2.4 Typical experimental plot of fabric shear properties (bending moment vs curvature), in this case of a fine, woven polyester blouse fabric, measured on the Kawabata KES-F system. (Courtesy G. Stylios, University of Bradford)

The fabric strength will clearly by increased by those factors that increase yarn strength, so increasing yarn linear density will increase fabric strength, while increasing yarn twist will be beneficial up to a point but will then reduce fabric strength. Increasing the number of threads per centimetre will also increase the strength of the fabric at first, but eventually the yarns become sufficiently densely packed to prevent the yarns under load from straightening, so their strength cannot be realized and the fabric strength falls.

Drape is one of the controlling factors in the visual aesthetics of a fabric. However, for many applications, the tactile aesthetics are at least as important as the visual properties. These are normally referred to as fabric 'handle' or 'hand'; this property is assessed by manipulating the fabric between the fingers. Consequently, the properties that are important in the subjective experience of hand are the tensile, shear and bending properties at low stress levels, plus the compressibility of the fabric in the thickness direction and the surface characteristics of the fabric.

Fabric compressibility depends on the compressibility of the yarns and the compressibility of the structure. Textured or bulked yarns are clearly more compressible than highly twisted yarns, and so produce more compressible fabrics. Fabrics that have space that yarns can move into under compression will be more compressible; thus fabrics with fewer interlacings, with less crimp or that are jammed, i.e. at the upper limit of density, will tend to be harder to compress. Finishing treatments can have a significant influence by raising a softer layer on the fabric surface or by imparting more freedom to the threads in the fabrics.

Surface characteristics also depend on both the yarns and the fabric structure. In terms of texture, monofilaments and untextured multi-filament yarns have a smooth surface, textured or bulked yarns have a rough surface, and twisted yarns are intermediate, with a level of smoothness that depends on the filament size and the level of twist. Similarly, densely woven fabrics with long floats made from fine yarns, such as sateens, give a smooth surface, whereas open fabrics, or fabrics with a large number of interlacings, such as plain weave, made from coarse yarns, will possess a rough surface. In terms of friction, smooth yarns and fabric surfaces will tend to have a higher coefficient of dynamic friction than rough surfaces, as smooth surfaces will have a higher area of contact with the fingers. The position is complicated, however, by any mechanical interaction that may occur between the fingerprint ridges and individual filaments or fibres.

2.3.3 Knitted fabrics

Knitted fabrics are constructed by interlacing loops formed from a single set of yarns. In warp knitting the yarns run in the machine direction and the loops are formed in the cross-machine direction; in weft knitting the yarns run across the fabric with the loops formed in the machine direction. In warp knitting especially, yarns may be 'laid-in' in either the machine direction or the cross-direction, or both; such laid-in yarns are comparatively straight and have a strong influence on the mechanical properties obtained.

Warp knitted fabrics with laid-in yarns are often intended to compete directly with woven fabrics, so their mechanical properties are similar to woven fabrics by design, with the laid-in yarns controlling the tensile properties, for example. Attention will be concentrated on knitted fabrics without laid-in yarns, and the ways in which their properties are distinct from those of woven fabrics.

In general, knitted fabrics are more isotropic than woven fabrics. As a consequence, the tensile, shear and bending moduli are all of the same order. This comes about because the mechanism in each case is similar, the deformation of the knitted loops. This is controlled, at first, by yarn bending, then by the yarns slipping past interlacings from one loop to another, and then by yarn extension. Yarn torsion also plays a part, as knitted loops adopt a three-dimensional shape.

Knitted fabrics are able to adopt smooth double curvatures, i.e. to drape well, although the mechanism of area change is different from shear in woven fabrics, involving distortion of the loops rather than the relative rotation of yarns.

In tension, knitted fabrics have low initial modulus, as the mechanism is that of the loops changing shape. This is controlled by the bending and torsional properties of the yarns. As fabric extension proceeds past the normally encountered, low stress region, the loop is straightened and the mechanism changes. If the yarns have relatively high modulus, and friction at the contact points between loops is relatively low, then yarn may slip past contact points into the stretched loops. This will be followed by extension of the yarns themselves. If the yarns have relatively low modulus and frictional constraints are high, then yarn extension will occur first. The first case would be more typical of staple fibre acrylic yarns in fine gauge knitwear, the second of textured continuous filament polyamide in hosiery, where some entanglement may occur at interlacings to increase frictional constraints. Fabric extensibility and recovery may be enhanced in the wales (cross-) direction of weft knitted fabric through the use of rib structures, such as may be used at the waist, cuffs or neck of knitwear. Rib structures produce a macroscopic crimp in the fabric, which can be deformed by a process of yarn bending.

Knitted fabrics characteristically have a soft handle, being easy to stretch, bend and shear, and knitted fabrics are usually more easily compressed than woven fabrics. In compression, the loop shape will be flattened, involving yarn bending, though yarn compression is also important in fabrics made from bulked or textured yarns.

Knitted fabrics may have a wide range of surface characteristics, depending on the yarn properties and fabric structure, for similar reasons to woven fabrics. Fine gauge warp knitted fabrics made from fine mono-filament or untextured continuous filament yarns will have a smooth surface, especially where long surface floats are introduced. At the other extreme, coarse, bulked, folded yarns in a plain weft knit will have a large-scale, textured surface.

2.3.4 Nonwoven fabrics

The term nonwoven encompasses a wide range of fabrics that are characterized by being made directly from fibres or filaments rather than yarns. The fibres or filaments may be held together by entanglement, by friction, by fusing at fibre-to-fibre contact points, by adhesive binder, or by fusing filaments together in small areas distributed in a pattern across the fabric, using heat and pressure. A detailed discussion of nonwovens and their methods of production will be given elsewhere in this book.

The mechanical properties of nonwovens are controlled by the properties of their constituent fibres or filaments and their structure. The principal structural parameters are the fibre or filament orientation distribution and the degree of bonding. The fibre orientation distribution can range from, at one extreme, all the fibres or filaments aligned in a single direction (machine or cross), to a completely random orientation in which no direction is preferred. The degree of bonding can range from no bonding, with the only constraint that of friction in a low density web of fibres, to a fully bonded web, where every fibre contact point is fused or glued, with densely packed fibres. The important difference that arises from these parameters is the freedom of individual fibres or filaments to move within the fabric.

In a web of fibres given cohesion only by friction, tensile loads result in extension through the mechanism of fibres slipping past one another. Extensions are permanent, as little or no strain energy is involved. As frictional constraints or the degree of bonding are increased, individual fibres or filaments are stretched between bond points. If the fibres are straight, the mechanism is one of fibre extension; if the fibres are curved, the mechanism is that of fibre bending, at least initially. The greater the degree of bonding, the greater the elastic component of extension and the ability of the nonwoven to recover. Similarly, the greater the proportion of fibre stretching, the higher the tensile modulus. Clearly, nonwovens in which fibres can slip past one another with little constraint will have low tensile strength. Again, with increasing bonding, tensile strength will increase but extension to break will decrease. Fibre orientation relative to the direction of extension is important; fibres must be aligned, or be able to align themselves, with the direction of applied tensile load for the nonwoven to achieve significant tensile strength.

Similar mechanisms apply in shear. Fibres will slip, bend or extend depending on the degree of bonding and the fibre orientation distribution. If the degree of bonding is high and fibres are distributed in all directions, the situation is similar to that in paper with high shear modulus. This prevents easy area reduction and hence imparts a 'papery' drape, i.e. such nonwovens do not drape well and are unable to adopt double curvatures smoothly.

Bending is also controlled to a large extent by the degree of bonding. If the fibres are free to move within the fabric and the fabric is thin, individual fibres will only suffer bending and torsion, as they will move to the neutral axis or will slip through the fabric to avoid fibre extension. In thicker fabrics, fibres will not be able to move to the neutral axis, but will again avoid fibre extension by slipping or will avoid axial compression by buckling, so that individual fibres only bend or twist. Consequently, fabrics with a low degree of bonding will have low bending moduli. In contrast, fibres in fabrics with high bonding and dense packing of fibres will not be free to move or slip, so fibres away from the neutral axis of fabric bending will be forced to stretch. If the fibres are straight initially, fibre extension becomes the dominant mechanism and the fabric bending moduli are high and increase rapidly with fabric thickness. One advantage of bonding a nonwoven in a pattern of small bond points distributed across the fabric is that it confines areas of high bending stiffness to the bond points, so that overall the fabric has low bending stiffness without sacrificing tensile strength.

Nonwoven fabrics are not normally noted for their handle characteristics. The compressibility of nonwovens depends on the fibre properties, the degree of bonding and the number of fibres per unit volume. The compressive modulus increases with the degree of bonding and the density of fibre packing. High bonding results in good recovery, though repeated compression of nonwovens, such as needlefelts, held together by friction and entanglement, results in a consolidation of the fabric to higher packing densities, reduced thickness and increased compressive modulus. In some cases this is undesirable, as with thick nonwovens used for thermal insulation, where the need is to trap as much air as possible per unit volume. Hollow fibres are often used here, as they have greater bending stiffness than solid fibres of the same linear density, giving better recovery from compression.

It is not possible to generalize about the surface friction and texture of nonwovens, as these depend on the fibre or filament linear densities and mechanical properties, the number of fibres per unit volume, the fibre orientation distribution, and the degree of bonding. In addition, finishing treatments are sometimes used to fuse the surface fibres to prevent fibre loss, and these alter the surface characteristics.

2.4 Fabric properties related to end-use performance

2.4.1 Use in garments

Garments (and similar items) are manufactured by joining together panels of fabric, usually by sewing. Problems can be caused at this stage by inappropriate fabric behaviour. It is convenient to begin by considering simple seams in which two pieces of fabric are joined in a flat seam.

Sewing requires the insertion of one or more sewing threads into the fabric structure without breaking or cutting yarns. In the case of lockstitch, two interlocking loops are trapped in the fabric at each needle insertion, so the structure has to accommodate four thread diameters. There is an interaction between the sewing operation and the fabric properties. In the case of woven fabrics, if the structure is open these extra threads can be accommodated easily, and, if the sewing thread tensions are not too high, the seam will lay flat. If the fabric is closely woven and too coarse a needle and sewing thread are used, the fabric will be unable to absorb the in-plane compression along the seam and seam pucker will result as the fabric buckles. The remedy then is to reduce the sizes of needle and thread used.

Even if appropriately sized needle and thread are used, if the tensions in the sewing thread are high, the fabric will be compressed along the seam line and the seam will pucker. The fabric mechanical properties at low stress are important here; if significant tensile extension is possible at low stress then significant in-plane compression will also be possible and pucker is less likely. Similarly, increased fabric bending stiffness will increase the critical force at which fabric buckling occurs and so will decrease the risk of pucker. If the fabric is jammed or nearly jammed, with adjacent yarns touching or nearly touching, the scope for in-plane compression is reduced and the risk of pucker is increased. However, bias extension corresponds to shear, with its low modulus, so a realignment of the seam away from the warp or weft direction by a few degrees increases the in-plane compressibility along the seam and so reduces the risk of pucker. Transverse fabric compressibility is also useful in seam formation, as it provides a mechanism for the relaxation of sewing thread tensions without pucker.

Knitted fabrics do not normally suffer from seam pucker, as in-plane compression is absorbed through distortion of the loops. However, it is necessary for yarns to slip at the contact points between loops to make room for the needle. If the frictional constraints are high and the needle strikes a part of the fabric where the length of yarn between contact points is short, the yarn segments may have insufficient extensibility to allow passage of the needle. When yarn segments break in this way, a hole is formed in the fabric at the point of needle penetration; this is referred to as sewing damage. Sewing damage is serious because most knitted structures are, topologically, simple knots, so that when a yarn is broken the structure loses its integrity and the hole may propagate in the well-known 'ladder'. The remedy to this problem is to reduce the internal friction in the fabric by finishing with a suitable lubricant or fabric softener. If the fabric has high friction with metal surfaces and enough layers of fabric are being sewn, friction with the needle during high speed sewing may result in needle heating. Under appropriate conditions, needle temperatures may rise quickly to the point where synthetic fibres melt on contact with the needle, producing holes in the fabric, also referred to as sewing damage.

Friction between layers of fabric, controlled by the surface characteristics of the fabrics, is normally relied upon to retain the two sides of the seam in the correct positional relationship. If this frictional constraint is insufficient, the upper layer may slip during sewing, resulting in differential extension between the layers that will cause pucker when the constraints on the seam are released. This problem requires special precautions to be taken during sewing, as it is not normally acceptable to modify the surface characteristics of these fabrics.

Some seams deliberately match two fabric edges of different lengths to produce a three-dimensional shape called fullness. This is particularly evident in the seam joining a sleeve to the shoulder in a tailored jacket. To avoid pucker in this situation the fabrics must have the ability to be stretched and compressed in the local plane of the fabric, at varying angles to the warp direction. If the fabric lacks extensibility, or if the bending stiffness is too low, pucker results. This has led to the definition of a number of 'formability' measures, based on tensile, bending and sometimes shear properties, to describe the ability of a fabric to form aesthetically pleasing seams under these conditions.

Internal friction, as measured by hysteresis in mechanical properties, has other effects than sewing damage. Excessive hysteresis in tensile properties can result in permanent extensions in a fabric. This may be manifested as unsightly 'bagging' or 'seating' in a garment, or as permanent distortions introduced as a result of the forces imposed on a garment during making-up. In bending, excessive hysteresis can also result in permanent distortions in the form of wrinkles or creases. Elastomeric fibres may be introduced into fabrics to provide greater restoring forces to overcome frictional constraints and prevent permanent distortions.

Insufficient hysteresis in shear can cause difficulties with some woven fabrics. If there is little friction at the yarn interlacings, the shear modulus will be low and the fabric will behave rather like a pin-jointed mechanism. In some garments, long fabric panels are secured by widely spaced side seams; low shear modulus can lead to the hem of such panels dropping. Low shear modulus also leads to dimensional instability in making fabric lays prior to cutting, during cutting and during making-up.

The importance of fabric properties at low stress to making-up performance has become recognized over the last two decades in the development of measurement systems designed to predict garment production problems. These have become known in general as 'fabric objective measurement' methods and their application is the subject of considerable current research.

2.4.2 In-service properties

It is possible to identify a number of secondary properties that are of considerable importance to the in-service behaviour of fabrics. As with the mechanical properties, these physical properties depend on both the fabric structure and the properties of the constituent yarns. Many of these properties are related to the ability of a fabric to resist a particular mode of failure.

One of the common failure modes of fabrics is tearing. Tearing occurs when the stresses imposed on a fabric result in a concentration of stress at a point, loading individual yarns to failure. The tear propagates when the failure of one yarn results in the load being transferred to the adjacent yarn, which then fails in turn. Frictional constraints at yarn interlacings and the extensibilities of the yarns control the resistance to tearing. If the yarns are inextensible and frictional constraints are high, the yarns cannot slip, so that several yarns resist the stress concentration at once, or align themselves with the direction of the tear. Greater extensibility or reduced frictional constraints allow the yarns to bunch at the point of the tear and so increase the tear resistance. Although the yarn strength is secondary to extensibility, rip-stop yarns of greater strength are effective in increasing tear resistance.

Pilling is a failure of fabric appearance consisting of unsightly balls of fibre attached to the surface of the fabric. These balls of fibre, or pills, are produced by rubbing of the fabric during use. Pilling requires loose fibre at the fabric surface, or fibres that can easily be loosened by rubbing, as shown in Fig. 2.5. The appearance of the pills is controlled by the different rates at which a series of processes occur. The rate of formation of pills

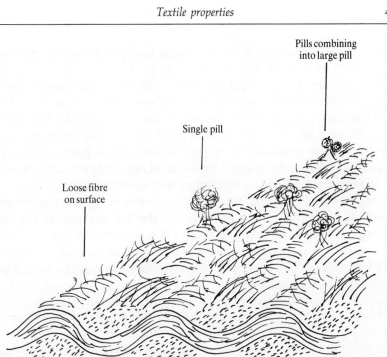

Pills combining
into large pill

Single pill

Loose fibre
on surface

Figure 2.5 The mechanism of pill formation, loose fibres at the surface of the fabric becoming twisted together to form small pills, and small pills combining to form large pills.

is controlled by the rate at which fibre can be loosened. This is a function of the yarn structure and fibre strength. Fibres can be loosened easily from staple yarns of relatively low twist, from multifilament yarns with little or no twist, and from textured or bulked yarns, especially if the filament is fine or of low strength.

The appearance of the pills is further controlled by the rate at which small pills coalesce to form fewer, larger pills, and by the rate at which pills break away and are lost. The rate of loss of pills is determined by the strength of the fibres that attach the pills to the surface. Because synthetic fibres tend to be stronger than natural fibres, pills tend to be lost at a slower rate with synthetic fibre fabrics. Controlling pill formation then becomes a matter of limiting the availability of loose fibre; using yarns with higher twist gives fabrics with enhanced pilling resistance.

If loosened fibre is lost from the surface before pills can form, the fabric is worn away by a process of abrasion. Factors that increase the resistance to abrasion include using high twist yarns, using coarser filaments and yarns of greater linear density, and employing a tighter fabric structure (i.e. more threads per centimetre) and a greater number of interlacings. These factors all assist in preventing fibres being pulled from the yarn and broken.

The range of pore sizes in a fabric is important to the permeability and filtration properties of fabrics. Woven fabrics tend to have well-defined pores in a narrow range of sizes as a consequence of the fabric structure. This can be changed by the yarn structure; bulked, textured or hairy yarns will have the effect of closing the pores to some extent. Knitted fabrics will also have well-defined pores, but the three-dimensional shape of the knitted loops will result in some pores passing through the fabric at an angle, increasing their length. Nonwoven fabrics, in general, will have a wide range of pore sizes, whose upper and lower limits will depend on the method of production and the fabric thickness. The permeability of a fabric obeys D'Arcy's law, so thinner fabrics and larger pores will lead to increased permeability. Air permeability can be reduced by using bulked or textured yarns and tighter fabric structures to reduce pore sizes. In filtration, pore size controls which sizes of particles are retained by the fabric and which are allowed to pass, so fabric pore size is matched to the expected particle size distribution to ensure efficient filtration.

Pore size is also important to the water repellency properties of fabrics. Water repellency is measured in terms of transmission of water through a fabric and absorption of water by a fabric. The wetting angle of water on the fabric is critical; if water can spread by wetting it will penetrate the fabric quickly. Most synthetic fibres are hydrophobic, so water remains on the surface in the form of drops. The fabric structure has an obvious effect; a tight structure results in small pores in the fabric which do not allow the drops of water to penetrate.

Further reading

It has not been possible in the space available to present more than a summary of the properties of textiles made from synthetic fibres. The following publications are suggested for further reading.

G A Carnaby, E J Wood, L F Story (eds), *The Application of Mathematics and Physics in the Wool Industry*, Wool Research Organisation of New Zealand Special Publications, Vol 6, Wool Research Organisation of New Zealand (Inc), Christchurch, New Zealand, 1988.

J W S Hearle, P Grosberg, S Backer *The Structural Mechanics of Fibres, Yarns and Fabrics*, Vol 1, Wiley-Interscience, Chichester, 1969.

J W S Hearle, B Lomas, W D Cooke, I J Duerdon *Fibre Failure and Wear of Materials: an Atlas of Fracture, Fatigue and Durability*, Polymer Science and Technology Series, Ellis Horwood, Chichester, 1989.

J W S Hearle, J J Thwaites, J Amirbayat (eds), *The Mechanics of Flexible Fibre Assemblies*, NATO Advanced Study Institute Series E: Applied Science No. 38, Sijthoff and Noordhoff, Alphen aan den Rijn, 1980.

S Kawabata, R Postle, M Niwa (eds), *Objective Specification of Fabric Quality, Mechanical Properties and Performance*, Proceedings of the Japan–Australia Joint Symposium, Kyoto, Textile Machinery Society of Japan, Osaka, 1982.

S Kawabata, R Postle, M Niwa, *Objective Measurement: Application to Product Design and Process Control*, Proceedings of the Third Japan–Australia Joint Symposium, Textile Machinery Society of Japan, Osaka, 1986.

R Postle, S Kawabata, M Niwa (eds), *Objective Evaluation of Apparel Fabrics*, Proceedings of the Second Australia–Japan Bilateral Science and Technology Symposium on Objective Evaluation of Apparel Fabrics, Parkville, Victoria, Textile Machinery Society of Japan, Osaka, 1983.

W J Shanahan, D W Lloyd, J W S Hearle, Characterising the elastic behaviour of textile fabrics in complex deformations, *Textile Res. J.*, **48**(9), 495 (1978).

CHAPTER 3

Dyeability 1: nylon fibres

D M LEWIS

3.1 Introduction

The term 'nylon' refers generally to any fibre-, bristle- or film-forming polymeric amide. A variety of polyamides are presently manufactured and are marketed under several different trade names; the two most important textile representatives being nylon 6,6 and nylon 6; other types such as nylon 6,10, 6,12, enjoy limited, speciality usage. Nylon 6,6, the first commercially available, wholly synthetic fibre, was introduced by du Pont in 1938 and is made by the polymerization of adipic acid and hexamethylene diamine (scheme 1).

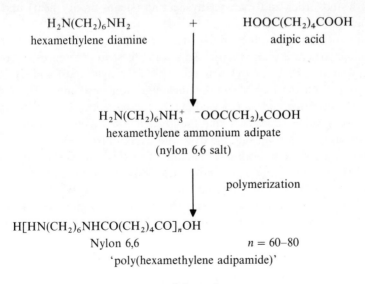

$$H_2N(CH_2)_6NH_2 \quad + \quad HOOC(CH_2)_4COOH$$
hexamethylene diamine   adipic acid

$$H_2N(CH_2)_6NH_3^+ \quad {}^-OOC(CH_2)_4COOH$$
hexamethylene ammonium adipate
(nylon 6,6 salt)

polymerization

$$H[HN(CH_2)_6NHCO(CH_2)_4CO]_nOH$$
Nylon 6,6 $n = 60–80$
'poly(hexamethylene adipamide)'

Scheme 1

Nylon 6, which was commercially introduced in 1939 by IG Farbenindustrie under the name Perlon L, is prepared by the polymerization of ε-caprolactam (scheme 2).

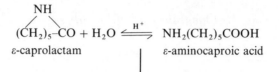

$$NH_2(CH_2)_5CONH(CH_2)_5CONH(CH_2)_5COOH$$

$$H[NH(CH_2)_5CO]_nOH$$

nylon 6 $\qquad n = 130-160$

Scheme 2

The two polyamides are thermoplastic and are melt spun. The fibres, formed by rapid cooling, are first drawn in order to develop polymer strength and order and then set by heating (steam or dry heat) under tension.

Although generally quite similar, nylon 6,6 has a more compact and crystalline structure than nylon 6. This is manifest in differences in both melting point (250°C for nylon 6 and 230°C for nylon 6,6) and dyeing behaviour: both the rate of uptake and migration (levelling) of a given dye are lower for nylon 6,6 and as a consequence the wash- and wet-fastness will be superior to that found for nylon 6. Thus the dyer must exercise care in selecting suitable dyes and dyeing processes.

Nylon fibres contain both terminal amino ($-NH_2$) and carboxylic acid ($-COOH$) groups. The amino end-groups impart substantivity to acid, metal complex, mordant, direct and reactive classes of dye; the fibre is also readily dyed with disperse dyes. In addition, cationic dyes may be applied to nylons with modified dyeing characteristics (i.e. basic-dyeable nylon).

3.2 Acid or anionic dyes

Acid dyes show high substantivity for polyamide fibres and give bright dyeings of relatively good light- and wet-fastness; however, the fastness to washing of many acid dyes on nylon leaves much to be desired.

Generally, the dyes are applied within the pH range 3–7 and in essence the dyeing mechanism involved closely resembles that which describes the application of acid dyes to wool. However, important differences must be noted. For wool, the point at which all the protonated amino groups $(-^+NH_3)$ are neutralized or occupied by adsorbed dye anions, which is known as the *saturation value* of the fibre, occurs at depths of shade in excess of 20 per cent on weight of fibre (owf). With nylons, however, the saturation value can in some cases occur with about 2 per cent owf acid dye.

The mechanism of dyeing with acid dyes can be represented as in scheme 3, where D represents a chromophoric residue (e.g. azo or anthraquinone).

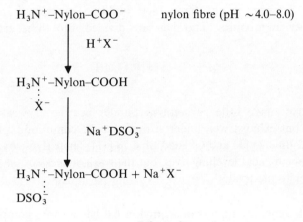

Scheme 3

In the pH range 4.0–8.0 nylon fibres exist mainly in the zwitterionic form; the carboxylic acid groups are dissociated and the liberated protons are taken up by the amino end-groups. In an acidic solution, dissociation of the carboxylic acid group is depressed, leaving the fibre with an overall positive charge, and the rapid-diffusing acid anions (X^-) are initially adsorbed onto the protonated amino groups. In the presence of an acid dye $(DSO_3^-Na^+)$ the slow-diffusing, high affinity dye anions gradually displace the acid anions to become adsorbed at the protonated amino end-groups.

The adsorption of simple acid levelling dyes is almost exclusively confined to protonated amine end-groups. Larger, more complex acid dyes that have substantivity for nylon at higher pH values (5–7) are adsorbed in excess of the saturation value of the nylon fibre (overdyeing); it is considered that adsorption of the dye of large molecular size at the protonated amino end-groups is augmented by forces such as Van der Waals and hydrophobic interactions.[1]

A characteristic feature of the acid dyeing of nylon is the phenomenon of barré dyeing, i.e. the appearance of stripes in the dyed goods. Barré arises from both chemical (amino and carboxyl end-group content) variations introduced during polymer manufacture and physical variations (crystallinity) introduced during melt spinning, drawing and setting processes. Nowadays variations in amino and carboxyl end-groups are minimal and full control of physical differences is required.

Modern fashion colours and fastness demands have recently exacerbated the barré problem because high molecular mass anionic dyes are required to meet wet-fastness requirements.

To help the dyer in dye selection, dye manufacturers market preferred groups of acid dyes for nylon based on their ability to cover physical and chemical differences in the fibre, their build-up properties and their compatibility in mixtures. The dyes are divided into three groups as follows.

Group 1

Group 1 dyes have little substantivity under neutral or weakly acid conditions but exhaust well under stronger acidic conditions (pH 3–4), e.g. CI Acid Blue 25 [I] and CI Acid Blue 78 [II]. These dyes cover barré nylon well, being acid levelling dyes, but the wash-fastness of the dyeings is only poor to moderate.

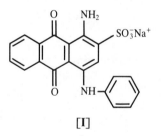

[I]

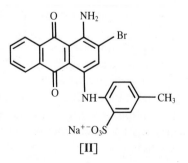

[II]

Group 2

These dyes exhaust well at pH 3–5 and give dyeings of higher wet-fastness than Group 1 dyes, e.g. CI Acid Blue 41 [III] and CI Acid Yellow 172 [IV].

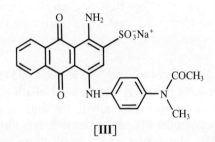

[III]

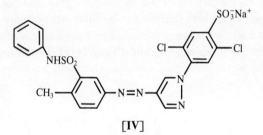

[IV]

Group 3

These dyes exhibit high 'neutral' (pH 5–7) substantivity and poor levelling properties. Their coverage of barré nylon is poor but the dyeings have high fastness to wet and washing treatments and good light fastness (5–6

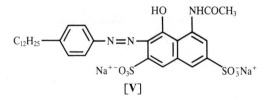

[V]

on the blue wool scale), e.g. CI Acid Red 138 [V] and CI Acid Blue 138 [VI].

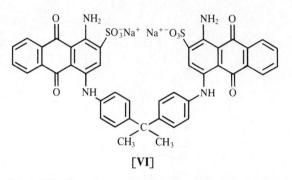

[VI]

Because of the difficulty of covering the physical and chemical variations in the fibre, it is usually necessary to employ dyebath additives to control exhaustion. Anionic agents compete with the anionic dyes for available sites within the polyamide, whereas cationic agents form complexes with the negatively charged dye that slowly break down and release the dye as the temperature is raised. Recently, the use of mixtures of anionic and cationic surfactants has been advocated. The anionic 'blocking' component is designed to be rapidly adsorbed by the substrate and become concentrated on those nylon fibres that tend to be dyed most rapidly. The cationic agent aids levelness by slowing dye adsorption due to complex formation.

A typical dyeing process on nylon 6,6 can be presented as in scheme 4.

Group 1 and 2 dyes

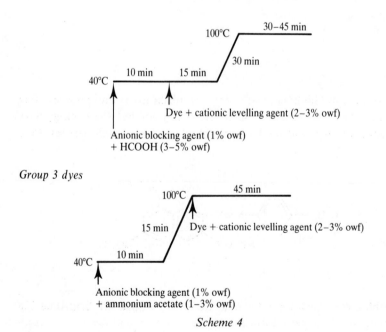

Scheme 4

The wash-fastness properties of acid dyes (especially Groups 1 and 2) on nylon leave much to be desired and it is thus common to aftertreat the dyed nylon with either natural or synthetic tanning agents to improve wash-fastness.

Natural tanning agents

The application of natural tanning agents is known as the 'full back-tan' and involves treating the dyed nylon with tannic acid (2 per cent owf), at pH 4.0 [CH₃COOH] for 20 minutes at 80°C, followed by a treatment in a separate bath containing potassium antimonyl tartrate (tartar emetic) (2 per cent owf) at pH 4.0 for 20 minutes at 80°C. It is considered that the full back-tan results in the formation of a sparingly water-soluble anionic complex of large molecular size, located at the surface of the dyed nylon, which restricts diffusion of anionic dye molecules out of the substrate during washing. Although the full back-tan often results in quite dramatic improvement in wash-fastness of acid dyes on nylon, the process suffers from the following disadvantages:

It is two-stage, therefore time consuming.
Tartar emetic is toxic.
Tannic acid discolours on exposure to light.
Back-tanning reduces the light-fastness of the dyes on nylon.
The process is expensive.
The effect can be partially lost on steam setting.

Major constituents of gallotannins are the polygalloylated glucoses. Shore[2] showed that the 1,2,4,6-tetragalloyl ester of D-glucose can exist as a remarkably flat stable structure [VII].

Synthetic tanning agents

A simpler aftertreatment for improving the wash-fastness of acid dyes on nylon was therefore developed employing synthetic tanning agents (Syntans). Typically, these Syntans are high molecular mass polycondensates of sulphonated phenol- or dihydroxydiphenyl- sulphone-formaldehyde. It is considered that, unlike the full back-tan, Syntans do not form a 'skin' as such at the fibre surface but instead are located within the periphery of the dyed substrate. Although not as effective as the full back-tan, Syntans can result in marked improvements in wash-fastness. They are generally applied at pH 3–5 (CH₃COOH or HCOOH) for 20 minutes at 80°C. Longer treatments promote diffusion of the Syntan from

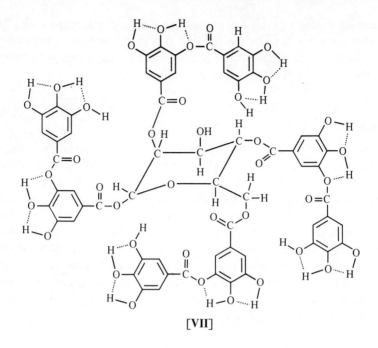

[VII]

the surface of the fibre and tend to lower wet-fastness, indicating that surface deposition of the anionic polymer is important.

Syntans are still widely used despite problems of reduction in light-fastness of aftertreated dyeings, slight changes in hue, firmer fabric handle and difficulties when used in foam laminating.

The improvements in wet-fastness achieved by the use of Syntans can be partially lost owing to the following factors:

Instability to dry heat
Post boarding sensitivity
Steam setting sensitivity.

3.3 Mordant dyes

These dyes are normally applied to nylon using the afterchrome technique shown in scheme 5. These dyes economically produce deep shades of excellent wash-fastness and very good light-fastness. However, the dyes tend to emphasize barré, the application procedure is rather long, and environmental problems related to chromium ions in dyehouse effluent are becoming pressing; a typical discharge level to be met for total chrome in effluents is 2 ppm.

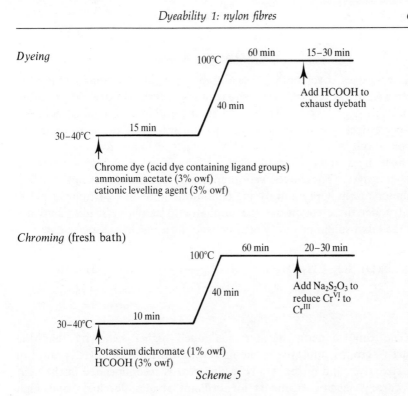

Dyeing

100°C — 60 min — 15–30 min

40 min

↑
Add HCOOH to
exhaust dyebath

30–40°C — 15 min

↑
Chrome dye (acid dye containing ligand groups)
ammonium acetate (3% owf)
cationic levelling agent (3% owf)

Chroming (fresh bath)

100°C — 60 min — 20–30 min

40 min

↑
Add $Na_2S_2O_3$ to
reduce Cr^{VI} to
Cr^{III}

30–40°C — 10 min

↑
Potassium dichromate (1% owf)
HCOOH (3% owf)

Scheme 5

3.4 Metal complex dyes

1:2 metal complex dyes are important for nylon dyeing especially where high light-fastness is required, e.g. automotive upholstery. The dyes exhibit excellent build-up, compatibility and very good wet- and wash-fastness characteristics on nylon. However, their ability to cover affinity variations in the fibre varies markedly and their application necessitates the use of blocking and (weakly cationic) complexing auxiliaries to control exhaustion.

The most common metals used in the 1:2 metal complex dyes are chromium and cobalt. Whilst chromium dyes are the commonest, cobalt dyes appear to offer the advantage of higher light-fastness. Symmetric premetallized dyes may be prepared from chromophores that contain sulphonate or carboxylic acid groups (e.g. Acidol M (BASF) and Elbelan N (Holliday Chemical Holdings)) or even from chromophores that contain no solubilizing groups (Veolan Fast (BASF)). The most level dyeings can be produced using the latter dyes but they are more expensive. Currently very popular metal complex dyes are asymmetric complexes of nonsulphonated and monosulphonated chromophores (e.g. Isolan S (Bayer), Lanasyn S (Sandoz), Lanacron S (Ciba Geigy) and Neutrichrome (formerly ICI but now Crompton & Knowles)).

3.5 Disperse dyes

Disperse dyes cover both physical and chemical variations in the fibre very well. However, set and wash-fastness properties of dyeings produced from this dye class on nylon, with the exception of 2:1 metal complex disperse dyes, are very poor and consequently their use is restricted to applications where high wet- and wash-fastness is not required.

Recently a study was made of the dyeing of nylon with vinyl sulphone disperse dyes.[3] These dyes were prepared *in situ* from the sulphatoethyl-sulphone form. Optimum dyeing conditions were 80°C, 1 hour at pH 8, and under these conditions the sulphatoethylsulphone readily activates to the vinyl sulphone dye, which behaves as a reactive disperse dye:

$$(D)-SO_2-CH_2CH_2OSO_3^- \ Na^+ \xrightarrow{\ OH^-\ } D-SO_2-CH=CH_2$$

water-soluble dye   vinyl sulphone
                                                     disperse dye

The vinyl sulphone residue has much higher reactivity than the reactive groups employed in the now withdrawn Procinyl (ICI) range of reactive dyes and further study in this area is recommended in the light of current fashion demands for brilliant shades, levelness and high wet-fastness.

The light-fastness of disperse dyeings on nylon is normally good.

3.6 Reactive dyes

Sulphonated reactive dyes

Reactive dyes contain fibre-reactive side-chains capable of forming covalent bonds with nucleophilic sites in appropriate fibres during the dyeing process; as a class they have been extremely successful on cotton and wool fibres, since they are capable of producing a very wide gamut of shades with excellent wet-fastness properties. However, on nylon fibres they have not been so successful, only giving sufficient covalent bonding in pale to moderate depths, in full depths the dye being present mainly as unfixed anionic dye which will exhibit the indifferent wet-fastness properties of acid dyes. In the light of renewed current demands for bright, full shades on nylon with good wet-fastness properties, the reasons for this unsatisfactory situation are worth further consideration.

Normally 1 kg of nylon 6,6 fibre contains 0.036 moles of primary amino groups and 0.090 moles of carboxylic acid groups situated at the ends of the polyamide chains. At ambient temperatures, the nylon fibre thus exists in the ionic states shown in scheme 6 according to the pH of the surrounding aqueous solution:

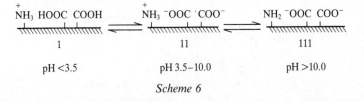

1	11	111
pH <3.5	pH 3.5–10.0	pH >10.0

Scheme 6

The situation pictured in scheme 6 is an approximation that assumes that the pK_a value of the carboxylic acid residues in nylon is about 3.5 and the pK_a value of the amine conjugate acid ($-CH_2-CH_2-CH_2-CH_2-CH_2-CH_2-{}^+NH_3$) is about 10. Since the amino residues are the only nucleophilic sites available to covalently anchor the reactive dye, it is apparent that the maximum depth of shade, the degree of dye fixation and hence the wet-fastness of the dyeing are very dependent on the available number of such residues in the fibre and the dyebath pH. Temperature also plays an important role; an increase in temperature from 25°C to 100°C lowers the pK_a values by about two units.[4] As nylon is usually dyed at or near the boil, it can be assumed that most of the nucleophilic primary amino sites will be available to react with the particular reactive dye only at dyebath pH values greater than neutral.

Reactive dyes currently available are all based on polysulphonated chromophores and are sold for the dyeing of wool and cotton. To achieve adequate substantivity of such anionic chromophores for the nylon fibre it is necessary to dye under weakly acidic conditions (pH 5.0–6.0). For the reasons discussed above, such conditions are inefficient in terms of achieving a high degree of dye–fibre covalent bonding, the nucleophilic amino residues being mainly in their nonreactive protonated form and as such tied up as zwitterions with anionic carboxylate groups.

Reactive disperse dyes

ICI developed a range of reactive disperse dyes, the Procinyls, specifically for nylon dyeing. These dyes were subsequently withdrawn from the market in the 1980s, but aspects of their chemistry and application to nylon are well worth recording. Scott and Vickerstaff[5] described the basic chemistry of this system.

Five dyes were marketed: Procinyl Yellow GS, Procinyl Scarlet GS, Procinyl Red GS, Procinyl Orange GS, Procinyl Blue RS. Structures were not widely published, but the patent literature indicates that a mixture of reactive groups was used, including chlorohydrin, chlorotriazine and chloroethylaminosulphone. A Russian paper on dyeing wool with Procinyl Blue RS confirmed its structure as [VIII].

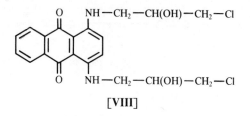

[VIII]

Using a nylon 6,6 fibre with an amino end-group content of 0.042 mol kg^{-1}, Scott and Vickerstaff showed that the above dye would fix to apparently higher levels than the maximum possible assuming monofunctional reaction at the primary amino residue. Dyeing was carried out initially at pH 4 at 95°C for 60 minutes followed by alkaline fixation at pH 10.5, again at 95°C for 60 minutes.

3.7 Nylons with modified dyeing properties

Chemically modified nylons find use predominantly in differential dyeing of yarn to produce multiple-colour or contrasting effects in piece dyeing of loop-pile tufted carpeting or woven and knitted cloths.

Since the dyeing behaviour of polyamide fibres is determined by the nature and number of end-groups present (–NH$_2$ and –COOH) and by the disposition of the composite polymer chains, fibres with dyeing properties that differ markedly from those of the standard products can be produced by modifying either or both of these characteristics.

Deep-dyeing nylon

Modification of the polymerization process to give a polymer with a higher amino end-group content and consequent increased substantivity towards anionic dyes can be achieved by the following methods.

1. Increasing the proportion of diamine in the original reaction mixture; however, this leads to difficulties in controlling the polymerization owing to loss of diamine from the system at an increasing rate with increasing excess of diamine.
2. Adding *p*-toluenesulphonic acid derivatives to the system, which protects the free amino end-groups during polymer and yarn manufacture, or by adding phosphonic acid or phosphonic acid derivatives, which prevents loss of diamine.

3. Producing a branched-chain polymer, for example, by copolymerization with trimesic acid [IX].

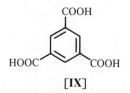

[IX]

4. Modifying the spinning process such that the disposition of the polymer chains favours penetration of dye molecules, e.g. by co-extruding a nylon 6,6 and and nylon 6 melt.

Basic-dyeable nylon

These fibres contain a significantly lower proportion of amino end-groups and have a reduced affinity for anionic dyes, but are readily dyed with cationic dyes. This can be achieved by the following means:

1. Addition to the polymerization system of, for example, acetylated caprolactam [X]

$$CH_3CON(CH_2)_5CO \longrightarrow \sim\!\sim NHCO(CH_2)_5NHCOCH_3$$

[X]

or butyrolactone [XI].

$$CH_2CH_2CH_2 \longrightarrow \sim\!\sim NHCO(CH_2)_3OH$$
$$\diagdown_{OCO}$$

[XI]

2. Introducing strongly acid (e.g. $-SO_3H$) groups into the fibre during the polymerization process. Such groups can be introduced either within a chain by copolymerization with, for example, 3,5-dicarboxy-benzene-1-sulphonic acid [XII]

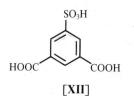

[XII]

or at the end of a chain by using, for example, 3,5-disulphobenzene-1-carboxylic acid [XIII].

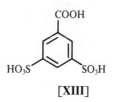

[**XIII**]

ICI Fibres Ltd marketed the following nylon fibres with modified dyeing properties:

Textured continuous filament yarns for carpets

Acid-dyeing	Deep Dyeing	(K 815)
	Ultra Deep Dyeing	(K 835)
Basic-dyeable		(K 825)

Continuous filament yarns for knitted and woven fabric

Acid-dyeing	Deep Dyeing	(Type 110)
Basic-dyeable		(Type 120)
Regular acid-dyeing		(Type 100)

The lack of attention paid to dyeing of unmodified nylon with cationic dyes is intriguing and deserves re-examination. On average, nylon contains twice as many carboxylate residues as amino residues and thus the fibre is overall negatively charged at pH values greater than about 5. This has been confirmed by Suzawa *et al.*,[6] who measured the ζ-potential of nylon 6 fibres of varying draw ratios at 25°C; the isoelectric points for fibres of draw ratio 1 (undrawn), 3 and 4 were found to be pH 5.2, pH 5.4 and pH 5.4, respectively.

In the 1940s and early 1950s when commercial methods for dyeing nylon were desperately being sought, the sort of cationic dyes available to the dyer were mainly those with delocalized cationic charges (e.g. Methylene Blue and Malachite Green). These dyes showed indifferent substantivity for nylon and, worse still, the dyeing produced exhibited very poor wash- and light-fastness properties.

Modern cationic dyes are often azo derivatives containing localized cationic charges (e.g. on pendent quaternized amino residues) and may be worthy of re-examination in terms of dyeing of nylon at pH 7; under these conditions the fibre amino residues are essentially deprotonated and the carboxylic acids, being fully ionized, are readily available as cationic dye sites.

The fully aromatic Nomex polyamides are difficult to dye and it is therefore of interest to note the proposal to dye these fibres from pyridine using cationic dyes.[7,8]

References

1. A N Derbyshire, R H Peters, *J. Soc. Dyers & Col.*, **71**, 530 (1988).
2. J Shore, *J. Soc. Dyers & Col.*, **87**, 3 (1971)
3. M Dohmyou, Y Shimizu, M Kimura, *J. Soc. Dyers & Col.*, **106**, 395 (1990).
4. U Altenhofen, H Baumann, H Zahn, *Proc. 5th Int. Wool Textile Res. Conf.*, Aachen, III, 1975, p. 529.
5. D F Scott, T Vickerstaff, *J. Soc. Dyers & Col.*, **60**, 104 (1960).
6. T Suzawa, T Saito, H Shinohara, *Bull. Chem. Soc. Japan*, **40**, 1596 (1967).
7. J Preston, W L Hofferbert, *Textile Res. J.*, **49**, 283 (1979).
8. K Silkstone, *Rev. Prog. Coloration*, **12**, 22 (1982).

CHAPTER 4

Dyeability 2: polyacrylonitrile and polyester fibres

S M BURKINSHAW

4.1 Polyacrylonitrile fibres: introduction

Copolymerization[1] of acrylonitrile reduces the structural regularity of the homopolymer, which improves both the processability and dyeability of the fibre. Neutral co-monomers reduce the compactness of the fibre and increase the permeability of the substrate to dyes, while the presence of polar co-monomers enhances nonionic interactions with dyes. Acidic co-monomers increase the number of anionic groups in the fibre and thus influence the rate and extent of uptake of cationic (basic) dyes whilst, in contrast, substantivity towards anionic dyes is imparted to the fibre by the use of basic co-monomers. Although both anionic- and cationic-dyeable polyacrylonitrile (PAN) fibres are available, the latter type predominates and, therefore, basic dyes are most widely employed in dyeing whilst anionic dyes enjoy very small usage; disperse dyes, which are applicable to both fibre types, are used to a much lower extent.

Acrylic fibres contain between 5 and 15 per cent of one or more co-monomers, whilst modacrylic fibres contain more than 15 per cent co-monomer. The latter fibres, the majority of which possess reduced flammability,[1,2] enjoy less textile usage than their acrylic counterparts; despite the similarity of anionic dye sites within acrylic and modacrylic fibres,[3] the latter fibres possess a greater propensity to unlevel dyeing with basic dyes[4] that accrues from a higher rate of dye uptake at low temperatures and a lower rate of dye diffusion within the denser structure of modacrylic polymers. Wet spinning, using aqueous solutions, is preferred to dry spinning for PAN fibres[3,5,6] since it provides superior dyeability arising from a more accessible microstructure.[7] Characteristically, the fibres possess desirable, all-round properties that vary according to the fibre's chemical composition and the spinning method employed. The properties of PAN fibres of relevance to dyeing are their high resistance to conditions typically encountered during dyeing, their hydrophobic nature, which contributes to the characteristic high light-fastness of cationic dyes on the substrates and, most importantly, their tendency to soften at temperatures above their T_g, which is responsible for the excellent wet-fastness properties of resultant dyeings as well as the marked temperature dependence of cationic dye adsorption.

4.2 Basic or cationic dyes

The chemistry of this class of dye has been discussed elsewhere;[8-11] the dyes can be divided into two chemical types, namely localized dyes, in which the positive charge is localized on one atom (commonly a nitrogen atom), e.g. CI Basic Orange 30:1 [I], and delocalized dyes, as typified by CI Basic Blue 41 [II], in which the positive charge is delocalized over the entire dye molecule.

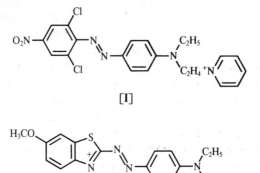

[I]

[II]

The majority of cationic dyes used for dyeing PAN fibres have been specifically developed for this purpose. Two types of basic dye are available, namely 'conventional' and 'migrating' cationic dyes, the latter type having been developed to overcome the generally low migrating power of 'conventional' cationic dyes on PAN fibres, which causes problems in achieving level dyeing. Typically, both types of cationic dye display brilliant hues of excellent all-round fastness properties on PAN fibres.

4.3 Dyeing with basic (cationic) dyes

Cationic dyes exhibit high substantivity towards basic-dyeable PAN fibres, this being primarily attributable to the large negative ζ-potential[12] present at the fibre surface in water. It is considered that in most cases dyeing occurs from an almost constant, saturated concentration of dye at the fibre surface.[13-15] The adsorption of cationic dyes on basic-dyeable PAN fibres follows a Langmuir mechanism,[16,17] denoting that the substrate possesses a saturation value (S) for cationic dye uptake, which is of great significance in dyeing. The Langmuir adsorption mechanism demonstrates that dye uptake occurs primarily by virtue of ion–ion interaction operating between the cationic dye and the anionic groups (e.g. SO_3^-, OSO_3^- and COO^-) in the fibre. However, interactions such as ion–dipole, dipole–dipole and dipole–induced dipole forces may also contribute to dye–fibre substantivity; these forces arise from the presence of the highly polar cyano groups in PAN fibres and, for example, polar substituents in cationic dyes.

The mechanism of cationic dye adsorption has been described in terms of ion exchange (scheme 1) in which the fibre exchanges cations (e.g. Na^+, H^+, K^+) associated with its component anionic groups with the dye cation.[14-20]

$$F \begin{cases} SO_3^- \\ OSO_3^- \\ COO^- \end{cases} \begin{matrix} H^+ \\ K^+ \\ Na^+ \end{matrix} + D^+ \leftrightarrow F \begin{cases} SO_3^- \\ OSO_3^- \\ COO^- \end{cases} D^+ + H^+, Na^+, K^+$$

D^+ is the cationic dye and F is the fibre

Scheme 1

In the case of exchange of the dye cation (D^+) for a proton (H^+) associated with the fibre, the process being subject to the condition that the saturation value of the fibre (S) is given by Eqn 4.1, an ion-exchange constant K_H^D (Eqn 4.2) can be defined that can be regarded as a measure of dye affinity.[21]

$$S = [D^+]_f + [H^+]_f \qquad [4.1]$$

where subscripts f and s represent fibre and solution, respectively.

$$K_H^D = \frac{[H^+]_s}{[D^+]_s} \frac{[D^+]_f}{S - [D^+]_f} \qquad [4.2]$$

At constant dyebath pH, rearrangement of Eqn 4.2 gives the Langmuir isotherm (Eqn 4.3).

$$\frac{1}{[D^+]_f} = \frac{1}{K'S} \frac{1}{[D^+]_s} + \frac{1}{S} \qquad [4.3]$$

where $K' = K_H^D/[H^+]_s$.

Using Eqn 4.3, a linear plot of $1/[D^+]_f$ versus $1/[D^+]_s$ at constant pH was obtained for several fibres; the values of S derived from the plots agreed well with those determined titrimetrically.[16]

4.3.1 Effect of pH on dyeing

It has been shown that equilibrium dye uptake increases with increasing pH of application,[16,17] this being attributed to increase of the dissociation of weakly acidic groups in the PAN fibres with increasing pH;[16] the effect of pH on dye uptake has also been explained in terms of the change in ζ-potential of the fibre.[28] Cationic dyes are commonly applied to PAN fibre in the pH range 3.5–6, this representing the range of highest stability for both the substrate and dye,[13,22,23] and preferably in the range pH 4.5–5.5. A buffer system (e.g. acetic acid/sodium acetate) is commonly used to counteract any undesirable rise in dyebath pH that may occur during dyeing.

4.3.2 Effect of electrolyte on dye adsorption

Electrolytes (e.g. sodium sulphate) reduce both the rate[16,23] and equilibrium uptake[16] of cationic dyes, the sodium ions competing with the dye cation for anionic sites in the fibre[16,23] and also displacing the ion-exchange equilibrium (scheme 1) to the left.[18] The retarding effect of electrolytes was shown to depend markedly on the nature of the electrolyte cation, anions having little effect on retardation of dye uptake;[24–26] it is considered that retardation in dye uptake will also result from the reduction in the ζ-potential of the fibre that occurs in the presence of electrolytes.[23,27] The retarding effect exerted on dye uptake is often employed in dyeing to promote level dyeing, sodium chloride or, preferably, sodium sulphate commonly being used.

The reduction in equilibrium dye uptake imparted by electrolytes is considered[21] to confirm the ion-exchange mechanism of dye adsorption depicted in scheme 1. To account for the retarding effect of sodium ions on dyeing equilibrium, two ion-exchange constants, K_H^D and K_H^{Na} (Eqns 4.4 and 4.5), have been defined.[17]

$$K_H^D = \frac{[H^+]_f [D^+]_s}{[D^+]_f [H^+]_s} \tag{4.4}$$

$$K_H^{Na} = \frac{[Na^+]_f [D^+]_s}{[D^+]_f [Na^+]_s} \tag{4.5}$$

In the presence of competing sodium ions, the ion-exchange process being subject to the condition that the saturation value of the fibre is given by Eqn 4.6, the Langmuir equations 4.7 and 4.8 are obtained.

$$S = [D^+]_f + [H^+]_f + [Na^+]_f \qquad [4.6]$$

$$S - [D^+]_f = \frac{[D^+]_f}{[D^+]_s} \frac{1}{K''} \qquad [4.7]$$

$$\frac{1}{[D^+]_f} = \frac{1}{K''S} \frac{1}{[D^+]_s} + \frac{1}{S} \qquad [4.8]$$

where $1/K'' = (K_H^D[H^+]_s + K_H^{Na}[Na^+]_s)$.

Rosenbaum[17] found that using Eqn 4.8 linear plots of $1/[D^+]_f$ versus $1/[D^+]_s$ at constant pH and sodium ion concentration were obtained, the slopes of which varied with pH and sodium ion concentration.

A Donnan approach has been applied by several authors to the equilibrium sorption of cationic dyes on PAN fibres[20,28–32], according to which, provided the fibre is electrically neutral, the concentration of all ions [X] on the substrate is related to their concentration in the internal solution phase $[X]_i$ by a distribution coefficient (K) and their concentration in the internal solution phase is related to their concentration in the external solution $[X]_s$ through a Donnan coefficient (λ). Equation 4.9 is applied to all ions present (e.g. D^+, H^+, Na^+, Cl^-) and the resulting set of distribution equations is combined to yield the exchange coefficient, K_H^D (Eqn 4.10), not necessarily assuming that an ion-exchange mechanism operates.

$$[X]_i = \lambda^z K[X]_s \qquad [4.9]$$

where $[X]_i$ and $[X]_s$ are the ion concentrations in the internal phase and external solution, respectively, z is the charge on the ion, K is the distribution coefficient of the ion and λ is the Donnan coefficient.

$$\frac{K_D}{K_H} = K_H^D = \frac{[D^+]_f [H^+]_s}{[D^+]_s [H^+]_f} \qquad [4.10]$$

A mechanism of dyeing of PAN fibres with basic dyes has been proposed based on this particular approach that considers all ions (Na^+, H^+, Cl^- and D^+) involved and two types of dye site (see section 4.3.3).[32–34]

4.3.3 Overdyeing

Although the Langmuir adsorption mechanism denotes that PAN fibres possess a saturation value for cationic dye uptake, this being determined by the number of acidic groups present in the substrate, several workers have observed 'overdyeing', or dye uptake greater than that predicted from the anionic group content of the fibre.[17,20,35] It has been proposed that this can be attributed to a small amount of the dye dissolving in the fibre by means of a simple partition mechanism[17,23,35] or by dye aggregation occurring in the fibre.[17,35] Harwood *et al.*[20] observed that an apparent saturation value (S_1) was obtained that was equal in magnitude for each of five dyes used but only approximately 80 per cent of the total acid group content of the PAN film employed. These workers also showed that, in some cases, dye uptake in excess of S_1 could occur, to an extent equal to the total acid group content of the substrate. Consequently, Harwood *et al.*[20] proposed that dye adsorption may be associated with two kinds of anionic site, namely the sulphonic acid end-groups, which corresponded to the quantity S_1, and the sulphate groups, which corresponded to the quantity S_2. It was suggested that dye uptake in excess of the apparent saturation value (S_1) could arise by virtue of a solution mechanism for dyes of low substantivity.

4.3.4 *Kinetics of dyeing*

Although the diffusional behaviour of cationic dyes in PAN fibres has been little studied owing to experimental difficulties,[18] dye diffusion is markedly non-Fickian; this is attributable to the high substantivity of the dyes for the limited number of anionic dye sites within the fibre. The sigmoidal concentration–distance profiles typically obtained[36] clearly demonstrate the pronounced concentration dependence of dye diffusion; values of diffusion coefficient increase with increase in dye concentration.

4.3.5 *Effect of temperature on dye adsorption*

As mentioned earlier, the T_g of PAN fibres plays a major role in dyeing, temperature having a dramatic effect on both the rate and, to a lesser degree, the extent of dye uptake. The T_g of PAN fibres will vary according to the composition of the copolymer; values between 56°C and 80°C have been reported by several workers[37–40] for various PAN fibres in water. Rosenbaum[14,15,41] first proposed a free volume model to describe the marked temperature dependence of dye diffusion within PAN fibres. At temperatures below T_g, little, if any, dye uptake occurs owing to the virtual absence of segmental mobility of the constituent polymer chains; at or about the T_g of the fibre, the rate of dye uptake increases markedly owing to the onset of segmental mobility. Further increase in temperature causes an exponential increase in free volume within the polymer which, in turn, results in an equally dramatic increase in both the accessibility of anionic dye sites and the rate of dye diffusion. The marked increase in dyeing rate above T_g occurs over a relatively small temperature range and saturation of the anionic dye sites is rapidly achieved. Temperature does not significantly affect the saturation value of PAN fibres;[13] although the number of accessible dye sites varies with temperature, the number remains reasonably constant at temperatures above T_g but decreases rapidly at temperatures below T_g.[37,42]

$$\log D_a = \log D_0 - E^*/RT \qquad [4.11]$$

Although according to Eqn 4.11 a linear relationship of slope E^*/R should be obtained for a plot of $\log D_a$ versus $1/T$, Rosenbaum[41] obtained a sigmoidal relationship for cationic dye diffusion into PAN fibre. Below T_g the plot of D_a against $1/T$ approached linearity, but at temperatures above T_g curvature was obtained; furthermore, at temperatures below the T_g of the substrate, the activation energy of diffusion (E^*) was almost constant, then increased markedly at T_g before decreasing exponentially with further increase in temperature. Thus T_g of the fibre presented the major barrier towards dye uptake. Using the WLF equation,[43] in the form expressed in Eqn 4.12, Rosenbaum[41] correlated the temperature dependence of both cationic dye diffusion and change in the physical properties of PAN fibres and, in doing so, demonstrated that dye diffusion within PAN fibres was governed by the segmental motion of the polymer chains.

$$\log \frac{D_T}{D_{T_g}} = -\log a_T = \frac{A(T - T_g)}{B + (T - T_g)} \qquad [4.12]$$

where A and B are constants, D_T and D_{T_g} are diffusion coefficients at ambient temperature T and at T_g, and $\log a_T$ is the shift factor.

However, several workers[44,45] have proposed that both free volume and porous matrix models operate in the diffusional process, the relative contribution of each model depending on the structure of the particular fibre employed.

4.3.6 Migrating cationic dyes

The characteristic rapid uptake and low migration power of conventional cationic dyes on PAN fibres, which can be attributed to the dyes' high substantivity arising mainly from their large molecular size and highly cationic and hydrophobic nature, causes problems in achieving level dyeing; to overcome this, recourse is made to the use of accurate pH and temperature control, compatible dyes and retarding agents. The smaller molecular size and greater hydrophilicity of migrating cationic dyes results in lower substantivity and greater migrating power on PAN fibres than with their conventional counterparts.[46] The dyes are applied to PAN fibres in a manner similar to that for conventional basic dyes, in the presence of electrolyte and a cationic retarding agent that is specifically intended for use with migrating cationic dyes and whose K-value is similar to that of the dyes; the mixture of migrating and conventional dyes is not normally recommended.

4.3.7 Carrier dyeing

Since Eqn 4.12 shows that the diffusion of basic dyes in PAN fibres is governed solely by the term $(T - T_g)$ and therefore that the rate of diffusion, and thus of dyeing, can be increased either by increasing T or reducing T_g, or both, several workers have investigated the reduction of T_g in order to expedite dyeing. This approach concerns both carrier dyeing and solvent-assisted dyeing, which are discussed in the section dealing with the dyeing of polyester fibres. Many workers have investigated the plasticizing action of a variety of compounds on the dyeability of PAN fibres with basic dyes, benzyl alcohol having been most widely used. Recent work[47–49] clearly demonstrated the plasticizing action of several compounds and also that the extent to which cationic dye uptake on Courtelle S was enhanced by several plasticizers was proportional to the amount of each compound adsorbed by the fibre, the maximum extent of uptake of each plasticizer being related to the association solubility parameter (δ_a) of the plasticizer.

4.3.8 Retarding agents

Owing to their high substantivity towards PAN fibres and their rapid rate of uptake over a small temperature range (10–15°C) above T_g, conventional cationic dyes exhibit very low migration on PAN fibres. Since temperature control and electrolyte addition are inadequate to expedite levelling, use is made of one of two types of retarding agent.

Anionic retarders,[5,6,13] which nowadays are little used, tend to be dye-specific[50] and can affect the compatibility of cationic dyes;[51] they interact with the cationic dye, forming a thermally unstable anionic dye–retarder complex of low substantivity for the substrate that does not diffuse into the fibre.[13] The anionic complex is loosely bound to the fibre surface at low dyeing temperatures, thus promoting dye migration, whereas, at high temperatures, the complex decomposes, releasing dye for adsorption onto the substrate. A surplus of retarder is usually required as well as the addition of a nonionic surfactant to the dyebath to prevent precipitation of the complex in the dyebath.

There are two types of cationic retarder: conventional and polymeric. Those of the former type are typically colourless, water-soluble quaternary ammonium compounds[5,6,13] of comparatively low relative molecular mass (M_r).[52] They operate by competing with the dye cations for anionic dye sites at both the surface and the interior of the fibre, thereby reducing the rate of dye uptake and promoting dye migration. The quantity of retarder required, which is calculated from information provided by the manufacturer, must not result in blocking of dye uptake[13] and is determined by the concentration of dye, the composition and physical form of the fibre as well as the dyeing machine employed.[52] The K-values of cationic retarders commonly lie in the range 2–3.5;[51] retarders of K-value slightly lower than those of the dyes being used provide optimum results.[50,51] Polymeric cationic retarders are typically quaternized polyamines of high M_r[52] that do not promote dye migration, since they are adsorbed only at the fibre surface. They reduce both the rate and extent of dye uptake by reducing the negative ζ-potential at the fibre surface and providing an electrical potential barrier at the surface of the substrate. It is proposed[13] that the compatibility of cationic dyes, especially low affinity ones, can be affected by certain polymeric retarders.

4.3.9 Dye–fibre characteristics

Since PAN fibres contain a limited number of anionic dye sites, the dyeing behaviour of individual dyes and fibres must be known in order to achieve a given hue and depth of shade when using mixtures of dyes. Such information is available in the form of three constants; methods for determining these have been devised by the Society of Dyers and Colourists (SDC).[53]

The fibre-specific *fibre saturation value* (A or D_f), has a value between 1 and 3 for most commercial fibres and represents the number of accessible anionic dye sites per unit mass of the fibre. The dye-specific *dye saturation factor* (f), which is related to the purity and M_r of the dye, represents the saturation characteristics of a given commercial dye, whilst the *saturation concentration* (C) is the quantity of commercial dye (expressed as per cent on mass of fibre (omf)) that will saturate a given fibre (i.e. $C = A/f$). The dye and fibre manufacturer provide values of these three constants.

The compatibility of cationic dyes when used in admixture has received much attention since incompatible dyes block the uptake of each other and also exhaust more rapidly than compatible dyes under the same application conditions. The compatibility of cationic dyes on PAN fibre is characterized by the compatibility or K-value (or C or CV value) of the dye, as first proposed by Beckmann,[54] which quantifies the adsorption of cationic dyes in admixtures, on a 1 to 5 scale, in terms of the diffusion coefficient and affinity of the dye. The K-values of cationic dyes are provided by the dye maker; in admixture, dyes of equal K-value exhaust virtually identically, dyes of low K-value exhausting before dyes of higher K-value. The standard dyeing time (e.g. T_{70}),[51] which relates to rate of exhaustion of individual dyes, also provides a means of quantifying the rates of exhaustion of dyes in mixtures. The migrating power of a cationic dye is described by the relative affinity, A_r, of the dye, this quantity being closely related to compatibility;[51] a high A_r value is given to a dye of low migrating power and denotes that although the dye's absorption behaviour is only slightly influenced by other cationic dyes, electrolytes or cationic retarders, its uptake is greatly influenced by anionic retarders.[13]

4.3.10 Gel dyeing

As previously mentioned, PAN fibres are commonly wet spun from aqueous solutions of either inorganic or organic solvents, residual solvent being removed from the extruded filaments by washing.[55] Gel- or producer-dyed PAN fibres are produced by passing the fibre in the gel state (i.e. after washing of the filaments), often in tow form, through an aqueous cationic dyebath at relatively low temperature, followed by rinsing, drawing and drying;[55] since the fibre is extremely absorbent in the gel state, dye diffusion within the water-filled substrate is very rapid.[6]

4.4 Disperse dyes

The chemistry of disperse dyes is discussed in the section dealing with the dyeing of polyester fibres. The dyes enjoy limited use on PAN fibres, primarily because of the excellent all-round fastness properties and brilliant hues furnished by cationic dyes on the fibres; furthermore, although disperse dyes exhibit excellent migration power on PAN fibres, they provide dullish shades, exhibit poor build-up and, in some cases, only moderate light-fastness on PAN fibres.

4.4.1 Dyeing with disperse dyes

Disperse dyes display low substantivity towards PAN fibres. This can be attributed to the hydrophilic nature of the substrate, in so far as, since it is likely that the anionic groups in the fibre will be ionized at the pH values used in dyeing with disperse dyes, the fibre will contain a reasonably high proportion of strongly hydrophilic groups. In addition, strong hydrophobic interaction between the disperse dye and the substrate cannot be envisaged, as the hydrophobicity of PAN fibres is provided mainly from aliphatic chain methylene groups.

The adsorption of disperse dyes on PAN fibres follows a Nernst (partition) mechanism[56,57] denoting that, in contrast to that of cationic dyes, disperse dye uptake occurs essentially by nonionic forces of interaction. It is suggested[35] that dipole–dipole interaction is the major contributor towards uptake of anthraquinone disperse dyes although dispersion and dipole–induced dipole forces could also contribute to dye–fibre substantivity; it can be considered that additional forces of interaction, such as hydrogen bonding, hydrophobic interaction as well as ion–dipole interaction, may also contribute towards dye–fibre substantivity.

4.5 Polyester fibres: introduction

This account concerns only poly(ethylene terephthalate) (PET) fibres. The hydrophobic fibres are predominantly dyed with disperse dyes. Owing to the compact and highly crystalline structure of PET fibres, recourse is made to the use of high temperatures (commonly 130°C) or carriers to achieve acceptable commercial dyeing rates. Copolymerization of terephthalic acid and ethylene glycol with a third co-monomer improves the dyeability of the fibre by reducing the structural regularity of the homopolymer and, in addition, enhances the pilling performance of the fibre; both the dyeing and pilling behaviours of the fibres are also affected by modifications employed in fibre production.[58] Nonionic co-monomers provide 'non-carrier-' on 'deep-' dyeing PET fibres,[59] whilst anionic co-monomers impart basic-dyeability to the fibres. The use of nitrogen-containing co-monomers confers substantivity towards anionic dyes.[60] Acid- and basic-dyeable PET fibres enjoy much less usage than their disperse-dyeable counterparts.

PET fibres possess excellent textile properties and generally high resistance to conditions encountered under typical dyeing and finishing processes; however, although the fibres undergo relatively little hydrolysis in the pH range 4.5–6,[59] even under high temperature dyeing conditions, they are slowly degraded by concentrated acids. Hot, concentrated alkalis hydrolyse the fibre, although this is limited mostly to surface saponification at temperatures up to the boil.[59]

4.6 Disperse dyes

Disperse dyes derive their name from the fact that they are of low water solubility and are applied to PET and other hydrophobic fibres (e.g. polyacrylonitrile, see PAN section) and also polyamide (see Chapter 3) from fine aqueous dispersion. The chemistry of disperse dyes has been discussed by many authors.[10,61–65] Owing to their low cohesive energy in the solid state, the dyes are volatile and can thus be applied in the vapour phase. The dyes were originally devised for the coloration of the secondary cellulose acetate fibres; with the subsequent introduction of the more hydrophobic cellulose triacetate and, more importantly, polyester fibres, disperse dyes of higher sublimation fastness were developed,[10,22,61–66] as typified by CI Disperse Blue 79 [III] and CI Disperse Red 60 [IV].

No universal system of classifying the dyeing behaviour of disperse

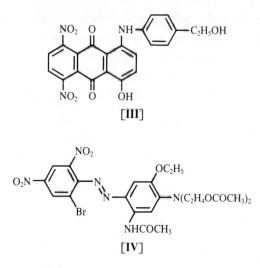

[III]

[IV]

dyes has been adopted by the dye makers;[67] some dye manufacturers classify their dyes for use on PET fibres in accordance with, for example, the dyes' sublimation character or the ability of the dyes to cover variations in texturized PET material. The SDC has proposed a series of tests to characterize the dyeing behaviour of disperse dyes on PET[68] under practical exhaust dyeing conditions. Generally, the dyes exhibit good build-up on polyester fibres and yield a wide variety of hues of very good fastness to wet treatments and light, although some dye combinations can exhibit anomalous light fading.

Some disperse dyes degrade during aqueous dyeing, giving rise to off-shade and pale dyeings, although in some cases such pH-induced changes are reversible. It has been concluded[69] that the colour yield obtained with disperse dyes decreases with increasing pH; in some cases, application at pH 8–10 destroys the dye. Dulling of shade of anthraquinone disperse dyes can sometimes occur owing to the presence of metals in the dyebath.[70] Commonly, aqueous dyeing is carried out in the pH range 4–5.5, the dyebath pH being adjusted using acetic acid or a suitable buffer system.

In aqueous dyeing, the sparingly water-soluble disperse dyes must be presented to the substrate as a fine, stable dispersion in order to avoid unlevel and weak dyeings. In order to achieve this, the synthesized disperse dye is usually milled in the presence of a dispersing agent to yield the required particle size and particle size distribution. The dispersing agent, commonly an anionic compound of relatively high M_r,[52,71,72] expedites milling, enables the dye to be prepared in powder or liquid forms, and also facilitates dispersion of the dye in water; the dispersing agent also maintains a stable dispersion during dyeing.

4.6.1 Barré effects

Barré effects in PET fibres arise mainly from variations in crystallinity introduced during extrusion and, more prevalently, during texturizing and heat-setting prior to dyeing. Increased crystallinity results in a reduction in both the rate and saturation value of dye uptake. In the case of exhaust dyeing with disperse dyes, good barré coverage can be achieved using temperatures up to 135°C and prolonged dyeing times, and also using suitable carriers at lower temperatures (commonly 110–125°C), although such conditions can degrade the handle, bulk and elasticity of the fibres. Generally, thermofixation using temperatures of about 210°C gives better coverage of barré effects than exhaust dyeing methods.

4.6.2 Oligomers

PET fibres contain, typically, between 1.5 and 3.5 per cent[73] of low M_r esters, the principal oligomer being cyclic tris(ethylene terephthalate). Although little migration of these esters occurs at temperatures below 110°C, the extent of migration increases with increase in temperature.[73] Although the crystalline oligomers are not dyed with disperse dyes, their deposition on the fibre surface can impair the spinning behaviour of yarns and the presence of the crystalline cyclic trimer in the dyebath may induce crystal growth of the dye or dye agglomeration under certain conditions. Deposition of the cyclic trimer on the dyeing vessel or fibre surface may occur during cooling of high-temperature (HT) dyebaths, such deposition being reduced by avoiding cooling of the dyebath prior to discharge, by lowering the dyeing temperature to 120°C, and by using the shortest possible dyeing time. Oligomer deposits on the fibre surface are removed by reduction clearing, more severe treatments often being necessary to remove the cyclic trimer from dyeing vessels.[59]

4.6.3 Dyeing with disperse dyes

As previously mentioned, disperse dyes are applied to hydrophobic fibres other than polyester; the mechanism of dye adsorption is generally considered to be identical for all such hydrophobic fibres. Soon after their introduction for use on cellulose diacetate fibres, Clavel[74] proposed that disperse dyes were adsorbed from a saturated aqueous solution of the dye and, as dyeing proceeded, this aqueous solution was replenished by dissolution of dye particles present in bulk dispersion. Burns and Wood[75] supported Clavel's proposal and also suggested that dye adsorption occurred at sites within the substrate; in contrast, however, Kartaschoff[76] suggested that the solid dye particles formed a surface layer on the secondary cellulose acetate fibre and the solid dye then diffused within the substrate by means of a solid-state or solid-solution mechanism. It is now generally accepted that the aqueous dyeing of hydrophobic fibres with disperse dyes occurs from true aqueous solution; the dyebath comprises a small amount of dye in aqueous solution with, in the early stages of dyeing, the majority of dye in the form of particles in bulk dispersion. Monomolecular dye adsorption occurs at the periphery of the fibre from the aqueous dye solution; as the adsorbed dye molecules diffuse from the periphery to the interior of the substrate, the depleted aqueous dye solution is replenished by dissolution of monomolecular dye particles from the bulk dispersion, thereby maintaining a saturated aqueous solution of dye from which adsorption onto the fibre surface can occur. This process continues until either the fibre is saturated with dye or the dyebath is exhausted.

The adsorption of disperse dyes onto PET fibres follows a Nernst or partition mechanism obeying Eqn 4.13, the isotherms being linear up to saturation of the fibre.[77] Linear partition isotherms have been obtained for the adsorption of disperse dyes on other hydrophobic fibres.[22]

$$K = D_f/D_s = S_f/S_s \qquad [4.13]$$

where K is the partition coefficient and D and S represent the concentration of dye and the saturation value of the dye, respectively, in the fibre (subscript f) and solution (subscript s).

The linearity of adsorption isotherms for disperse dyes led many authors to conclude that the disperse dye interacts with PET and other hydrophobic fibres by means of the solid-solution mechanism as proposed by Kartaschoff. However, other workers argue that disperse dye–PET interaction cannot be described as solid-solution and that dye adsorption onto hydrophobic fibres occurs via adsorption-at-sites as suggested by Burns and Wood. Clearly, the precise mechanism of interaction between the dye and the hydrophobic fibre is unclear, no direct experimental evidence having been obtained. However, as Vickerstaff[78] discussed, there is little, if any, difference between these two mechanisms, in so far as the sites with which the dye interacts according to the adsorption-at-sites mechanism can be considered as being identical to those with which the dye interacts if it dissolved in the substrate according to the solid-solution mechanism.

It has been proposed that the disperse dye is monomolecularly dispersed in both the PET fibre and dyebath phases,[77,79] although other workers suggest that some degree of aggregation of the dye in both the fibre and dyebath may occur.[80] Debate still attends this matter, owing to the experimental difficulties involved in such examinations.

Several workers[77,81] suggest that disperse dyes interact with PET fibres by means of hydrogen bonding, although other authors[80,82] propose that the adsorption of disperse dyes on all hydrophobic fibres occurs by virtue of the combination of strong hydrogen bonding and weak dispersion forces. It has been proposed that dipole–dipole and dipole–induced dipole forces of interaction may also contribute towards disperse dye–hydrophobic fibre interaction.[83,84] The solubility parameter concept,[85,86] which was originally developed to predict the ease of mixing of nonpolar liquids, has been extended to the mixing of fibres (semicrystalline polymers) with dyes and also carriers. In the case of interaction between PET fibre and typical disperse dyes, three contributions to the cohesive energy density can be expected for each component, namely dispersion forces, hydrogen bonding and polar forces; the total solubility parameter (δ_t) can thus be partitioned into contributions from the various forces involved as given by Eqn 4.14.

$$\delta_t^2 = \delta_d^2 + \delta_p^2 + \delta_h^2 \qquad [4.14]$$

where δ_d, δ_p and δ_h are the contributions from dispersion forces, polar forces and hydrogen bonding, respectively.

Equation 4.14 more simply takes the form of Eqn 4.15:

$$\delta_t^2 = \delta_d^2 + \delta_a^2 \qquad [4.15]$$

where δ_a, the association solubility parameter, represents all polar forces that contribute to cohesive energy density.

With regard to the substantivity of disperse dyes towards PET fibres, Gerber[87] found that, from a total of 60 monoazo disperse dyes, only a few whose δ_t solubility parameter was within 1 (cal cm^{-3})$^{0.5}$ of that of polyester showed good substantivity for the substrate. Ingamells and Thomas[88] found that the equilibrium uptake of 12 related monoazo disperse dyes on PET at 100°C increased with increasing δ_d value of the dye; maximum dye uptake occurred for the dye whose association solubility parameter was closest to that of the fibre. It was also demonstrated[88] that the relative magnitude of the dispersion (δ_d) and polar (δ_a) cohesive forces involved in dye–fibre interaction, as given by δ_d^2/δ_a^2, influenced dye affinity for PET, since maximum equilibrium dye uptake occurred when δ_d^2/δ_a^2 of the dye was closest to the value of δ_d^2/δ_a^2 for PET. Hence, the solubility parameter concept has provided some information concerning the nature of the interaction between the dyes and PET.

4.6.4 Effect of temperature on dyeing

For both PET[77] and diacetate fibres,[89,90] it has been demonstrated that an increase in application temperature decreases the partition coefficient,

K, in Eqn 4.13, but increases the saturation value (S_F) of disperse dye uptake. The reduction in K that accompanies an increase in temperature can be attributed to the exothermic nature of and the negative entropy change involved in disperse dye adsorption together with an increase in solubility of the dye in the aqueous phase.[21,22] The increase in saturation value of dye uptake that attends an increase in application temperature can also be attributed to an increase in aqueous solubility of the dye.

4.6.5 *Effect of particle size on dyeing*

As previously mentioned, disperse dyes are commonly milled in order to achieve the desired particle size and particle size distribution. Owing to the range of particle sizes present in a typical dye dispersion and the higher solubility of the smaller particles, the dye solution may become super-saturated with respect to the large particles, leading to crystal growth on the surface of the larger particles, which in turn will reduce the mean solubility of the dispersion and thus reduce dye uptake. Such crystal growth, which can occur during cooling of the dyebath during HT dyeing and shading operations, can be promoted by nonionic levelling agents; anionic dispersants stabilize a disperse dye dispersion against crystal growth.[91]

4.6.6 *Effect of dispersing agent on dyeing*

Dispersion failure during exhaust dyeing results in dye agglomeration and possibly dye precipitation, giving rise to unlevel dyeing; dispersion failure is most commonly encountered during HT dyeing.[72] Although additional anionic dispersing agent can be added by the dyer to maintain dispersion stability under such conditions,[72] certain dispersing agents (e.g. lignin-sulphonates) promote reduction of azo disperse dyes during HT dyeing[92] and partial reduction of anthraquinone dyes.[93] Dispersing agents increase the aqueous solubility of disperse dyes, thereby affecting both the rate and extent of uptake of disperse dyes on hydrophobic fibres.

4.6.7 *Effect of levelling agents on dyeing*

Although dispersing agents enhance the migration and levelling of disperse dyes on polyester fibres during HT dyeing, the addition of nonionic or anionic or an anionic/nonionic blend surfactant enhances levelling during exhaust dyeing. Nonionic levelling agents increase the solubility of the dye and reduce the rate of dyeing[72] but can restrain dye uptake, leading to a reduction in colour yield.[72] Anionic surfactants increase the cloud point of the nonionic compound to avoid dispersion failure and concomitant unlevel dyeing. Anionic levelling agents also increase dye solubility but exert little restraint on dye uptake.[72]

4.6.8 Effect of crystal form of dye on dyeing

Different crystal forms of a chemically identical disperse dye were found to exhibit different saturation values on cellulose diacetate[94] and also PET,[95] this being considered[94,96] to be due to a difference in the solubility of the different crystallographic forms of the dye in the water and fibre. Modification of a particular crystal form of a disperse dye can occur during dyeing.[94] Such transformations, which can be influenced by surfactants,[94,96] can result in dyeings of reduced colour yield[95] as well as low rub fastness.[94]

4.6.9 Kinetics of dyeing

Several workers have demonstrated the concentration independence of dye diffusion within PET fibres at 100°C using infinite dyebaths.[22,84,97–99] Authors also have demonstrated that the diffusion coefficient of disperse dyes in PET is independent of dye concentration at temperatures up to 210°C.[98,99] The fact that similar results have been obtained by other workers for disperse dyes on several hydrophobic fibres has led to the general conclusion[66] that the diffusion of disperse dyes within all hydrophobic fibres is characterized by a constant diffusion coefficient.

4.6.10 Isomorphism

Evidence has been found of isomorphism during admixture dyeing with disperse dyes in so far as, with dyes of similar structure, less dye was adsorbed onto PET in admixture than when the dyes were applied separately.[77] Similar findings were made on cellulose diacetate fibres.[100] This has been attributed to such isomorphous pairs of dyes interacting in the dyebath with the effect that the aqueous solubility of the dyes is reduced.[100]

4.6.11 Carrier dyeing

The term 'carrier' was first ascribed by Waters[101] to compounds that, when added to a disperse dyebath, accelerated the rate of dyeing. Although carriers are nowadays mostly employed in the dyeing of PET and triacetate fibres, their use is declining owing to environmental concerns. Carriers are, typically, aromatic compounds of small molecular size, such as *o*-phenylphenol, diphenyl or chlorinated benzenes, which, when added to the dyebath in solution or emulsion form, enable commercially acceptable dyeing rates to be achieved at 98°C rather than at high temperature (commonly 130°C). Carriers are adsorbed onto PET fibre by an identical mechanism to that of disperse dyes,[102–104] linear (Nernst) isotherms having been obtained for the adsorption of various carriers on PET fibres.[103,106] It is considered that the substantivity of carriers for PET can be primarily attributed to the operation of nonpolar forces[104,107] although polar forces may also contribute to adsorption. The solubility parameter has also been applied to the adsorption of carriers; although little[108] or no[109] correlation was observed between the δ_t values of several carriers and PET in terms of carrier action[108] and solubility of carriers in PET,[109] and also little[109] or no[107] correlation between the δ_a values of several carriers and PET in terms of PET fibre plasticization[107] and solubility of carriers in PET,[109] the finding[107] that the extent of PET fibre plasticization was related to coincidence of the δ_d values of the carrier and substrate demonstrated that nonpolar forces predominate in the interaction between carriers and PET.

Many theories have been proposed to explain carrier action.[105,110] Of the various mechanisms proposed, only one, namely fibre plasticization, is now generally accepted. Salvin[105,111] established that in order for a carrier to function it must be adsorbed by the fibre. Several workers[102,112] proposed that carriers, once adsorbed onto the fibre, reduce the inter-molecular forces operating within the substrate, thereby increasing the segmental mobility of the macromolecular chains, and therefore that carrier action entails disruption of the fibre structure; the plasticization theory of carrier action has received considerable support from the results of many workers using various hydrophobic fibres.[110]

4.6.12 High temperature dyeing

Owing to the highly compact and crystalline structure of PET fibres, the use of high temperatures (in the region of 125–135°C) enables commercially acceptable dyeing rates to be achieved. Characteristically, HT dyeing enables excellent barré coverage and high colour yield to be achieved in short dyeing times, and therefore offers advantages over carrier dyeing at 98°C. However, carriers can also be used to advantage in HT dyeing to provide superior coverage of barré in textured PET fabrics as well as lower temperatures (typically 110–125°C) so as to reduce impairment in handle and bulk of texturized PET and liberation of oligomers. Carriers that promote dye migration but impart low shrinkage to PET under HT dyeing conditions are commonly employed.[59,72]

'Rapid dyeing' HT methods, which are primarily intended to shorten the HT dyeing process but at the same time maximize dye migration, rely upon optimizing the rate of dyeing. In such methods, careful temperature control enables uniform dye uptake to be achieved during the heating-up phase, which thereby reduces the time required at maximum temperature (e.g. 130°C). In view of the characteristic poor dispersion stability of disperse dyes under HT dyeing conditions, the use of dispersant-free dyes dissolved in the presence of surfactants has been proposed,[113] the benefits of this system being[114] possibly shorter dyeing cycles, high dispersion stability and on-tone build-up over a range of shades, although high liquor circulation would be required to ensure level dyeing.

4.6.13 Thermofixation

In this continuous pad–bake dyeing process, a high rate of dyeing is achieved using temperatures in the region of 210°C.[73] The method offers benefits such as the fact that dyeing and heat-setting can be combined, pressurized equipment and carriers are not required, and a high throughput of material and high dye fixation can be achieved. After impregnation with dye dispersion, the PET fibre is dried and then baked for between 5 and 90 seconds in the region of 175–225°C. On 100 per cent PET, thermofixation is restricted mostly[115] to the dyeing of narrow fabrics with disperse dyes and the dyeing of texturized PET fabrics. Migration inhibitors are required to reduce dye migration during drying[59,115] and temperatures between 160 and 180°C may be used to reduce stiffness and loss of bulk of the goods.[59]

4.6.14 Solvent-assisted dyeing

This technique, in which organic solvents such as *n*-butanol,[116] 2-phenoxyethanol and *o*-dichlorobenzene[117] are added to the disperse dyebath to accelerate the rate of dyeing of PET fibre, can be considered[73] to be a variant of carrier dyeing.

4.6.15 Solvent dyeing

There has been much interest in the use of organic solvents as alternatives to water in the dyeing of PET with disperse dyes, chlorinated hydrocarbons and predominantly perchlorethylene (PER) having received most attention. Several advantages are considered to be associated with the use of PER rather than water: the fibre swells to a lower extent in PER[118] so that mechanical deformation is lower and heating costs are lower for PER dyeing.[118] PET fibres are rapidly wetted by PER,[118] removal of the solvent being effected by treatment with steam or hot air.[119] The adsorption of disperse dyes onto PET follows a Nernst mechanism,[118] as observed for aqueous dyeing. However, evidence suggests that the solvent from which dyeing occurs may be involved in dye–fibre interaction.[120] Owing to the greater solubility of disperse dyes in PER than in water, low partition coefficients are obtained using PER,[118] resulting in low colour yields.

4.6.16 Afterclearing

To remove surface dye and auxiliaries, pale and medium depth dyeings are scoured (e.g. detergent $1 \, g \, l^{-1}$; 40–50°C; 30 min).[61] Medium and heavy depths are often reduction cleared (e.g. detergent $2 \, g \, l^{-1}$; $Na_2S_2O_4$ $2 \, g \, l^{-1}$; ammonia (880) $2 \, cm^3 \, l^{-1}$; 50–60°C; 30 min),[61] followed by rinsing and, if necessary, neutralization with, for instance, dilute acetic acid. Reduction clearing is carried out on certain forms of PET materials, regardless of depth of shade, to remove oligomers.[73] Residues of anthraquinone dyes that may remain after reduction clearing can be removed by an oxidative treatment.[73]

4.7 Vat dyes

Vat dyes are nowadays mostly used in the dyeing of PET–cellulosic fibre blends.

4.8 Azoic colorants

The use of azoic combinations is mostly restricted to the production of black shades owing to the generally high fastness and wide range of hues secured on PET with disperse dyes, the 'concurrent' and 'reversed' azoic dyeing processes being used,[61] mostly for diazotized and developed black disperse dyes.[73,115]

References

1. D J Poynton, *Textile Progr.*, **8**(1), 51 (1976).
2. A R Horrocks, M Tunc, D Price, *Textile Prog.*, **18**(1,2,3), 151 (1989).
3. K Silkstone, *Rev. Prog. Col.*, **12**, 22 (1982).
4. A M Jowett, *Int. Dyer*, **154**, 437 (1975).
5. I Holme, *Chimia*, **34**, 110 (1980).
6. I Holme, *Rev. Prog. Col.*, **13**, 10 (1983).
7. J P Craig, J P Knudsen, V F Holland, *Textile Res. J.*, **32**, 465 (1962).
8. R Raue, *Rev. Prog. Col. Rel. Topics*, **14**, 187 (1984).
9. U Mayer, E. Seipmann, *Rev. Prog. Col. Rel. Topics*, **5**, 65 (1974).
10. P Gregory in *The Chemistry and Application of Dyes*, ed. D Waring, G Hallas, Plenum, New York, 1990.

11. D R Baer in *The Chemistry of Synthetic Dyes*, **IV**, ed. K Venkataraman, Academic Press, New York, 1971.
12. G Weston, unpublished results, see T Vickerstaff, *Hexagon Digest*, **9**, 3 (1954).
13. W Beckmann in *The Dyeing of Synthetic-Polymer and Acetate Fibres*, ed. D M Nunn, Dyers Co. Pub. Trust, Bradford, 1979.
14. S Rosenbaum, *J. Appl. Polymer Sci.*, **7**, 1225 (1963).
15. S Rosenbaum, *Textile Res. J.*, **34**, 52, 291 (1964).
16. D Balmforth, C A Bowers, T H Guion, *J. Soc. Dyers & Col.*, **80**, 577 (1964).
17. S Rosenbaum, *Textile Res. J.*, **33**, 899 (1963).
18. J Cegarra, *J. Soc. Dyers & Col.*, **87**, 149 (1971).
19. W R Remington, H E Schroeder, *Textile Res. J.*, **27**, 177 (1957).
20. R J Harwood, R McGregor, R H Peters, *J. Soc. Dyers & Col.*, **88**, 216 (1972).
21. H H Sumner in *The Theory of Coloration of Textiles*, 2nd edn, ed. A Johnson, Society of Dyers and Colourists, Bradford, 1989.
22. R H Peters, *Textile Chemistry* 3, Elsevier, Amsterdam, 1975.
23. W Beckmann, *J. Soc. Dyers & Col.*, **77**, 616 (1961).
24. M Bouche, *Teintex*, **33**, 519, 585 (1968).
25. M L Longo, D Sciotto, M Torre, *Textile Res. J.*, **52**, 233 (1982).
26. G Alberti, A Cerniani, M R De Giorgi, *Ann. Chim.*, **73**, 283 (1983).
27. S Rosenbaum, *Textile Res. J.*, **34**, 159 (1964).
28. T H Guion, R McGregor, *Textile Res. J.*, **44**, 439 (1974).
29. R McGregor, *Textile Res. J.*, **42**, 172 (1972).
30. R McGregor, *Textile Res. J.*, **42**, 536 (1972).
31. R McGregor, T H Guion, *Textile Res. J.*, **44**, 433 (1974).
32. G Alberghina, S-L Chen, S Fischella, T Iijima, R McGregor, R M Rohner, H Zollinger, *Textile Res. J.*, **58**, 345 (1988).
33. G Alberghina, M E Amato, S Fisichella, *J. Soc. Dyers & Col.*, **104**, 279 (1988).
34. G Alberghina, M E Amato, S Fisichella, *J. Soc. Dyers & Col.*, **105**, 163 (1989).
35. F Feichtmayer, A Wurz, *J. Soc. Dyers & Col.*, **77**, 626 (1961).
36. R J Harwood, R McGregor, R H Peters, *J. Soc. Dyers & Col.*, **88**, 288 (1972).
37. R Asquith, H S Blair, N Spence, *J. Soc. Dyers & Col.*, **94**, 49 (1978).
38. G M Bryant, A T Walker, *Textile Res. J.*, **29**, 211 (1959).
39. Z Gur-Arieh, W C Ingamells, *J. Soc. Dyers & Col.*, **90**, 8 (1974).
40. D Aitken, S M Burkinshaw, J Catherall, R Cox, R E Litchfield, D Price, N G Todd, *J. Appl. Polymer Sci., Appl. Polymer Symp.*, **47**, 263 (1991).
41. S Rosenbaum, *J. Polymer Sci., Part A*, **3**, 1949 (1965).
42. R Asquith, H S Blair, N Spence, *Textile Res. J.*, **47**, 446 (1977).
43. M L Williams, R F Landel, J D Ferry, *J. Am. Chem. Soc.*, **77**, 3701 (1955).
44. R M Rohner, H Zollinger, *Textile Res. J.*, **56**, 1 (1986).
45. T Hori, H S Zhang, T Shimuzi, H Zollinger, *Textile Res. J.*, **58**, 227 (1988).
46. W Biedermann, *Rev. Prog. Col.*, **10**, 1 (1979).
47. D Aitken, S M Burkinshaw, J Catherall, R Cox, D Price, *J. Appl. Polymer Sci., Appl. Polymer Symp.*, **47**, 270 (1991).
48. D Aitken, S M Burkinshaw, D Price, *Dyes and Pigments*, **18**, 23 (1992).
49. D Aitken, S M Burkinshaw, *J. Soc. Dyers & Col.*, **108**, 219 (1992).
50. J Park, *A Practical Introduction to Yarn Dyeing*, Society of Dyers and Colourists, Bradford, 1981.
51. F Hoffman, *Rev. Prog. Col.*, **18**, 56 (1988).

52. T M Baldwinson, in *Colorants and Auxiliaries*, Vol 2, ed. J. Shore, Society of Dyers and Colourists, Bradford, 1990.
53. D G Evans, *J. Soc. Dyers & Col.*, **89**, 292 (1973).
54. W Beckmann, *11th Int. Text. Seminar*, Kingston, Ontario (1968).
55. P Ackroyd, *Rev. Prog. Col.*, **5**, 86 (1974).
56. H R Hadfield, W M Sokol, *J. Soc. Dyers & Col.*, **74**, 629 (1958).
57. R C D Kaushik, S D Deshpande, *Ind. J. Textile Res.*, **14**, 119 (1989).
58. J S Ward, *Rev. Prog. Col.*, **14**, 98 (1984).
59. *Dyeing and Finishing of Polyester Fibres*, B 363e, BASF, Germany (1975).
60. K W Hillier in *Man-made Fibres Science and Technology*, ed. H F Mark, S M Atlas, E Cernia, **3**, 1, Interscience, New York (1968).
61. S M Burkinshaw in *The Chemistry and Application of Dyes*, ed. D R Waring, G Hallas, Plenum, New York, 1990, pp. 237–379.
62. L A Shuttleworth, M A Weaver in *The Chemistry and Application of Dyes*, ed. D R Waring, G Hallas, Plenum, New York, 1990, pp. 107–163.
63. J F Dawson, *Rev. Prog. Col.*, **3**, 18 (1972); **9**, 25 (1978); **14**, 90 (1984).
64. O Annen, R Egli, R Hasler, B Henzi, H Jacob, P Matzinger, *Rev. Prog. Col.*, **17**, 72 (1987).
65. C V Stead in *The Chemistry of Synthetic Dyes*, **III**, ed. K. Venkataraman, Academic Press, New York, 1970, pp. 385–462.
66. S R Iiveraja Iyer in *The Chemistry of Synthetic Dyes*, **VII**, ed. K. Venkataraman, Academic Press, New York, 1974.
67. D Blackburn in *The Dyeing of Synthetic-polymer and Acetate Fibres*, ed. D M Nunn, Dyers Co. Pub. Trust, Bradford, 1979.
68. A N Derbyshire, *J. Soc. Dyers & Col.*, **93**, 228 (1977).
69. C H A Schmitt, *Am. Dyest. Rep.*, **51**, 974 (1962).
70. J F Leuck, *Am. Dyest. Rep.*, **68**(11), 49 (1979).
71. S Heimann, *Rev. Prog. Col.*, **11**, 1 (1981).
72. A N Derbyshire, W P Mills, J Shore, *J. Soc. Dyers & Col.*, **88**, 389 (1972).
73. R Broadhurst, in *The Dyeing of Synthetic-polymer and Acetate Fibres*, ed. D M Nunn, Dyers Co. Pub. Trust, Bradford (1979).
74. R Clavel, *Rev. Gen. Matiere Col.*, **28**, 145, 167 (1923); **29**, 95 (1924).
75. H H Burns, J K Wood, *J. Soc. Dyers & Col.*, **45**, 12 (1929).
76. V Kartaschoff, *Helv. Chim. Acta*, **8**, 928 (1925).
77. M J Schuler, W R Remington, *The Faraday Society Discussion on Dyeing and Tanning*, 201 (1954).
78. T Vickerstaff, *The Physical Chemistry of Dyeing*, Oliver and Boyd, London, 1954.
79. K Hoffmann, W McDowell, R Weingarten, *J. Soc. Dyers & Col.*, **84**, 306 (1968).
80. C H Giles in *The Theory of Coloration of Textiles*, 2nd edn, ed. A Johnson, Dyers Co. Educational Trust, Bradford, 1989.
81. M M Allingham, C H Giles, E L L Neustadter, *Disc. Faraday Soc.*, **16**, 248 (1954).
82. J Shore in *Colorants and Auxiliaries*, ed. J. Shore, Society of Dyers and Colourists, Bradford, 1989, Ch. 3.
83. T G Majury, *J. Soc. Dyers & Col.*, **70**, 442, 445 (1954).
84. O Glenz, W Beckmann, W Wunder, *J. Soc. Dyers & Col.*, **75**, 141 (1959).
85. J H Hildebrand, R L Scott, *The Solubility of Nonelectrolytes*, 3rd edn, Reinhold, New York, 1950.

86. G Scatchard, *Chem. Rev.*, **8**, 321 (1931).
87. H Gerber, *J. Soc. Dyers & Col.*, **94**, 298 (1978).
88. W C Ingamells, W C Thomas, *Textile Chem. Col.*, **16**(3), 55 (1984).
89. C L Bird, H K Partovi, G Tabbron, *J. Soc. Dyers & Col.*, **75**, 600 (1959).
90. H J White, *Textile Res. J.*, **30**, 329 (1960).
91. H Leube, *Textile Chem. Col.*, **11**, 106 (1979).
92. A Murray, K Mortimer, *J. Soc. Dyers & Col.*, **87**, 173 (1971).
93. G Prazak, P Dilling, *AATCC Nat. Tech. Conf. Books of Papers*, 283 (1979).
94. W Biedermann, *J. Soc. Dyers & Col.*, **87**, 105 (1971).
95. J A Shenai, M C Sadhu, *J. Appl. Polymer Sci.*, **20**, 3141 (1976).
96. H Braun, *Rev. Prog. Col.*, **13**, 62 (1983).
97. J Cegarra, P Puenta, *Textile Res. J.*, **37**, 343 (1967).
98. K V Datye, R Rajendran, *Indian J. Technol.*, **4**, 101 (1966).
99. K V Datye, S C Pitkar, R Rajendran, *Indian J Technol.*, **4**(7), 202 (1966).
100. C L Bird, P Rhyner, *J. Soc. Dyers & Col.*, **77**, 12 (1961).
101. E Waters, *J. Soc. Dyers & Col.*, **66**, 609 (1950).
102. F Fortess, V S Salvin, *Textile Res. J.*, **28**, 1009 (1958).
103. D Balmforth, C A Bowers, J W Bullington, T H Guion, T S Roberts, *J. Soc. Dyers & Col.*, **82**, 405 (1966).
104. D M Koenhen, C A Smolders, *J. Appl. Polymer Sci.*, **19**, 163 (1975).
105. V S Salvin *et al.*, *AATCC Piedmont Section, Am. Dyest. Rep.*, **48**(22), 23 (1959).
106. C R Jin, D M Cates, *Am. Dyest. Rep.*, **53**, 64 (1964).
107. W C Ingamells, A Yabani, *J. Soc. Dyers & Col.*, **93**, 417 (1977).
108. A H Brown, A T Peters, *Am. Dyest. Rep.*, **57**, 284 (1968).
109. E C Ibe, *J. Appl. Polymer Sci.*, **14**, 837 (1970).
110. W C Ingamells in *The Theory of Coloration of Textiles*, 2nd edn, ed. A. Johnson, Society of Dyers and Colourists, Bradford, 1989.
111. V S Salvin, *Am. Dyest. Rep.*, **49**, 600 (1960).
112. M J Schuler, *Textile Res. J.*, **27**, 352 (1957).
113. J Navratil, *J. Soc. Dyers & Col.*, **106**, 283 (1990).
114. J Navratil, *J. Soc. Dyers & Col.*, **106**, 327 (1990).
115. A N Derbyshire, *J. Soc. Dyers & Col.*, **90**, 273 (1974).
116. A N Derbyshire, E D Harvey, D Parr, *J. Soc. Dyers & Col.*, **91**, 106 (1975).
117. J K Skelly, *J. Soc. Dyers & Col.*, **91**, 177 (1975).
118. B Milicevic, *Textile Chem. Col.*, **2**, 87 (1970).
119. B Milicevic, *J. Soc. Dyers & Col.*, **87**, 503 (1971).
120. C Heit, M Moncrief-Yeates, A Palin, M Stevens, H J White, *Textile Res. J.*, **29**, 6 (1959).

CHAPTER 5

Clothing and fabrics

I HOLME

5.1 Introduction

The introduction of synthetic fibres has created greater changes in the textile and clothing industries in the last half century than had previously occurred from 1760, the dawn of the industrial revolution.[1] In the nineteenth century, and even as late as 1920, many types of fabrics based upon the natural fibres, cotton, silk, wool, linen, etc., had become virtually standardized for certain purposes. A style of fabric was developed at a price suitable for the end-use and change, if any, occurred only slowly by a combination of fashion, social change and technological diffusion. In the nineteenth century, social class and standing in the community, and even the type of occupation, could be clearly perceived by the clothes that a person wore and by the types of fabrics from which the clothing was manufactured.

However, with the introduction of synthetic fibres a new era of mass production and a new technological age began.[1,2] Imaginative research and development led to the introduction of innovations that fundamentally transformed the nature of textile processing and enabled novel methods to be introduced into both textile and clothing manufacture. As the range of synthetic fibres expanded, the limitations imposed on the design, construction, properties and performance of both textiles and clothing by the restricted range and variety of natural fibres available were largely overcome.

The concept of fibre engineering introduced progressively from the 1960s onwards has enabled synthetic fibre producers to engineer the properties of fibres, filaments, yarns, fabrics and garments to be suitable for both established and new end-uses.[3-6] In particular, the range of performance criteria that can now be met using synthetic fibres in garment form by the use of the appropriate combination of fibre engineering and dry and wet textile processing has been greatly expanded.[3-7] The continual introduction of novel varieties of synthetic fibres with improved appearance, handle and performance is a constant stimulus to the fabric and clothing design teams. In this way the engineering limits are extended to provide greater comfort in wear, more luxurious handle and aesthetics, and improved appearance and performance.

In virtually every sector of the textile and clothing industries the impact of synthetic fibres upon the direction of technology has been profound. It has led to streamlined methods of production, higher processing speeds, and improved standards of quality in the final article, whether it be a garment or textiles designed, for example, for upholstery or carpet end-uses. The growth in fibre consumption and in fibre consumption per capita in many countries, which could not be achieved through the use of natural fibres alone, has been met by the use of synthetic fibres. Particular mention may be made of the growth rate for polyester fibres, forecast to grow at 3.6 per cent per annum (1986–1995), the highest for all textile fibres.[8,9] Polyester fibres in staple form blended with cellulosic fibres or wool have been a major success story from the 1970s through to the 1990s, with a projected annual growth rate for polyester/cellulose blends of 5.3 per cent over the same period.[8,9]

Of particular importance in the field of clothing has been the dominating tendency to provision of easy care performance that stemmed from the ability to heat-set synthetic fibres and fabrics to provide dimensional stability during washing and drip drying, and largely eliminating the need for ironing.[1,7] The ability to dye and print synthetic fibres to a wide range of bright, deep colours with high standards of colour-fastness to machine washing, light and rubbing, has been a major factor in the success of synthetic fibres.[1,10] In the United Kingdom in 1985, for example, 83 per cent of households possessed some form of washing machine and 80 per cent of garments in the wash load were coloured.[1] In Europe some 500 million garments are washed daily, and the growth in synthetic fibres and blends has aided in decreasing wash temperatures.[1,11]

The use of synthetic fibres in fabrics and garments has therefore been widely accepted, their success stemming from their unique combination of dependability and versatility. Thus, in contrast to natural fibres, which are produced in restricted ranges of fibre fineness and staple length, synthetic fibres may now be engineered from microfibres to relatively coarse fibres, and produced as staple fibre of the appropriate length or as continuous filament. Moreover, the type of fibre crimp and the crimp frequency can be varied to modify the spinnability and the resultant yarn bulk, texture and handle. The fibre lustre, dyeability, and colour may be varied at will, and the fibres can often be modified to simulate the handle and appearance of natural fibre fabrics.[4,6,7] The handle can also be modified by the application of appropriate softening agents.[12]

5.2 Fabric structure

Synthetic fibres are used in the construction of all types of textile fabrics. The major types of fabrics produced include woven, knitted (both weft knitted and warp knitted), raschel, lace and nonwoven fabrics.[13] Apart from the many types of nonwoven fabrics that are produced by a variety of novel methods[14] (e.g. spun-laid fabrics, thermobonding of synthetic fibres or filaments, needle punching, adhesive bonding, etc.), all the major fabric production routes normally involve the use of yarns. Yarns are also used in braided and tufted fabric constructions.[13] It is axiomatic that yarns composed of staple fibres bound together by the increased frictional forces imparted by the yarn twist and also continuous filament yarns must possess an appropriate strength and flexibility to withstand the fabric manufacturing process.

5.3 Functional design of textiles

For most textile fabrics used in clothing, two sets of requirements are essential, namely the aesthetic and the functional requirements.[15] The aesthetic requirements encompass properties such as colour, pattern, lustre, texture, handle and drape, and these properties must appeal mainly to the visual and tactile senses of the wearer. In all fabric end-uses, however, the functional requirements are also of importance, and indeed in the industrial textile sector the functional requirements are paramount, for the aesthetic requirements are often of little importance.

Textile fabrics must have the appropriate physical and mechanical properties that will enable them to maintain their structural integrity during garment making and during end-use; for clothing, for example, this means for the wear life of a garment.[15,16] Textile fabric must be serviceable and will therefore require some minimum level of tensile strength, tear strength or bursting strength to maintain the structural integrity during use and provide a satisfactory level of durability appropriate for the intended wear life of the garment or fabric.

During the wear life of a garment the textile fabric will be subjected to mechanisms that decrease the fabric strength. Abrasion and flexing of the fabric will occur and the necessity to launder clothing to maintain a clean appearance introduces further mechanical and chemical actions that can lead to a decrease in fabric strength.[15] Synthetic fibres are, however, relatively hydrophobic and do not swell markedly in water, nor do they exhibit a markedly inferior wet strength, in contrast to the natural fibres like silk and wool.[17] Fabrics and garments made from synthetic fibres can therefore be made more durable to the action of mechanical and chemical agencies.

In many cases the fabric may be exposed to sunlight (e.g. curtains, awnings, automotive fabrics) and to weathering (tarpaulins, sails for yachting, etc.). Careful selection of the fibre type is required for such end-uses. For protection against sunlight it is particularly important to incorporate stabilizers in synthetic fibres against the action of ultraviolet radiation, which causes polymer degradation.[5,7] The synthetic fibres are generally more resistant to weathering than the natural fibres because of their hydrophobicity. The presence of water in textile fibres can greatly hasten the degradative processes caused by photochemical attack.[18]

Biological organisms such as fungi and bacteria may also attack textile fibres, particularly under warm, moist conditions, and this can lead to significant strength losses in natural fibre fabrics.[18,19] Synthetic fibres are more durable to the action of such organisms and their compact physicochemical structure and greater hydrophobicity compared with natural fibres generally ensure a greater level of protection against biological attack.

The functional design of a fabric implies a target lifetime, or wear life, which must be generally acceptable to the consumer.[15] Although the destructive action of various physical, chemical, photochemical and biological agencies must be understood and taken into account, the product lifetime will in addition depend upon other factors. Clothing, for example, will not be worn to the point of physical disintegration but may be discarded long before this point for aesthetic reasons such as loss of appearance, fading, loss of lustre or the the development of shine. Fashion changes can also lead to the effective lifetime of a garment being terminated as the garment is discarded in favour of a more fashionable item.

A major advantage of the synthetic fibres is the way in which the properties of the fibres or filaments may be modified physically or chemically to engineer their properties to be appropriate for the functional design of a fabric.[5-7] This process takes into account many fibre parameters depending upon the specific end-use. The importance of specific synthetic fibre parameters in the design of fabrics and clothing will now be considered in more detail.

5.4 Fibre properties and fabric performance

The relationship between individual component fibre properties and the resulting properties of a textile fabric is very complex. The behaviour of individual fibres is modified in both a cumulative and an interactive manner when they are combined into a yarn and/or into a fabric structure.[20] Some general properties, however, will normally persist from the fibre stage through to the yarn or fabric stage. Thus, high tenacity synthetic filaments yield strong tear-resistant fabrics, while antistatic nylon or polyester fibres retain their antistat performance in lingerie and specialist protective clothing fabrics. The suitability of a particular synthetic fibre to satisfy a given end-use thus demands the correct balance between quantified fibre characteristics or properties and the necessary end-use performance specifications.

Matching the end-use performance specifications may be a difficult task where these are expressed not only in objective but also in subjective terms. The translation of subjective terms such as handle, resilience and comfort into objectively defined fibre characteristics is a complex task. Nevertheless, research studies are making progress in this field and this has been greatly aided by the ability of synthetic fibres and filaments to be engineered into different variants designed to provide specific performance properties.

One approach has been the use of a fibre identity or performance diagram to illustrate the balance of fibre properties.[20] A number of fibre parameters of interest are selected and are plotted on axes, one for each fibre parameter, in a radial fashion. This is depicted in Fig. 5.1, in which 12 physical and chemical properties are shown. The selection of the fibre properties can clearly be specific to a particular end-use, but in Fig. 5.1 the properties have been arranged in four approximately equivalent sectors representing tensile properties, moisture and electrical characteristics, the fibre behaviour when subjected to heat and light energy, and

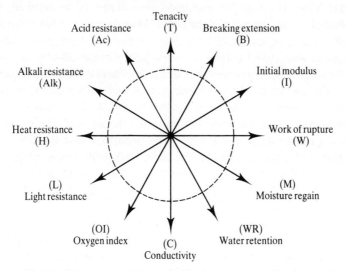

Figure 5.1 Radial representation of fibre parameters. [Redrawn from A. R. Horrocks[20] by courtesy of Benjamin Dent Publications Ltd]

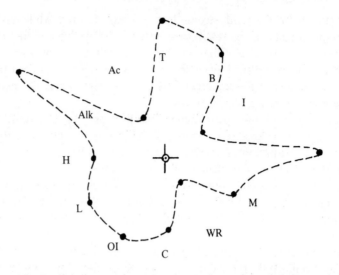

Figure 5.2   Radial diagram for nylon 6,6. Redrawn from A R Horrocks[20]
by courtesy of Benjamin Dent Publications Ltd]

factors associated with chemical resistance. These major fibre properties
are also summarized in Table 5.1 in which average fibre properties have
also been calculated that could be imagined to be acceptable for a
so-called 'ideal' fibre for use in normal textile end-uses.[20]

If the radial axis length of each parameter is plotted for synthetic fibres,
e.g. nylon 6,6 (see Fig. 5.2) or for natural fibres, the balance of advantageous
properties and of any limitations may be clearly established. Such fibre
identity profiles may be utilized by synthetic fibre producers to simulate
the balance of fibre properties that impart natural fibre characteristics,
e.g. cotton-like or wool-like characteristics, to their synthetic fibres.

Polyester fibres, for example, demonstrate excellent performance in
many respects, but the very low moisture regain and the related low
electrical conductivity can create static problems, and the fibre is also
sensitive to alkaline treatments. The superposition of fibre identity profiles
may also be utilized for fibres with different properties to provide blended
yarns and fibres with a more favourable overall balance of properties for
a particular end-use.

In the area of specialty and high performance fibres, the fibre identity
diagram will clearly demonstrate the nature of the modified property.
Epitropic polyester fibres, for example, exhibit electrical conductivity
values of the order of 0.2, at least twice the value for normal polyester,
while flame-retardant synthetic fibres such as modacrylic fibres[20] possess
high oxygen index values in the range 27–30. In addition, the fibre identity

Table 5.1. Typical fibre properties

Fibre	Tenacity (N/tex)	Breaking extension (%)	Modulus (N/tex)	Work of rupture (mN/tex)	Moisture regain (%)	Water retention (%)	Electrical conductivity (log Rs⁻¹)	Flammability OI (%)	Light resistance (months)	Heat resistance (°C)	Alkali resistance (hours)	Acid resistance (hours)
Cotton	0.3	5	5	10	7	50	0.14	18	3.5	105	10	0.5
Wool	0.14	40	2.5	40	17	44	0.10	25	3.5	90	0.5	4
Viscose	0.20	15	6	20	12.5	100	0.10	18	2.5	105	10	0.5
Polyamide	0.45	20	2.5	70	4.3	15	0.13	21	3*	100	4	1
Polyester	0.45	20	10	60	0.4	20	0.08	22	3.5	140	2	50
Acrylic (e.g. Orlon)	0.27	30	5	50	1.3	10	0.07	19	19*	120	5	100
Polypropylene	0.65	15	7	70	0	5	0.06	17	4‡	105	10	10
Average or 'ideal' fibre	0.3	15	5	40	5	50	0.10	21	3	120	2	2

Note: * represents semi-dull fibres; ‡ indicates UV stabilized fibre.
Electrical conductivity – Reciprocal log of resistance Rs of fibre at 65% rh; Heat resistance – Max. temp. for continual use, °C; Light resistance – Time to lose 50% strength in Florida sunlight; Alkali resistance – Time to lose 50% strength in 10% NaOH at 100°C; Acid resistance – Time to lose 50% strength in 10% HCl at 100°C. OI – Oxygen Index.
[Reprinted from A R Horrocks²⁰ by permission of Benjamin Dent Publications Ltd]

diagrams may be utilized to compare different variants of the same generic fibre.

5.4.1 Fibre cross-section

The ability to design and to manufacture markedly different fibre and filament cross-sections out of the same polymeric material has been one of the most significant developments in synthetic fibre engineering.[5,7] This represents a considerable advance over the developments in natural fibres, which have been restricted, in the main, to limited modifications in the fibre length, fineness, crimp and strength. All these parameters may be varied for synthetic fibres over much wider limits by modifying the fibre production stages. Synthetic fibre production is thus more flexible and more responsive to the demands of the consumer market for greater variety in the materials used in clothing and fabrics.

Acrylic fibres may be conventionally spun by wet spinning to give a circular cross-section, or dry spun to give a dog-bone cross-section.[5,7] Nylon filaments may be produced in trilobal and pentalobal bright versions to enhance the lustre for hosiery, clothing or carpet end-uses.[5,7] Hollow fibre cross-sections are utilized to improve thermal insulation, and nylon filaments with four hollow channels can be produced to provide internal reflection and light scattering and therefore greater soil-hiding capabilities in carpet pile fibres. The lustrous appearance of bright fibres may of course be subtly modified by the incorporation of titanium dioxide as a delustrant to produce duller, semi-matt and matt fibres.[5,7] Fibre blending or filament intermingling can thus be used with synthetic fibres to provide the widest range of appearance, texture and handle in textile materials.

The ability to produce polyblend manufactured fibres using two distinct polymer components has led to the introduction of side-by-side bicomponent fibres or conjugate fibres in which a helical crimp may be developed through the differential shrinkage of the components.[5,7,21,22]

Core–sheath bicomponent fibres in the symmetrical or asymmetrical form may be produced to take advantage of particular physical properties such as strength or conductivity of the inner component, or alternatively of the aesthetic, textile, adhesive or other properties of the outer component.[21,22] Crimp may also be generated in the asymmetrical form of core–sheath bicomponent fibres.

In another fibre variant, 'islands-in-the-sea' bicomponent fibres have uses similar to those of the core–sheath type or may be used as an intermediate in the production of other specialized types of fibre.[21,22] In addition, matrix–fibril bicomponent fibres and fibril–matrix bicomponent fibres (also known as biconstituent fibres) may be produced. In these fibres, fine fibrils of one component are embedded in a matrix of the other. The individual fibrils are, however, invariably of a very restricted length and do not extend along the full length of the fibre.[21]

The synthetic fibre producer may select polymers that utilize these novel fibre configurations to combine the properties of the two components. The polymer components may also be selected to adhere strongly together or, alternatively, to have poor adhesion so that separation of the components occurs during subsequent processing. In the latter case, finer filaments are thereby created. If the differing polymers are in alternate segments of a conjugate fibre system, then the filament will split into as many segments of finer filaments. Thus a 3 decitex/fil filament comprising six equal segments would split into six 0.5 dtex microfilaments where the decitex refers to the mass in grammes of 10 kilometres of filament. This approach has been utilized in the production of very fine filaments of 0.17–0.33 dtex/fil used in ultrasuede products, e.g. Ecsaine polyester (Toray) and Hilake polyamide/polyester (Teijin) ultrasuede products.[23,24] Ultra-fine fibres of a fineness as low as 0.001–0.01 dtex can be produced by the mixed melt islands-in-the-sea route by removal of one of the components by extraction. Alternatively, extraction of one of the components can be utilized to provide porous fibres.[24,25]

5.4.2 Fibre density

The fibre densities of dry fibres generally range from polypropylene (0.91 g cm^{-3}) to glass (2.5 g cm^{-3}) with most fibres in the range 1.1–1.6 g cm^{-3} and with only small changes in density when the fibres are immersed in water.[26] Fibres with low fibre density like polypropylene (0.91 g cm^{-3}), nylon 6,6 and nylon 6 (1.14 g cm^{-3}) and acrylic (orlon 1.19 g cm^{-3}) can be employed in end-uses where lightweight bulky materials are required. Acrylic fibres are widely used in high bulk knitwear and in blankets where the softness, warmth and easy washability of acrylic fibres are combined in a lighter weight product for greater comfort in use.[27,28]

Problems can sometimes be experienced when wet processing poly-propylene materials because the fibre density is less than that of water and because of the lack of absorption of water by the fibre. This may give rise to the material floating on the water and not being fully immersed, a problem that has also been noted with polyamide microfilament fabrics.[29] The higher fibre density of polyester filaments (1.39 g cm^{-3}) compared with polyamide fibres (1.14 g cm^{-3}) helps to prevent this problem in wet processing of polyester fibres.

Because polypropylene fibres have the lowest fibre density of all fibres, they have a much larger covering power, a factor of key importance in

Table 5.2. Covering power of fibres

	Density (g cm^{-3})	Specific volume (cm^3 g^{-1})	Cross-sectional area* (μm^2 tex^{-1})	Covering power (%)
Isotactic polypropylene	0.92	1.09	1087	100
Nylon	1.14	0.88	877	81
Acrylic	1.18	0.85	847	78
Polyester	1.38	0.72	725	67
Wool	1.32	0.76	756	70
Cotton	1.50	0.67	667	65
Viscose	1.52	0.66	658	60

* Calculated for 0.11 tex and circular cross-section.
[Reprinted from O Pajgrt, B Reichstädter and F. Ševčik[32] p. 51, by permission of Elsevier]

carpets.[30] Compared with a nylon fibre of equivalent linear density, polypropylene would have a 25 per cent greater number of fibres per kilogram. Polypropylene is a relatively inexpensive fibre and thus for the same weight of pile fibre in a carpet a much more luxurious looking carpet may be manufactured. Alternatively, a very favourable weight reduction is possible if the same degree of cover is required. The inertness, low price and low fibre density of polypropylene are also important factors in the use of polypropylene in geotextile applications because of the high coverage per unit mass.[31] The greater covering power of isotactic polypropylene is illustrated in Table 5.2.[32]

5.4.3 Fibre linear density

The fibre linear density is the mass per unit length of the fibre, normally expressed in dtex/fil (1 dtex refers to the mass in grammes of 10 kilometres of fibre).[33] Microfibres (and microfilaments) are generally defined as fibres or filaments of linear density approximately 1.0 dtex or less, although some commercial products may in practice be as coarse as 1.3 dtex.[34] In Western Europe microfibres are spun by conventional spinning methods to 0.5–0.7 dtex. However, even finer fibres, referred to as supermicrofibres, may be produced by spinning bicomponent fibres with a matrix/fibril structure. This can yield fibres in the 0.1–0.3 dtex/fil range, primarily used to obtain suede or silk-type effects. The production of microfibres and microfilaments has been particularly established with synthetic fibre technology capable of extruding such fine fibres and filaments, but commercially impossible to attain using natural fibres.[23–25]

Polyester and polyamide microfilaments have now been widely introduced into three major end-use categories since the mid-1980s.[35] These end-uses may be discussed in terms of their design for specific functions

Table 5.3. European consumption of microfabrics by end-uses (%)

End-uses of microfabrics	Polyester	Polyamide
Active sportswear	14.6	32.0
Leisurewear	30.2	33.0
Heavyweight fashionwear	33.6	16.0
Lightwear fashionwear	15.1	11.9
Lingerie and others	6.5	7.1

[Reprinted from I Heidenreich and H Ninow[35] by permission of Textiles Intelligence Ltd]

and for fashion purposes, e.g.

Function	Sports, e.g. skiing, walking, trekking, and jogging
Fashion plus function	Coats, blousons, jackets, parkas
Fashion	Blouses, dresses, trousers, skirts, women's suits, blazers, lounge suits

In 1990 the West European supply of microfilaments was estimated to be 8000 tonnes of polyester and some 11 000 tonnes of polyamide (of which only limited quantities of filament are below 1.0 dtex). Some 85 per cent of polyester microfilament is used in weaving, with 11 per cent used in warp knitting. For polyamide microfilaments the weaving sector consumes 95 per cent and warp knitting 5 per cent of the total. The production of microfabrics used in the clothing sector was estimated to be 95 million m^2 for polyester and 130 million m^2 for polyamide.[36] The end-uses of these two types of microfabrics are given in Table 5.3.

Functional clothing usually has to be waterproof, windproof, permeable to water vapour, durable and possess easy care properties.[35–37] By contrast, in the fashion plus function category greater importance is attached to appearance, handle and drape, although the functional properties are not ignored. In the fashion sector, functional properties are of minor importance and the fabric aesthetics, i.e. appearance, handle and drape, are the major criteria considered by consumers. The importance of various properties for so-called microfabrics made from microfibres is illustrated in Table 5.4.[35]

Morton and Hearle have demonstrated[38] that the flexural rigidity or stiffness R of a filament is given by

$$R = \frac{1}{4\pi} \frac{\zeta E T^2}{\rho} \qquad [5.1]$$

where ζ is the shape factor of the filament, E is the tensile modulus of the

Table 5.4. Importance of micro fabric properties in relation to end-use

Property	Function	Fashion plus function	Fashion
Waterproof	Indispensable	Important	Unimportant
Windproof	Indispensable	Important	Unimportant
Permeable to water vapour	Indispensable	Indispensable	Unimportant
Handle	Less important	Important	Indispensable
Appearance	Important	Important	Indispensable
Drape	Less important	Important	Indispensable
Wear resistance	Indispensable	Important	Less important
Easy-care properties	Indispensable	Indispensable	Important

Source: Akzo.

filament, T is the linear density of the filament and ρ is the filament density. For filaments produced from the same polymer, e.g. polyester, spun with the same cross-sectional shape (i.e. same shape factor), tensile modulus and density, it is clear that the filament stiffness or bending resistance is dependent upon the square of the filament linear density. Thus the flexibility of the filament increases markedly as the linear density is decreased to the microfilament level. The effect of decreasing the filament linear density is to decrease both the filament bending stiffness and the crimp contraction of textured filament yarns, as illustrated in Fig. 5.3.[35]

   Accompanying the lower bending stiffness there is a decrease in the individual filament strength of microfilaments, making the resultant microfabric ideal for emerizing.[35,39] Emerizing (or sueding) is a process in which fabric is passed over a series of rotating emery-covered rollers to produce a suede-like finish.[39,40] The increased number of filaments in

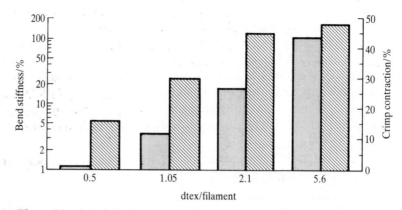

Figure 5.3   Relationship between filament bend stiffness (left axis) and crimp contraction (right axis). [Redrawn from reference 35 by permission of Textiles Intelligence Ltd]

microfibre yarns compared with standard fibre yarns results in the total yarn strength remaining high. However, one consequence of decreasing the filament linear density is to increase the specific surface, i.e., the surface area per unit volume of the filament. This has the effect of increasing the quantity of size required in the sizing of warp yarns for weaving and greatly increases the surface area for the absorption and desorption of dyestuffs.[41–44] Thus a polyester fibre of 3.5 dtex had a dyeing rate of one-fifth the rate of a polyester microfibre with a fineness of 0.55 dtex, and accordingly precautions must be taken in practical dyeings to ensure that level dyeing is obtained.[41] Considerably more dyestuff is also required to attain the same depth of colour on microfilaments, and there can be problems with colour-fastness to washing and on exposure to light.[41–44]

The effect of decreasing the filament linear density in polyester and polyamide filament fabrics is to produce a softer, more drapable fabric, because of the lower bending stiffness of each individual filament, as shown in Fig. 5.4. However, as the number of filaments in a given count of textured yarn increases, the yarns become bulkier because yarn crimping causes fabrics to contract. The use of finer filaments decreases the extent of this contraction leading to a lower resilience and a lower fabric bulk.[35]

Changing the filament linear density of polyester also introduces other changes in fabric properties, as shown in Fig. 5.5. The creasing tendency of the fabric increases on changing from coarse to fine filaments, and there are accompanying decreases in fabric abrasion resistance and resilience because of the use of finer filaments of lower strength and lower bending stiffness.[35] The resistance to slippage also falls, but this can be countered by increasing the density of the fabric construction to avoid seam slippage in garments.

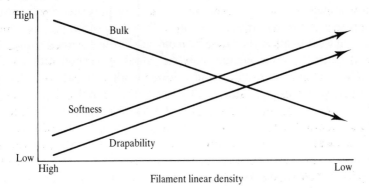

Figure 5.4   Effect of filament linear density on fabric properties. [Redrawn from reference 35 by permission of Textiles Intelligence Ltd]

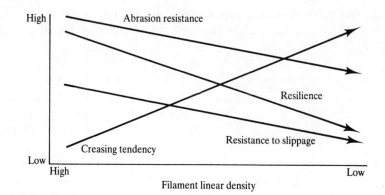

Figure 5.5   Effect of filament linear density on end-use properties. [Redrawn from reference 35 by permission of Textiles Intelligence Ltd]

In contrast to microfilaments, microstaple fibres in spun yarns are of less importance, but nevertheless can be used to provide improved handle, appearance and drape in a wide range of fibre blends. High spinning speeds have been claimed to be possible spinning microstaple fibres of <1.0 dtex into Nm 100 rotor yarns of high evenness and cleanliness as well as adequate strength.[35] The increasing fibre fineness, however, creates greater difficulty in opening the fibre stock, yields a lower carding performance and may offer a higher sensitivity to unfavourable yarn spinning conditions.

A major problem area is that of the propensity to pilling in fabric form, which has been shown generally to increase as the fibres become finer and as the staple length decreases. This can decrease the hard wearing qualities of a fabric and create an unfavourable appearance during wear.[45]

In relation to pilling of conventional polyester fibres, the problem is mainly caused by the relatively high tenacity of polyester fibres, which ensures that fibres that anchor the pill to the fabric surface do not break in normal wear to allow the pill to be removed. The use of lower molecular mass polymer, however, enables polyester fibres to be spun with a lower tenacity, a higher extensibility and a lower work to break at a defined extensibility.[46] This is illustrated by comparing low pill polyester staple with a limiting viscosity number of 0.38–0.48, a tenacity of 26 cN tex^{-1} and an extensibility of 40 per cent with regular polyester filament yarn of 0.55–0.65 limiting viscosity number, tenacity 50 cN tex^{-1} and extensibility 15 per cent. The lower the molecular mass of the polymer, the more readily are pills abraded from the fabric surface. Pill resistance for polyester improves as the limiting viscosity number falls below 0.48 and improves further as it falls to about 0.4.[46]

5.4.4 Fibre thermal conductivity

The thermal conductivity of fibres is higher than that of air, and it is generally recognized that where thermal insulation is required in fabrics and garments the most important factor is the proportion of still air incorporated in the material.[47–49] The thermal conductivity values for air and various fibres are given in Table 5.5.[47]

It can be seen that polypropylene fibres, polyurethane, aramid and polyester fibres all have lower thermal conductivities than natural fibres such as wool. Polypropylene fibres have the lowest thermal conductivity of all textile fibres and have therefore been termed the warmest.[48,50,51] This inherent fibre property is used to good advantage in the carpet sector, in which this feeling of warmth is utilized together with the low fibre density to produce luxurious, warm carpets for use in bedrooms where there is likely to be contact with the carpet pile fibre.[30]

Heat transfer through a textile material takes place by a combination of conduction through the air and fibres, and to a lesser extent by infrared radiation.[47,49] Radiation may pass from fibre to fibre, resulting in a very low rate of heat transfer. Alternatively, in an open construction, the radiation may pass directly through the interstices of the material without absorption and thus contribute significantly to the heat transfer. The interchange of heat between layers of clothing takes place by conduction through the air gap between the fabric layers, and by unrestricted infrared radiation. The heat transfer from layer to layer is also aided by the natural convection in the air trapped between the layers of clothing. Forced convection, as in conditions of high wind velocity, also influences the heat flow in both the air gaps and the fabric layers.

The thermal resistance of a mixture of fibres and air is approximately

Table 5.5. Thermal conductivities of air and various fibres

Material	Thermal conductivity/$W\ m^{-2}\ K^{-1}$
Air	0.026
Polypropylene	0.117
Polyurethane	0.126
Aramid	0.130
Polyester	0.141
Poly(vinyl chloride)	0.167
Wool	0.193
Polyacrylonitrile	0.200
Cellulose acetate	0.226
Polyamide	0.243
Viscose	0.289
Cotton	0.461

equivalent to the thermal resistance derived from a layer of still air of the same thickness less an amount due to the heat conduction in the fibres. This component is dependent upon the amount of fibres present, the packing factor of the material and the fibre thermal conductivity. The packing factor can vary from about 0.2 for a densely woven, cropped fabric, to 0.05 for loosely knitted underwear, or as low as 0.01 for fibre batting.[47]

Undoubtedly the thickness of a textile material is the most important factor determining the thermal resistance, with thicker materials providing greater thermal insulation.[47,49] In general there is a linear relationship between thermal resistance and fabric thickness, but in clothing this does not take into account the major influence of additional air layers trapped between successive layers of clothing. The thermal resistance of the fabric used in a cotton shirting fabric, at 0.91 K m^2 W^{-1}, may thus be compared with that of an anorak lining (0.20 K m^2 W^{-1}) and with a thick sliver knit jacket (0.41 K m^2 W^{-1}), illustrating the effect of fabric thickness on thermal insulation.[47]

The technique of raising or brushing of textile fabrics during fabric finishing is one method of decreasing the fabric density and increasing the fabric thickness, thereby increasing the thermal insulation. Such a technique is utilized in blanket manufacture, where both sides of the fabric are raised to provide a thick, thermally insulative material. The apparent warmth to the touch is also increased by this mechanical finishing treatment, because the projecting pile surface fibres keep the skin from contacting the more thermally conductive cores of the yarns, decreasing the rate of heat loss, a factor also used to advantage in underwear.[47]

The ability of high shrink and low shrink acrylic staple fibres to be blended and heat treated to provide high bulk acrylic yarns is one method of fibre engineering that leads to a greater yarn and fabric bulk, and hence to a greater thermal insulation value. This factor has been exploited widely in weft knitted sweaters to produce lightweight, bulky and warm clothing. Both polyester and polyamide continuous filaments may be textured by a variety of methods to provide a higher degree of bulk, stretch and texture. The greater fabric thickness and bulk again provides a higher level of thermal insulation compared with conventional low twist continuous filament fabrics.[49]

For most clothing systems the human body is clothed in a combination of garments, and the thermal insulation of the garment combination is greater than the sum of the individual thermal resistance values of each item.[47] Additional air layers are trapped between each layer and the next (or the skin), because each layer of clothing does not necessarily follow the contours of the neighbouring layers exactly. The thickness of the air layer depends upon the pattern and fit of the clothing and with body-hugging elastic knitwear or wet fabric adhering to the skin the thickness may effectively be zero. In addition to the air layers trapped within the fabric there is a boundary layer of still air over the outer surface of the clothing, the thermal resistance of which varies from $0.12 \text{ K m}^2 \text{ W}^{-1}$ down to $0.02 \text{ K m}^2 \text{ W}^{-1}$, the exact value depending upon the velocity of air movement around the body.[49,52]

Studies using an articulated manikin have demonstrated that the thermal insulation of a man's normal business suit, including underwear, is only 17 per cent of the total thermal insulation sensed by the body under stationary conditions, and about 25 per cent when walking.[53] Thus the thermal insulation provided by the layers of clothing alone is small compared with the thermal insulation provided by the air layers between garments and the air layer on the outer surface of the clothing. However, under polar conditions, the fraction of the total insulation contributed by the textile may rise to 80–90 per cent.[47,53]

Synthetic fibres and filaments can be engineered to provide a high level of thermal insulation, not only by bulking or texturing the yarn but also by introducing a modified fibre cross-section. Some polyester fibres have been produced with a hollow core or channel that simulates the medulla in wool.[54] The hollow core of the synthetic fibre thus creates a natural surface for still air and has the additional benefit of decreasing the weight of the fibre. This is of importance where thermally insulative textile materials may have to be carried for long times (e.g. sleeping bags).

Synthetic fibre fillings (termed fibrefill), principally made from polyester fibres, have been used very widely as filling materials for quilts and pillows, and some remarkably sophisticated fibres have been developed to simulate the bulk and thermal insulation of down (feathers from the breasts of geese and ducks).[55] The snowflake structure of natural down provides high bulk, trapping air and providing thermal insulation. The novel polyester fibrefill fibres contain one or more parallel air channels in the fibre cross-section, which increase the fibre bulk and entrap more air. Finer fibres provide greater bulk while heavier fibre versions are used as fibrefill in washable, quick drying pillows. Such fibres are used in pillows, quilts and sleeping bags, and in contrast to some other materials are odourless, mildew- and moth-resistant, and do not cause dust allergies.

Du Pont, for example, manufacture Dacron Fiberfill (one hole), Dacron Hollofill 4 (four holes) and Dacron Quallofil 7 (a combination of four-hole and seven-hole fibre structures). Dacron Quallofil 7 has a sawtooth crimp and improved recovery from compression, imparting improved drape and wear durability, as well as good thermal insulation.[56] The humidity that is given off by the human body during sleep is dissipated gradually because of the unique mix of fibre constructions that makes the filling light and airy. Such specialty fibres are usually coated with silicone lubricants to help the fibres slip over each other;[55] this helps to prevent matting, crushing and lumping and aids the task of shaking the quilt into shape. Flame-retardant versions of polyester fibrefill have also been developed for use where this property is required.

Another unique form of synthetic fibre filling was introduced by Du Pont in 1986. Comforel is a polyester fibre product that consists of large numbers of 5-mm balls of fibre, which closely simulate the appearance of down.[54-56] The great advantage of such a product over all other synthetic and animal fibrefills is that it is air transportable, and can be handled by blowing techniques. Another product simulating down is Kuraray Sanitar Duckfill (Kuraray).[55] This takes the form of short staple hollow fibres bound together into the shape of down and treated with an antimicrobial finish.

An alternative approach to providing increased thermal insulation is that of giving the air more surfaces to which it may cling. This technique is only available to man-made fibre production techniques and has been utilized in Thinsulate, a microfibre produced by 3M.[49,54] The microfibres are many times finer than a human hair and their use is claimed to cause the resulting fabric to come closer to the ideal thermal insulator, namely a vacuum, than other competitive insulation. An improvement of up to twice the warmth of an equivalent thickness of other insulation materials has been claimed for Thinsulate fabrics.

The end-uses for lightweight high loft insulation materials continue to expand. Some typical end-uses for Thinsulate insulation materials made from synthetic fibres are given in Table 5.6, the insulation being engineered for the specific end-use.[57]

The fibre composition of Thinsulate Type C, CS and CDS is 65/35 polypropylene/polyester, with the Thinsulate ceramic insulation Type YCS consisting of polypropylene/polyester/ceramic loaded fibres (65/26.2/8.8). In Thinsulate Type YCDS the scrim layers are sonically quilted in 152-mm widths. This material may be free hung in edge-stabilized constructions with no restrictions on panel size, and hence Type CDS can be put into a garment as a free-hanging liner. This is widely used in Scandinavian countries.

The thermal resistance of clothing is a measure of the insulation against heat loss.[58] The practical SI unit that is used to quantify the thermal insulation of garments, bedding, etc., is the tog which is defined as $0.1 \text{ m}^2 \text{ K W}^{-1}$. One tog is equivalent to 0.645 clo, a unit introduced in the United States in about 1940, with one clo being the amount of clothing required to keep a sitting man of average metabolic rate comfortable in the average indoor atmosphere at 21°C.

Some typical thermal resistance values (in togs) for clothing and

Table 5.6. Uses for Thinsulate (3M)

Thinsulate (3M)	Properties	End-uses
Thermal insulation Type CS	Low thickness, softness and drape for use with soft drapable outer shell fabrics in temperatures down to −40°C	Sportswear, bedding, window insulation, general outerwear, skiwear, gloves
Ceramic insulation Type YCS	Ceramic particles incorporated in the insulation fibres are heated by sunlight, decreasing heat loss from the body	Skiwear, outerwear
Compressed insulation Type B	Resists compression, maintaining warmth under moderate pressure	Footwear, diving suits
Light insulation Type KS	Very thin, drapable insulation for use in lightweight garments where other insulation materials would provide too much warmth or bulk	Windshells, golf jackets, running suits, slacks, shirts
Liteloft insulation Type THL	New generation of lightweight resilient high loft insulation. The material recovers its loft after compression, retaining its thickness and insulation value after several washes	Skiwear, general outerwear, bedding, sleeping bags

Table 5.7. Typical thermal resistance values of clothing and bedding

Product	Tog value
Clothing	
Shirting	0.1
Underwear	0.2–0.4
Thermal underwear	0.4–0.8
Suitings	1.0
Sweaters	1.0
Bedding	
Blankets	1–2
Continental quilts	7–14

[Modified from E Clulow[58] by permission of Benjamin Dent Publications Ltd]

bedding are illustrated in Table 5.7. Thermal underwear usually has a thermal resistance in the range of 0.4–0.6 togs, but this can be increased to 0.8 togs if the material has been raised to produce a surface pile. This increases the fabric thickness and entraps more air, increasing the thermal insulation. It has been demonstrated that the thermal resistance of

Table 5.8. Heat production rates and insulation for various activities

	Required for thermal neutrality		
	Heat production/W	Required insulation/K m^2 W^{-1}	Clothing thickness/cm
Sleeping	80	2.0	8.0
Sitting	110	1.45	5.8
Standing	175	0.91	3.6
Walking with pack and snowshoes	250–450	0.64–0.36	2.6–1.4
Pulling toboggan	500–1000	0.32–0.16	1.3–0.6

Skin temperature: 33°C; Air temperature: -40°C; Clothing conductivity: 0.4 K^{-1} m^{-2} W.
[Reprinted from G Holmes[60] by permission of Benjamin Dent Publications Ltd]

underwear fabrics is linearly related to the fabric thickness measured under conditions where the fabric is virtually uncompressed.[59]

In relation to bedding materials, the total thermal resistance of a sheet plus two blankets and a bedspread is normally just under 5 togs.[58] In contrast, the use of polyester hollow fibre filled continental quilts can enable the tog value to be varied from 7 to 14 togs using only one-third to one-half the weight of textile fibre because of the high thermal resistance of the still air trapped within the bulk of the quilt.

Most clothing designed for extremely cold climates utilizes the layer principle of thermal insulation.[49,60] The metabolic heat production depends upon the level of human activity and ideally, if the person is to remain thermally neutral, dry and comfortable, this heat production must be balanced by an equal rate of heat loss throughout the clothing without the evaporation of sweat being required. It is clear from the thermal insulation and clothing thickness values required for different activities[60] that they are too great to permit the wearing of the same environmental protective clothing systems for all activities, as shown in Table 5.8.[60] Generally the individual wears clothing designed for moderate activity levels, with a separate system for sleeping. High levels of activity lead to sweating and the condensation of sweat but its subsequent evaporation when the activity level drops to sedentary can result in excessive cooling.

5.5 Clothing

The conversion of a fabric into a garment takes place in a number of stages, illustrated in Fig. 5.6, in which the clothing manufacturer produces shell structures out of flat fabrics to match the shape of the human body.[61] The creation of the garment design and construction of patterns for the components of the design are followed by pattern grading to enlarge or diminish a style pattern to ensure a proper fit for all sizes.

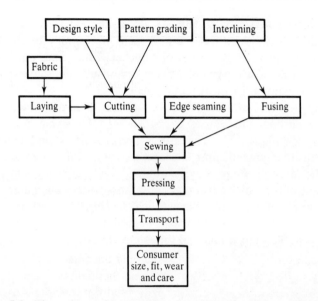

Figure 5.6 Flow chart of a garment manufacturing process. [Redrawn from R L Shishoo[61] by courtesy of Business Press Ltd]

The shape and size of the garment relative to the body is known as the fit and this will be strongly influenced by many factors that have to be taken into account when drafting patterns for a garment. Any tendency of the fabric to shrink, stretch, distort and drape because of stresses introduced by static or dynamic situations can influence the shape and size, and hence the garment fit.

In laying/spreading and cutting operations it is important that the fabric width is uniform and that the weft line is at right angles to the warp. Heat setting treatments on fabrics containing synthetic fibres like polyester and polyamide during fabric finishing have greatly aided this process. Uniformity of fabric width is essential when cutting out the garment components from superimposed fabric layers in order to achieve maximum utilization of the fabric during cutting.

The fabric tailorability is the ability and ease with which the fabric components can qualitatively and quantitatively be sewn together to form a garment.[61] In this respect good formability and good sewability are required. Fabric formability has been related to the maximum fabric compression that can be sustained before the onset of buckling. Good sewability relates to the ease of formation of shell structures and styles and the absence of fabric distortion and seam damage.

Multilayer structures are often needed in specific garment elements to produce the desired bending stiffness, drape and extension stiffness, and fusible interlinings, often based on synthetic fibres, may be employed.[62] The collars and cuffs of shirts, and jacket fronts and waistbands in trousers and skirts are particular examples of such multilayer structures that fulfil functional garment requirements.

Flat apparel fabrics must be converted into three-dimensional shapes either by cutting and sewing or by shaping by steam pressing and moulding techniques.[61,62] The use of heat, moisture and pressure on thermosensitive synthetic fibre fabrics can strongly influence the ease with which the resulting shape may be attained, and its subsequent stability. Indeed one of the major advantages of synthetic fibres compared with the natural fibres is the relative ease with which thermoplastic synthetic fibre and filament fabrics may be shaped or moulded by the use of heat, to provide a stable shaped material for use in clothing.

5.5.1 Clothing

Comfort in clothing is generally assessed as a neutral sensation, when the wearer of the clothing is physiologically and psychologically unaware of the clothing being worn. Positive comfort sensations tend to be more individualistic and less frequently noticed compared with the sensations of discomfort produced by wearing clothing.[63–66]

Pyschological discomfort arises when the clothing being worn is considered to be inappropriate for the person or for the occasion. The aesthetics of the clothing can lead to discomfort sensations from an inappropriate choice of colour or fashion, fabric construction and finishes, or unsuitability for an occasion, poor garment style (e.g. it does not flatter the figure nor fit properly) or from prejudice.[63,64]

Physiological discomfort can be subdivided into sensorial discomfort arising from what the fabric or garment feels like when it is worn next to the skin, from thermophysiological discomfort and from garment fit.[63,64] Sensorial discomfort may be initiated through an allergic response, a feeling of itchiness or prickle, or an initial cold feel to the fabric. In addition, other discomfort sensations can include skin abrasion, the shedding of loose fibres and the clinging of wet fabric to the skin. From the thermophysiological viewpoint, a fabric may feel too warm, too cold, or transport sweat away from the skin. In respect of garment fit, discomfort may be caused by a tight fit overall, or through localized tight areas (e.g. waistbands).[63,64,66]

Man-made fibres are generally more versatile in their aesthetic and intrinsic fibre properties compared with natural fibres. However, discomfort can arise from a variety of sources connected with a garment. For example, the use of fabric care labels sewn into garments can lead to discomfort where the heat-sealed edge may be inaccurately cut, or a heat-fused edge may be liable to crack at a fold, which can prove very uncomfortable to wear because of the skin prickle sensation that is generated.[63,64]

The sensorial feeling of clothing, i.e. the comfort or discomfort associated with how a fabric or garment feels next to the skin, is a very difficult quantity to assess and define scientifically. Nevertheless, the type of skin sensations produced by clothing worn next to the skin is a major factor in determining the overall comfort of a garment. It is known from research studies that fabric handle, as assessed by feeling with the hands, has proved to be unreliable as a guide in the prediction of sensorial comfort.[63,64] Differences in fabric structure, drape and fabric finish may be detected, but not in fabric hairiness, which is a major factor in determining sensorial comfort.[67]

The fingertips have a higher density of nerve endings than the general body surface. Moreover, the threshold for touch over the forearm is 33 g mm^{-2} and on the abdomen is 26 g mm^{-2}, whereas on the fingertips it is only 2 g mm^{-2}. During the handling of a fabric using the fingers and thumb, the thumb normally flattens any protruding surface fibres to give the impression of a smooth, soft, resilient fabric, thereby allowing an accurate profile of the fabric surface to be registered. However, when a garment is donned, the change in the conditions will be registered for the first few minutes. After this time an equilibrium condition will be reached unless there are no major discomfort sensations.

Major discomfort sensations experienced by the wearer of next-to-skin apparel may include a number of problem areas.[63-67] Local tightness or excessive looseness caused by poor garment fit is often an overriding discomfort sensation. Tickle is caused by fabric hairiness and influenced by garment fit.[67] Coarse or stiff fibres protruding from the fabric surface may cause a prickly sensation. Prickle is particularly noted with fabrics and garments containing wool.[68,69] Attention to the appropriate selection of synthetic fibre linear density, and yarn and fabric engineering, allied with fabric cutting and/or singeing treatments in textile finishing can minimize tickle and prickle discomfort sensations. The uniformity of the linear density of synthetic fibres compared with the variability in linear density often experienced when using natural fibres is here clearly an advantage. Local irritation caused by sewn-in garment labels and to a lesser degree by abrasion associated with seams can, however, still take place in garments composed of any fibre type, both natural or man-made.[63,64] Skin and nasal irritation may be caused by loose fibres that have been released from the fabric surface, and fibres that become attached to other garments can also cause annoyance.

For comfort, the perspiration given off by the human body must be considered, as well as the dry heat.[58,70,71] The human body constantly loses moisture as vapour, and under conditions where the level of activity or the external temperature is sufficiently raised the body also gives off liquid sweat. Where beads of perspiration may form on the skin a liquid film of sweat may be formed on the surface of the skin. Under adverse conditions 'wet cling' or 'tacky cling' may occur. Wet cling is caused by sweating, and tacky cling is associated with the presence of damp and sticky sweat residues on the skin.[58] The combination of sweat, sebum and damp skin gives a very tacky surface. A major factor influencing the amount of fabric cling is the area of the fabric in contact with the skin, which in turn is influenced by the fabric structure. The presence of sweat on the skin can also aggravate skin abrasion, mainly attributed to physical activity when the relative movement between the fabric and skin is frequent.

Another discomfort sensation is associated with the initial cold feel when a garment is first donned. This is primarily experienced with cold-weather clothing and is mainly influenced by the fabric surface contact areas with the skin.[58]

Static electrical effects have generally been considered to be a minor discomfort sensation, where fabrics may visibly spark or cling to the body. This can be unpleasant and detrimental to a person's comfort and the low moisture regain synthetic fibres are normally given an antistatic treatment on the fibre surface during textile finishing to avoid this problem. Body cling is particularly undesirable in ladies' wear, especially in lingerie, and to overcome this problem synthetic fibre producers have introduced antistatic versions of synthetic fibres that incorporate durable antistats such as in antistatic nylon fibres.[72] Synthetic fibres can also be made electrically conductive, as in the ICI 'epitropic' polyester fibres in which carbon is incorporated.[72-74] Such epitropic fibres can be engineered to form clothing that can be heated electrically using battery packs to protect against arduous climatic conditions where the windchill factor induces very low equivalent ambient temperature conditions, such as in polar environments. In other working situations where protection against incendiary sparks is essential, the use of durable antistatic finishes is clearly essential for synthetic fibre fabrics that, in general, would possess high electrical resistivity without suitable chemical modification or application of surface finishes.[72,73]

Allergies related to synthetic fibres are rare, and the number of cases of dermatological problems recorded over the years is very small.[75,76] In almost all cases the dermatitis appears to have arisen as a result of the presence of certain dyes, finishes or chemicals (e.g. antioxidants) present in the fibre, rather than from the synthetic fibre material used.

5.5.2 *Improving comfort in clothing*

Comfort in clothing has been shown to be difficult to define and physical, psychological and physiological factors all have to be taken into account. Certainly the transmission of heat, moisture and air through the textile material are very important factors. In addition a number of subjective elements such as the size and fit of the garment, the softness and the handle of the material, and the possibility of static generation must be considered.[63–67,71,77]

Consumer impressions of synthetic fibre materials have not always been favourable. Principal limitations are often associated with an unnatural handle, unpleasant thermal sensations, fabric clamminess in contact with the skin, a lack of moisture absorbency, unfamiliar skin contact sensations and static-related problems.[78] A considerable amount of research and development work has been carried out to attempt to overcome such criticisms, and some notable advances have been made, particularly for polyester fibres.[77]

Garment comfort may be improved by many techniques; for example, by fibre modification, by fibre blending, and by topical finishing (i.e. the application of chemical finishes to the fibre surface). The yarn and fabric construction may be varied and the design of the garment may be optimized for the appropriate end-use. All of these methods may be utilized individually or in combination to achieve the desired effect.[77]

The fibre engineering approach has been used commercially to modify polyester fibres to improve the garment comfort. A mixed cut staple length polyester fibre with a length distribution similar to that of combed cotton formed the basis for Comfort Fiber (Celanese), a polyester fibre that was also chemically modified to enhance tactile sensations.[79] Superior wicking and moisture evaporation have been achieved by producing a fibre cross-section with the appearance of a double scallop. This four-channel fibre was used by Du Pont to produce CoolMax fabric.[80]

Porous polyester fibres have also been produced. Wellkey (Teijin) contains a central canal and a porous wall with pore radii mainly in the range 5–1500 nm, enabling the fibre to absorb water and dry rapidly.[81] The incorporation of up to 10 per cent of a metallic oxalate (e.g. $K_3Al(C_2O_4)_3$) into a polyester prior to fibre spinning, followed by hydrosetting the drawn filaments under pressure at 120–140°C, generates a pore structure provided there is water present in the liquid state for a time sufficient to build up and stabilize the pore system.[82] With pore radii in the 5–40 nm range, the moisture regain of the porous polyester fibre was 1 per cent (at 20°C/65 per cent relative humidity (RH)) compared with 0.3 per cent for normal polyester. At 34°C/92 per cent RH the porous polyester exhibited a moisture regain of 8 per cent, considerably above that for normal polyester (0.5 per cent) and very close to that for cotton (10 per cent).

Treatment of polyester fabrics with aqueous sodium hydroxide (caustic soda) has been practised for many years and has been variously termed causticization, deweighting, or alkalizing of polyester.[83] A high fabric lustre can be obtained without a paper-like handle by a calender/heat set/caustic hydrolysis treatment, and on fabric made from untextured yarn a softer tactility and a less synthetic-like handle result from alkaline hydrolysis.[77,83] The treatment decreases the fibre diameter and produces a loss in fabric weight. It is one of the very few types of subtractive finishing treatments still used.

In a heat-set woven polyester filament fabric, the decrease in filament diameter leads to an increase in air permeability and increased yarn movement at interyarn crossover points. The latter is reflected by accompanying improvements in fabric drape and suppleness. Other benefits that may be attained are better fibre cohesion, increased water wetting and wicking, reduced wet cling properties and a decreased tendency to be soiled and to retain soil. The chain scission of the long chain molecules at the carbonyl groups results in the production of hydroxyl and carboxylate end-groups, with an improvement in hydrophilicity that may be utilized to provide better adhesion of topical (i.e. surface) finishes to the fibre surface.[77,83] The changes in fabric properties are generally associated with an improvement in the comfort properties.

Silk-like properties in textured polyester woven fabrics are generally associated with weight losses in the range of 10–30 per cent.[84] Bright fibres with round cross-sections lose weight more slowly than delustred fibres with multilobal cross-sections.[85] Textured yarns appear to be more reactive than flat yarns and cationic-dyeable polyester fibres (e.g. polyester containing the sodium salt of the 5-sulphoisophthaloyl group) react very rapidly with caustic soda.

Caustic soda treatment is a topochemical reaction in which the reaction is confined to the fibre surface.[77,83] After chain scission, the reaction products dissolve in the reagent solution, revealing a fresh surface, and further reaction then proceeds, with the fibres becoming progressively thinner as well as exhibiting surface pitting. The pitting is enhanced on delustred fibres and the increased surface roughness may be a contributory factor in decreasing the contact angle with water, as well as altering the reflection of light to give a silk-like lustre. Hydrolysis may increase the accessibility of hydrophilic groups present on the fibre exterior, and the number of hydrophilic groups will also be increased by the treatment, thereby leading to a lower contact angle with water, and hence to improved wettability and wicking properties.[83,86] This is particularly important in improving garment comfort under hot, humid conditions or under conditions of high metabolic workload where the transport of perspiration away from the skin greatly improves the sensation of garment comfort.

Comfort in garments is dependent upon the fit to the human body, and garments have been likened to a second skin. The skin is highly extensible and is able to extend and contract according to the dictates of the movement of the limbs and joints. Under normal conditions this extensibility or percentage stretch is around 15–50 per cent.[87]

These levels of percentage stretch in fabric are not attainable using conventional woven fabrics, which are relatively inextensible to an extent depending on the yarn crimp in the fabric. This in turn is determined by the yarn properties and the weave structure. Stretch woven fabrics are obtainable using slack mercerized cotton, but a much more convenient method is to use textured woven and textured weft knitted fabrics composed of synthetic filament yarns.

Air texturing, which involves the entanglement of synthetic filaments using high velocity air emerging from a vortex air jet, provides satisfactory levels of stretch suitable for use in woven fabrics for apparel and other end-uses.[88] Much higher levels of stretch, combined with higher bulk and texture, are obtained using false-twist textured synthetic filament yarns in which the degree of stretch and power of recovery can be utilized to provide body-hugging characteristics, accentuating the fashion shape and contours of the human body. These textured yarns combine high stretch at low loads with considerable powers of recovery from deformation.[89] The ability of synthetic filament yarns to be heat set, steam set, or hydroset ensures that the dimensional stability of such stretch fabrics and textured fabrics will withstand domestic laundering without shrinkage taking place.

The major use of false-twist textured yarns, which can be draw-textured continuously during synthetic filament production, is in weft knitted structures, although some textured synthetic filament yarns are also used in warp knitted and in woven fabric structures. In general the fabric extensibility is in the order weft knitted > warp knitted > woven fabric. The ease of distortion of the weft knitted yarn loops in the weft knitted structure is thus a major factor determining fabric extensibility. The warp knitted loop structure and the warp/weft crossover systems of woven fabrics are, in general, less extensible in the warp and weft directions.[90]

Textured synthetic filament yarns have been employed widely in weft knitted constructions for both outerwear and underwear. Textured yarns have been particularly utilized in knitting half-hose (socks) and in ladies' hosiery (stockings and tights). The world consumption of socks in 1990 was 10.1 billion pairs and of tights 8 billion pairs per year.[91] The very large extensions required around the knee joint in stretching and the requirement of immediate and complete return to the original shape on recovery from extension have favoured nylon ladies' hosiery at the expense of polyester and other synthetic filaments. The elastic recovery of nylon is superior to that of all other conventional synthetic filaments,[92] with the exception of elastane filaments. Thus nylon stockings and tights do not 'knee' or sag around the knee when the leg is bent. The durability of nylon and its ability to be textured in fine yarns has opened up new levels of comfort in ladies' lightweight hosiery. The ability of nylon to be dyed and printed to a wide range of bright, fast colours has also stimulated the ladies' hosiery sector in fashion terms.

However, in lightweight sheer stockings and tights, the power of recovery of even lightweight nylon yarns is not as high as that of higher yarns. A solution to this problem has been achieved by laying-in or plaiting elastane filaments such as Lycra (Du Pont) whilst knitting the textured nylon filament yarns. This improves the ease of deformation of the fabric and greatly increases the power of recovery from deformation.[93] This provides lightweight sheer stockings and tights that possess form-fitting characteristics with a very high level of comfort engineered into the garment.

In the field of sportswear there has been considerable growth in the use of synthetic fibres in energy wear or active sportswear.[94] The initial use of elastane filaments such as Lycra in this field was developed for competitive swimwear, providing form-fitting garments to reduce hydrodynamic drag and eliminate wet sag. Lycra for foundation garments originally had a high load power, a high unload power and a large hysteresis at point of wear. For swimwear, however, Lycra was developed in stretch garments with a lower power, both load and unload, and a low hysteresis. Thus the swimwear is designed so that as the muscles expand the amount of work that has to be done against the garment is minimized, while the necessary support is still provided—an important design consideration for all energy wear garments. Good chlorine and mildew resistance are built into the elastane fibre by modifying the fibre chemistry or incorporating suitable additives.

Energy wear garments are now being tailored to fit the requirements of specific sports, maximizing physical performance and providing protection. Garments are ergonomically designed as a second skin, conforming to the body and providing both freedom of movement and absence of irritation. Form-fitting garments may have to stretch by 15 to 50 per cent because of the limb movement of the human body when stretching in athletic activities.[87,95] Form-fitting garments enhance performance because of the reduction in drag and the slowing of the onset of fatigue.

Athletes wearing form-fitting Lycra-containing tricot garments have demonstrated a 5–8 per cent increase in leg power during multiple jumps compared with the wearing of loose-fitting shorts.[94] The microclimate against the skin decreases the negative impact of heat and cold on the body. Compression garments are thus claimed to give improved muscle performance, delaying the point at which the athlete tires. In addition, the muscle-warming effect can help to prevent injury.[94]

A wide variety of woven or knitted synthetic fibre fabrics containing elastane filaments are now worn in garments for skiing, ski jumping, riding, cycling, training, jogging, running, climbing, basketball, football, volleyball and tennis. Other garments are designed for aerobic dancing and ballet, to emphasize body movement, for aesthetic effect, and to provide comfort.

5.5.3 Synthetic fibres in leisurewear and sportswear

Within Western Europe the changes in consumer lifestyle over the past decade have seen the introduction of a more casual fashion look at the expense of formal wear.[9] This has often increased the consumption of the more extensible weft knitted fabrics compared with the traditional stiffer woven structures. Coupled with these trends has been the greater interest in personal health and fitness, and greater involvement with sports and leisure activities, both for indoor and outdoor activities. The latter has increased the need for foul-weather clothing for use in recreations such as sailing, golf, climbing and hiking, sportswear for athletics, training and jogging.[96]

It has been estimated that by 1992 the leisurewear market (40 per cent) and the sportswear market (14 per cent) would together dominate the outerwear market in Western Europe, with formal wear decreasing to 46 per cent of the market share.[8] The sports and leisure clothing and footwear markets are expected to rise in Europe according to a recent forecast.[97,98] The demand for specialist clothing and footwear designed and engineered for particular sports and leisure sectors seems set to rise, and this will introduce further market opportunities for synthetic fibres that are well suited for many sports and leisure activities. The sizes of the markets for each sector are illustrated in Table 5.9.

It is interesting to note that in 1990 Americans spent $1.5 billion on ski hardware (skis, boots, binders), protective clothing and accessories, and that worldwide skiing is a $28 billion industry.[99] In the United States, cycling ($3.3 billion) and golf equipment ($1.2 billion) are also large consumer markets.

The requirements for sportswear fabrics and garments are usually more severe than for leisurewear, for in many sporting activities the relative metabolic rate (RMR) may be considerably higher than in leisure activities. This is illustrated in Table 5.10, which illustrates the very wide range of RMR values for various sports.[100]

Table 5.9. The European market for sports and leisure clothing and footwear, 1990 and 1996

Sector	Sales value 1990/US$ bn	Estimated sales value 1996/US$ bn
Cycling/keep-fit (running, walking, gym and aerobic wear)	4.81	6.5
Sports footwear	3.33	4.69
Water sports/swimwear	2.59	3.39
Ski, horse-riding and mountain wear	2.55	3.29
Racket and ball sports (golf, football)	1.89	2.49

[Reprinted from reference 97 by permission of Frost and Sullivan]

Table 5.10. Quantity of perspiration and relative metabolic rate (RMR) generated by various sports and activities

Sport/activity	Perspiration/g/m^{-2} per 24 h	RMR
Golf	850	2
Tennis	2250	7
Hiking	850	2
Climbing	2530	8
Baseball		
Pitcher	1700	5.8
Fielder	870	2
Rugby	3400	11.1
Basketball	2650	12
Rowing	7000	24
Marathon	4300	14.3

[Reprinted from S Yagihara and T Furuta[100] by permission of Sterling Publications Ltd]

Under normal indoor conditions the loss of moisture by evaporation from the human body will rarely cause the clothing to be wet with liquid sweat (sensible perspiration). The transport of water vapour will thus occur almost entirely through the holes between the yarns in the clothing. It has been estimated for underwear fabrics under indoor conditions that 99 per cent of the moisture is passed through the interstices of the fabric and only 1 per cent through the fibres.[58] Thus polyester and poly(vinyl chloride) fibres with low moisture absorbency may be used satisfactorily in underwear.

However, where the wearer is active and may wear absorbent fibres like cotton or viscose, the higher quantity of liquid sweat produced will be absorbed by the fibres. Under such conditions, the wet fibres will be slow drying and will tend to cling more to the wet skin, possibly giving rise to sensations of discomfort. The use of synthetic fibres like polyester and nylon with low moisture sorption values provides conditions where moisture may be wicked away rapidly from the skin and evaporated from the fabric without necessarily generating wet cling conditions.

For active sportswear, Enka developed a special fabric (Golden Tower) composed of an inner fabric layer knitted from Diolen fibres that are provided with microscopically minute grooves on their surfaces.[101] These grooves act to transport liquid body moisture (perspiration) from the skin surface to an outer layer of moisture-absorbing cotton. Diolen spun yarns or Diolen FE filament yarns may be used for moisture transport, the latter being a special spun-like filament yarn that shows free protruding fibre ends. These fine fibre ends act as spacers between the fabric and skin just as with conventional spun yarns. The finer the fibres, the greater the number of channels along which moisture can be transported, and hence the greater the forwarding action away from the skin. This type of fabric has been used in tennis shirts and trackwear and the Diolen contributes not only to garment comfort but also to the dimensional stability and washability of the garment.[101] Washability is an important factor in active sportswear, which can rapidly become heavily soiled depending on the type of activity.

In the Du Pont Comforteam fabric similar principles are utilized in lightweight double-face knitted or woven fabrics in which the inner fabric is made from type D-241 polyester and the outer fabric from an absorbent fibre-like cotton. Type D-241 has a unique scalloped oval cross-section forming four channels along its length, and has 30 per cent more surface area than conventional circular cross-section fibres. Perspiration is drawn along the channels by capillary action away from the skin, and the fibres possess microscopic irregularities in the fibre surface that contribute to a soft handle. In other work on active sportswear, comfort polypropylene was demonstrated to have excellent properties for the inner fabric with its very low moisture regain, coupled with its excellent wicking properties.[50] It should be noted that polypropylene is used in babies' disposable napkins (diapers), where a layer of polypropylene is used to wick the liquid into the absorbent layer to keep the babies' skin dry.

Dunova acrylic fibre (Bayer) has been specifically designed for use in skin-contact clothing such as sportswear, leisurewear and underwear.[102] This fibre has a porous core-and-sheath structure that allows moisture to pass rapidly through the fibres, thereby keeping the area next to the skin dry and promoting a greater feeling of comfort. The Dunova fibres thus absorb moisture rapidly and permit the passage of moisture through to the outside of the fabric, where it is evaporated. This avoids the damp, sticky, uncomfortable sensation that could otherwise be generated after exertion or sporting activities. Moreover, Dunova does not swell, because the water is absorbed into the porous fibre centre. This permits un-diminished ventilation because the air permeability and the moisture

Table 5.11. Important factors in the selection of workwear

Order of importance*	Wearer	Employer	Rental operator
1	Comfort	Durability	Ease of maintenance
2	Protection	Protection	Durability
3	Appearance retention	Price	Protection
4	Durability	Comfort	Price
5	Ease of maintenance	Appearance retention	Appearance retention
6	Price	Ease of maintenance	Comfort

* 1 = most important   6 = least important
[Reprinted from I Holme[104] by permission of World Textile Publications Ltd]

diffusion properties are unimpeded. The swelling of natural fibres like cotton as they absorb moisture can lead to an appreciable decrease in the air permeability (17–28 per cent) in plain knit fabrics, whereas the comparable decrease in air permeability for Dunova is only 2 per cent. Dunova is approximately 40 per cent lighter than cotton and is claimed not to give rise to post-exercise chill.[102]

5.5.4 Synthetic fibres in workwear

The important factors to be considered in the selection of textile material for workwear vary according to the viewpoint of the wearer, the employer, and the rental operator[103,104] and are summarized in Table 5.11.

From the selection point of view, synthetic fibres in workwear can impart high standards of durability, protection, appearance retention and ease of maintenance, while the level of comfort can be attained through appropriate fibre, fabric and garment engineering.

Workwear is worn for protection and also for identification.[105] It may show the status, skills and responsibilities of individuals and identify them as belonging to a specific group or company. Workwear must protect the workers and their personal clothing from damage and soiling that may arise from the type of work done or the external environment, e.g. the weather. Workwear may also serve to protect the product or the immediate environment from the worker, e.g. from contamination. Suitably chosen workwear may also advertise a service or a company, and enables standardization of the type of clothing worn regularly by the individual or by a group of individuals.

Because workwear is worn day-to-day it is clear that it will be subject to extremes of wear and tear, and moreover it must be worn and washed much more frequently than fashion garments.[105] The high tensile strength, tear strength and abrasion resistance of clothing made from synthetic fibres, combined with a built-in high performance level in terms of the resistance to acids, alkalis and a wide range of organic solvents, often 44offers many advantages over the natural fibres. Fabrics made from natural fibres can suffer losses in mechanical properties as a result of chemical finishing to improve the degree of protection.

The qualities that are demanded of washable workwear are clearly dependent upon the fibre and fabric properties that influence garment performance.[105] Fabric strength, wear resistance, dimensional stability, colour fastness and pilling or snagging resistance are major parameters that must be considered, and synthetic fibres such as nylon not only have a very high wear resistance but also may be heat set to provide superior dimensional stability compared with natural fibres. Even in blend form, such as in polyester/cotton blends, the high strength, durability and dimensional stability of the polyester component are inherent in the fibre. In contrast, the properties of the cotton must be modified by chemical cross-linking to upgrade the performance to provide the satisfactory level of easycare performance that can be attained simply by heat setting polyester.

The response of fibres on exposure to chemicals, the moisture absorbency, antistatic properties, strength, and reaction on exposure to heat and/or flame are also of great importance for many workwear fabrics. The selection of fibres suitable for the end-use is often complex and compromises have to be reached in terms of performance, comfort and durability.

Synthetic fibres generally provide good resistance against chemicals, and suitable fibres may be selected to protect against specific reagents. Where this is not possible using fibres alone, the fabric may be coated with an appropriate impervious protective coating. The moisture absorbency of synthetic fibres is generally low, but can be increased by fibre engineering.[5] The antistatic properties of synthetic fibres are generally significantly improved by copolymerization, by the use of additives, or by application of surface finishes. The ability of synthetic fibres to be produced in high strength versions, by selection of an appropriate physico-chemical structure in high performance fibres or by drawing of conventional synthetic fibres to increase the chain molecular orientation, can also be important in many textile end-uses.

In respect of their heat resistance, there is normally a limited temperature range below the softening point within which satisfactory physical properties are maintained. In this respect aramid fibres such as Nomex (Du Pont) and PBI (polybenzimidazole; Celanese Corporation) exhibit superior high temperature resistance.[106]

Most synthetic fibres, in common with the natural fibres, will burn on exposure to a flame. Modacrylic fibres that contain chlorine-containing

Table 5.12. UK practice
Fabric types in UK domestic washing (per cent of total)

	1955	1965	1975	1985
Cotton	80	74	55	44
Nylon	5	18	30	15
Polyester–cotton	0	+	7	25
Polyester	0	+	2	5
Wool	15	6	4	4
Acrylic	0	+	2	7
Coloured materials	15	30	62	80

[Reprinted from I Holme[109] by permission of World Textile Publications Ltd]

co-monomers and aramid fibres and polybenzimidazole fibres can provide higher levels of flame retardancy.[107] In some fire scenarios the ability of synthetic fibres to self-extinguish by melt-dripping can be beneficial. However, in blends with natural fibres, the thermal decomposition of the natural fibre can form a carbonized skeleton that imparts a scaffolding effect that prevents melt-dripping from taking place.[108] The other problem is that of skin injuries caused by the melting of the synthetic fibre and the subsequent release of heat to the skin on re-solidification of the molten polymer on cooling.

Flame-retardant varieties of synthetic fibres can be produced using suitable co-monomers, incorporation of flame-retardant additives, or application of appropriate flame-retardant chemicals in fabric finishing. Thus the range of options available for modifying synthetic fibres is far wider than that for natural fibres, for which recourse can only be made to the application of flame-retardant chemicals in textile finishing.[108]

5.5.5 Garments containing synthetic fibres and aftercare treatments

The introduction of synthetic fibres has exerted a seminal influence upon the types of fabrics to be found in an average household wash load.[109] The changes in the United Kingdom, for example, are highlighted in Table 5.12, in which it can be seen that the decline in the importance of cotton is matched by an accompanying growth in the market share of synthetic fibres, the growth varying according to the vagaries of fashion. Particularly notable is the rise in the percentage of coloured materials in domestic washing, rising from 15 per cent in 1955 to 80 per cent in 1985. Some 500 million garments are washed every day in domestic washing machines in Western Europe.[109]

The ability to dye and print synthetic fibres and their blends with bright colours that possess a high colour-fastness to washing, light and rubbing has been a major factor in the success of synthetic fibres in the apparel market, as well as a stimulus to the development of novel synthetic dyestuffs.[1] Synthetic fibres like nylon and polyester combine a high strength with a light weight and the fibres have a low moisture absorbency and moisture retention. High temperatures in domestic washing can render synthetic fibres vulnerable to damage in washing or pressing. Thus the wash temperatures recommended for synthetic fibre garments are usually in the 40–50°C range.[110] This has been a distinct advantage from the viewpoint of energy conservation and the environment, because less energy is consumed per wash cycle and the discharge temperature of the waste water is considerably lower than in the 95°C boil wash cycle often used on cotton materials.

Garments made from synthetic fibres even for workwear do not usually require commercial laundering, but can be washed domestically in a machine or by hand, at low temperatures. The garments can usually be given a short spin (centrifugal hydroextraction cycle) to remove the capillary water in the structure, which rapidly ensures that the garments are taken out of the washing machine in a state of low moisture content. As a result, the subsequent drying of the garment may be rapidly achieved by line drying or appropriate tumble drying. Alternatively, the garments may be allowed to drip-dry overnight. Synthetic fibre materials may be tunnel washed and dried, but the high moisture absorbency of cotton materials is a disadvantage, often rendering tunnel drying completely unsuitable. Usually, synthetic fibre garments in which the fibres have been heat set retain their dimensions through domestic laundering and may require no ironing or only minimum-iron conditions to retain their fresh, smart and clean appearance.

It is to be noted that in the United Kingdom washing machine ownership during the period 1955 to 1985 increased from 15 to 83 per cent, with 64 per cent of washing machines in 1985 being of the drum automatic type.[109] The standardized wash cycles recommended in fabric care labelling schemes are thus implemented using microprocessor control to give precise control over all aspects of the wash cycle. Synthetic fibres can be engineered to minimize pilling during wear and do not felt like wool, or shrink like cotton that has not been compressively shrunk. Garments from synthetic fibres can thus be washed safely in multiple washing treatments for a large number of wash cycles before the effective wear life is achieved. Even under hospital washing conditions, where sterilization is required, the wear life of 100 per cent polyester bed-sheets is considerably greater than for all-cotton or for polyester–cotton blended bed sheets. Chemical degradation of fibres can occur during domestic laundering owing to the action of modern washing powders, which are highly alkaline and contain peroxygen compounds with oxidizing powers.

This degradation is much higher in cotton and other natural fibres than in fabrics and garments composed of synthetic fibres such as polyester, nylon and acrylic fibres, which are more durable to the action of such agents.

References

1. I Holme, *Chem. Br.*, **27**, 627 (1991).
2. G Loasby, *J. Textile Inst.*, **42**, P411 (1951).
3. I Holme, *Rev. Prog. Col.*, **1**, 31 (1970).
4. I Holme, *Rev. Prog. Col.*, **7**, 1 (1976).
5. A J Hughes, J E McIntyre, G Clayton, P Wright, D J Poynton, J Atkinson, P E Morgan, L Rose, P A Stevenson, A A Mohajer, W J Ferguson, in *The Production of Man-made Fibres*, ed. P J Alvey, *Textile Prog.*, **8**(1), The Textile Institute, Manchester, 1976.
6. S K Mukhopadhyay, in *The Structure and Properties of Typical Melt-spun Fibres*, ed. P W Harrison, *Textile Prog.*, **18**(4), The Textile Institute, Manchester, 1989.
7. M Lewin, E M Pearce (eds) *Fiber Chemistry, Handbook of Fiber Science and Technology*, Vol IV, Marcel Dekker, New York, 1985.
8. P W Leadbetter, A T Leaver, *15th IFATCC Congress, Book of Lectures*, Lucerne, June 1990.
9. I Holme, *Rev. Prog. Col.*, **22**, 1 (1992).
10. D M Nunn (ed.) *The Dyeing of Synthetic-polymer and Acetate Fibres*, Dyers Company Publications Trust, Bradford, 1979.
11. D R Karsa, *Rev. Prog. Col.*, **20**, 70 (1990).
12. I Holme, *Chemspec 92 BACS Symposium*, Manchester, March 31–April 1, 1992, p. 51.
13. M A Taylor, *Technology of Textile Properties: An Introduction*, 2nd edn, Forbes Publications, London, 1981, p. 64.
14. A T Purdy, *Developments in Non-woven Fabrics*, ed. P W Harrison, *Textile Prog.*, **12**(4), The The Textile Institute, Manchester, 1983.
15. H M Taylor, *Textiles*, **13**(1), 23 (1984).
16. M A Taylor, *Technology of Textile Properties: An Introduction*, 2nd edn, Forbes Publications, London, 1981, p. 144.
17. W E Morton, J W S Hearle, *Physical Properties of Textile Fibres*, 2nd edn, The Textile Institute and Heinemann, London, 1975, Chapter 13, p. 265.
18. K Slater, *Textile Degradation*, ed. P W Harrison, *Textile Prog.*, **21**(1/2), The Textile Institute, Manchester, 1991.
19. T L Vigo, in *Handbook of Fiber Science and Technology*, Vol II, *Chemical Processing of Fibers and Fabrics Functional Finishes*, Part A, ed. M Lewin, S B Sello, Marcel Dekker, New York, 1983, Chapter 4, p. 367.
20. A R Horrocks, *Textile Horizons*, **2**(1), 38 (1983).
21. M C Tubbs, P N Daniels (eds) *Textile Terms and Definitions*, 9th edn, The Textile Institute, Manchester, 1991, p. 23.
22. S P Hersh, in *Handbook of Fiber Science and Technology*, Vol III, *High Technology Fibers*, Part A, ed. M Lewin, J Preston, Marcel Dekker, New York, 1985, Chapter 1, p. 1.

23. J D Geerdes, *Int. Fiber J.*, **5**(3), 24 (1990).
24. M Okamoto, in *Tomorrows Ideas and Profits: Polyester 50 Years of Achieve-ment*, ed. D Brunnschweiler, J W S Hearle, The Textile Institute, Manchester, 1993, p. 108.
25. T Yasui, J Matsuura, in *Tomorrows Ideas and Profits: Polyester 50 Years of Achievement*, ed. D Brunnschweiler, J W S Hearle. The Textile Institute, Manchester, 1993, p. 210.
26. W E Morton, J W S Hearle, *Physical Properties of Textile Fibres*, 2nd edn, The Textile Institute and Heinemann, London, 1975, Chapter 6, p. 154.
27. A Cavallaro, G Cazzaro, *Opportunities for Man-Made Fibres*, Shirley Institute Publication **S36**, 1979, p. 67.
28. B G Frushour, R S Knorr, in *Fiber Chemistry, Handbook of Fiber Science and Technology*, Vol IV, ed. M Lewin, E M Pearce, Marcel Dekker, New York, 1985, p. 171.
29. J C Dupeuble, *The Indian Textile Journal*, **101**(11), 92 (1991).
30. P T Slack, *Textile Technology International*, Sterling Publications, London, 1988, p. 32.
31. J B Young, in *Polypropylene Textiles*, Shirley Institute, Publication **S44**, 1982, p. 47.
32. O Pajgrt, B Reichstädter, F Ševčik, *Production and Applications of Poly-propylene Textiles*, Textile Science and Technology, Vol 6, Elsevier, Amsterdam, 1983, p. 51.
33. M C Tubbs, P N Daniels (eds) *Textile Terms and Definitions*, 9th edn, The Textile Institute, Manchester, 1991, p. 364.
34. *Ibid* p. 193.
35. I Heidenreich, H Ninow, *E.I.U. Textile Outlook International* No. 40, 37 (March 1992).
36. D & K Consulting S.A. *Textile Month.*, 23 (January 1992).
37. S Davies, *E.I.U. Textile Outlook International* No. 27, 65 (January 1990).
38. W E Morton, J W S Hearle, *Physical Properties of Textile Fibres*, 2nd edn, The Textile Institute and Heinemann, London, 1975, p. 401.
39. H B Goldstein, *Text. Chem. Col.*, **25**(2), 16 (1993).
40. M C Tubbs, P N Daniels (eds) *Textile Terms and Definitions*, 9th edn, The Textile Institute, Manchester, 1991, p. 105.
41. J Hilden, *International Textile Bulletin, Dyeing/Printing/Finishing*, No. 3, 19 (1991).
42. P Richter, *Colourage Annual*, 1991, p. 55.
43. P Leadbetter, S Dervan, *J. Soc. Dyers & Col.*, **108**, 369 (1992).
44. *Polyester Microfibres*, ICI Colours, April 1991.
45. J Stryckman, *Wool Science Review*, No. 42, 32 (1972).
46. J E McIntyre in *Fiber Chemistry, Handbook of Fiber Science and Technology*, Vol IV, ed. M Lewin, E M Pearce, Marcel Dekker, New York, 1985, p. 1.
47. B Holcombe, *Wool Science Review* No. 60, 12 (1984).
48. T A Hardman in *Polypropylene Textiles*, Shirley Institute Publication **S44**, 1982, p. 33.
49. S M Watkins, *Clothing the Portable Environment*, Iowa State University Press, Ames, Iowa, 1984, Chapter 2, p. 16.
50. B Piller, *Textile Technology International*, Sterling Publications, London, 1991, p. 179.

51. Reference 32, p. 54.
52. E T Renbourn, W H Rees, *Materials and Clothing in Health and Disease*, Lewis, London, 1972.
53. K-H Umbach, Verbesserung des Trageokomforts von Herrenbekleidung durch optimierte Kleidungsventilation, *Hohensteiner Forschungsbericht* (January 1981).
54. D Gintis, in *Tomorrows Ideas and Profits: Polyester 50 Years of Achievement*, ed. D Brunnschweiler, J W S Hearle, The Textile Institute, Manchester, 1993, p. 232.
55. G Crawshaw, *EIU Textile Outlook International* No. 35, 41 (July 1987).
56. DuPont Technical Information.
57. 3M Technical Information.
58. E Clulow, *Textile Horizons*, **4**(9), 20 (1984).
59. B V Holcombe, B N Hoschke, *Textile Res. J.*, **53**(6), 38 (1983).
60. G Holmes, *Textile Horizons*, **5**(11), 25 (1985).
61. R L Shishoo, *Textile Asia*, **20**(2), 66 (1989).
62. M A Taylor, *Technology of Textile Properties, An Introduction*, 2nd edn, Forbes Publications, London, 1985, p. 139.
63. J E Smith, *Textiles*, **15**(1), 23 (1986).
64. J E Smith, *Textiles*, **22**(1), 18 (1993).
65. K Slater, Comfort Properties of Textiles, *Textile Progress*, **9**(4), The Textile Institute, Manchester, 1977.
66. M J Denton, Textiles for Comfort, *Third Shirley International Seminar*, Didsbury, June 15–17 (1971).
67. J Smith, *Textile Horizons*, **5**(8), 35 (1985).
68. R J Mayfield, *Textile Horizons*, **7**(11), 35 (1987).
69. G R S Naylor, C J Veitch, R J Mayfield, R Kettlewell, *Textile Res. J.*, **62**(8), 487 (1992).
70. P Mehta, *Wool Science Review* No. 60, 23, April 1984.
71. L Fourt, N R S Hollies, *Clothing Comfort and Function*, Marcel Dekker, New York, 1970.
72. N Wilson, *8th Shirley International Seminar*, Shirley Publication S24, 1976.
73. D M Brown, M T Pailthorpe, *Rev. Prog. Col.*, **16**, 8 (1986).
74. J E McIntyre, in *Fiber Chemistry: Handbook of Fiber Science and Technology, Volume IV*, ed. M Lewin and E M Pearce, Marcel Dekker, New York, 1985, p. 40.
75. K L Hatch, *Textile Res. J.*, **54**, 664 (1984).
76. K L Hatch, *Textile Res. J.*, **54**, 721 (1984).
77. S H Zeronian, M J Collins, *Text. Chem. Colorist*, **20**(4), 25 (1984).
78. B M Latta, *Clothing Comfort: Interaction of Thermal, Ventilation, Construction and Assessment Factors*, Ann Arbor Science, Ann Arbor, Michigan, 1977, p. 33.
79. J P Casey, *F.I.T. Review*, **1**, 29 (1985).
80. *Textile Horizons*, **6**(9), 40 (1986).
81. T Suzuki, *Microporous Polyester Fibers and Textile Use*, Textile Research Institute, 56th Annual Conference, Washington, DC, April 1986.
82. N Mathes, W Lange, K Gerlach, US Patent 4371485, February 1 (1983).
83. S H Zeronian, M J Collins, *Textile Progress*, **20**(2), (1989).
84. A A-M Gorrafa, *Text. Chem. Colorist*, **12**, 83 (1980).
85. B M Latta, *Textile Res. J.*, **54**, 766 (1984).

86. E M Sanders, S H Zeronian, *J. Appl. Poly. Sci.*, **27**, 4477 (1982).
87. P Harnett, *Wool Science Review* No. 60, 3, April (1984).
88. K Wilson, *Textile Technology International*, Sterling Publications, London, 1990, p. 115.
89. D K Wilson, T Kollu, *Textile Progress*, **21**(3), 22 (1991).
90. M A Taylor, *Technology of Textile Properties: An Introduction*, 2nd edn, Forbes Publications, London, 1981, p. 88.
91. M Newbury, S Schlehuser, *Knitting International*, **98**(1166), 27 (1991).
92. R Meredith, *J. Text. Inst.*, **36**, T147 (1945).
93. Du Pont Technical Information, Lycra Superfit Hosiery, Bulletin L-527.
94. C W Ericson, Energy Wear, Text of a lecture delivered at Leeds University, UK, 29 September 1992.
95. W Kirk Jr, S M Ibrahim, *Textile Res. J.*, **36**, 37 (1966).
96. R Hill, *Textiles* **14**(2), 30 (1985).
97. The European Market for Sports and Leisure, Clothing and Footwear, 1990 and 1996, Frost and Sullivan Report E1627/31.
98. *International Dyer* **177**(10), 6 (1992).
99. C A Michaels, *The British Association for Chemical Specialities Symposium*, 31 March–1 April (1992), p. 37.
100. S Yagihara, T Furuta, *Textile Technology International*, Sterling Publications, London, 1991, p. 214.
101. *Textile Horizons*, **5**(8), 37 (1985).
102. Dunova Technical Literature (Bayer).
103. M Elmasri, *Survival 90 Conference*, University of Leeds, 1990.
104. I Holme, *Textile Month*, June, 56 (1990).
105. K Edwards, E Klouda, *Textiles*, **16**(3), 70 (1987).
106. M Jaffe, R S Jones, *Handbook of Fiber Science and Technology: Volume III*, High Technology Fibers Part A, ed. M Lewin and J Preston, Marcel Dekker, New York, 1985, p. 349.
107. A R Horrocks, M Tunc, D. Price, *Textile Progress*, **18**(1)/(2)/(3) (1989), ed. L. Cegielka.
108. I Holme, *Colourage Annual*, 1991, p. 85.
109. I Holme, *International Dyer* **170**(11), 6 (1985).
110. M A Taylor, *Technology of Textile Properties: An Introduction*, 2nd edn, Forbes Publications, London, 1981, p. 183.

CHAPTER 6

Carpets

E D WILLIAMS

6.1 Introduction

The use of textiles as floor coverings undoubtedly grew out of the use of fibrous materials such as the woody stems of plants to cover the earthen floors of caves and tents. The consolidation of these by interlacing, as in braiding or weaving, was a logical next step, and from such methods woven mats were probably the first textile floor coverings.

As outlined by Robinson,[1] the art of carpet weaving probably originated in the southern part of the territory of ancient Persia about 4000 BC. There the shepherds spun coarse wool with distaff and spindle, dyed this yarn with juices of plants and wove these into rugs on a frame loom to serve as floor coverings as well as to provide soft materials to sit and sleep on.

The oldest carpet seems to be one found by a Russian archaeologist at Pazyryk in Southern Siberia. This carpet of the 'Persian' style dates from 2400 BC. It is made of wool pile yarn knotted to a warp of smaller yarns using a type of knot still employed in making hand-knotted 'Oriental' carpets today.

Robinson outlines the ebb and flow of the use of carpets and the art of carpet making westward through the Roman empire to be lost again in Western Europe with the fall of Rome. The Saracens reintroduced the skill in about AD 711, and carpet making was practised in Italy and southern France in the fifteenth century. By the end of the sixteenth century there were carpet weavers at Aubusson and Beauvais in France.

Carpet making spread widely during the sixteenth and seventeenth centuries, initially using the hand knotting of individual tufts of yarn but gradually introducing long yarn lengths by weaving to make loops in the pile and finally cutting these to make single tuft cut-pile carpets. Power was applied to carpet looms in 1840–1850 when E. P. Bigelow of the United States obtained a patent for this process. This heralded the change of carpet making from a cottage industry to a factory style of manufacture.

Other important methods of producing textile floor coverings include bonding a corrugated web of fibres to a preformed backing fabric, which was introduced in the 1930s; the insertion of pile yarns into a preformed backing by a sewing or 'tufting' process, which was first done on wide width material in 1949; the warp knitting of loop pile materials; and the production of felts by punching randomly oriented fibre batts into a backing material on a needle loom.

Worldwide the tufting process is the most important manufacturing method today, followed by weaving, bonding, needle-loom and knitting.

6.2 Carpet types: methods of manufacture

6.2.1 Hand knotted

The classical hand knotted carpet consists of predyed pile yarns that are individually knotted by hand around the warp yarn using one of the two traditional knots shown in Figs 6.1 and 6.2 and are then cut to the desired pile height. The Persian knot makes a more uniform and denser pile and permits a sharper definition of the pattern if all other factors are equal, but the quality of such a carpet is determined by many other factors such as the quality of the yarn, the density of the knots, the quality of the dyeing, etc., so that the value cannot be determined by the knot type alone.

6.2.2 Loom woven

Carpets made on the machine loom have the basic construction shown in Figs 6.3 and 6.4. The carpet pile and backing fabrics are produced at the same time. The warp and weft yarns are woven as in a plain weave while the predyed pile yarns are woven over pile wires that extend the

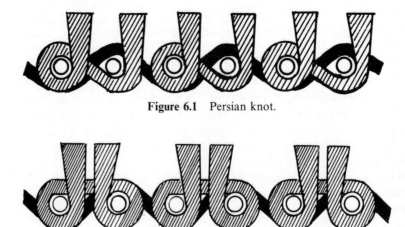

Figure 6.1   Persian knot.

Figure 6.2   Turkish knot.

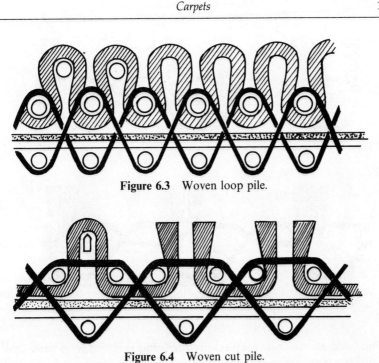

Figure 6.3   Woven loop pile.

Figure 6.4   Woven cut pile.

width of the loom and serve to create the loops of the pile yarn that extend above the carpet surface.

If these wires are provided with a knife edge at the end a cut pile is made as they are withdrawn during the weaving process; otherwise a loop pile is made. It is also possible to use wires that have varying width along their length. In this way, the pile height varies across the carpet width and patterns may be produced by varying the wire profiles in series.

By using more than one pile warp in a Jacquard loom, patterns of multiple colours may be made. This is done by weaving the colours that are not desired on the face into the backing and causing only the desired colour to loop over the pile wire at each insertion. The number of frames used corresponds to the number of such coloured warps.

Such carpets were originally called Brussels carpets if loop pile and Wilton carpets if cut pile. Cut-pile carpets of single colour are often known as plush or velvet carpets and may be woven in a single frame construction.

The Axminster carpet construction shown in Figs 6.5 and 6.6 avoids the use of the buried pile yarn that uses much of the expensive pile merely as filler in the backing by employing a set of pincer-like grippers, one for each warp yarn row, which extend across the entire width of the carpet loom and can withdraw lengths of yarn of the proper colour from a yarn carrier under the control of a Jacquard mechanism and place it into the weave shed so as to allow it to be secured with the weft shot yarns.

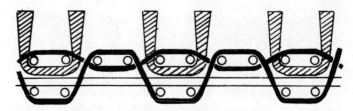

Figure 6.5 Axminister woven conventional.

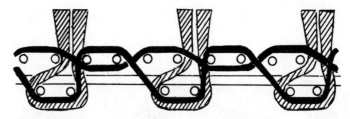

Figure 6.6 Axminster woven through-the-back.

The hand woven constructions are predominantly made of wool fibres. Although synthetic fibres are used in machine woven carpets, wool accounts for a major part of the production of these styles also. The carpet construction methods discussed below are the ones in which the synthetic face fibres have become dominant.

6.2.3 Tufted

Tufted carpets are made by inserting the pile yarn into a backing fabric using needles similar to those of a conventional sewing machine. The needles penetrate the backing from above and the pile yarn is held below the backing fabric by a looper. For the production of loop pile carpets this looper merely withdraws to release the loop thus formed as the needle descends to make the next switch as shown in Fig. 6.7.

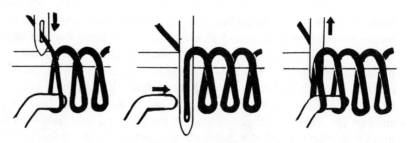

Figure 6.7 Tufted loop pile (level).

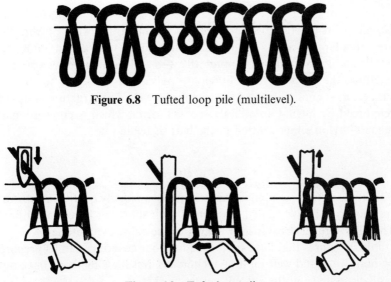

Figure 6.8 Tufted loop pile (multilevel).

Figure 6.9 Tufted cut pile.

The height of the loop formed is controlled by the rate of feeding of the pile yarn, which is regulated by a metering device. If a constant rate of feed is used then the carpet has a uniform pile height often termed 'level loop', while if the feeding rate is intentionally varied in a regular manner a pattern of loops of various heights may be produced, as shown in Fig. 6.8.

If the looper has a sharp knife attached, the pile produced is a cut pile, as illustrated in Fig. 6.9. Many variations of this basic pattern such as mixtures of cut and loop pile as well as sideways displacement of the tuft rows between stitches are used to create styling effects in the tufted carpet industry.

6.2.4 Bonded

Bonded carpets are made by folding a warp of face yarns that is delivered from a creel into an accordion-folded sheet that is then bonded on each side to a backing material precoated with an adhesive. The backing material is usually a nonwoven fibreglass material; the adhesive is commonly poly(vinyl chloride); and heat is used to effect the bonding. The resulting sandwich structure, which has the face yarns bonded at both ends of the folds, is then separated by cutting parallel to the backings to make two carpets in which the cut-pile face yarn is secured to the backing by being bonded with the vinyl resin.

6.2.5 Needle punched

Needle punched structures are prepared from carded webs of fibre that are cross-lapped and are then consolidated by the action of barbed needles, which are pushed through the web and then withdrawn so as to interlock the batts as some fibres are pulled perpendicular to the web surface, thus entangling and densifying the web. The structure is usually reinforced by having an adhesive-coated scrim, which is conventionally made of nylon either centred in the batt or behind it.

6.2.6 Other

Structures similar to bonded carpets may be made by coating a substrate such as an adhesive-coated vinyl sheet with short cut fibres called flock. The fibres are electrostatically drawn to the substrate material.

Pile structures that have been used for floor coverings can be prepared by both warp and weft knitting techniques, but these methods have only a small presence in the current carpet market.

Machine braided spun yarns, usually containing a core of waste staple fibres as filler and to add bulkiness, are sewn together in a spiral pattern to make floor coverings that are intended to simulate similar structures that were previously hand made as a home industry.

Carpet constructions

Carpet constructions may be categorized depending on whether the pile yarn is in uncut loops or in cut lengths and what twist levels and ply structures are used in the yarn constructions. The styles described below include only the major types prevalent in today's carpet markets. The structures of the major types are shown in Fig. 6.10.

6.2.7 Loop pile single level

These styles constitute an important proportion of the commercial styles used in relatively heavily trafficked areas. They are given variety in visual appearance through the use of multicoloured yarns such as heather yarns and of colour patterns created by using precoloured yarns or yarns with different dyeing characteristics that create different colour effects during finishing.

6.2.8 Loop pile multilevel

These effects are created during the carpet preparation and may be arranged at random or in a regular pattern to produce visual variety.

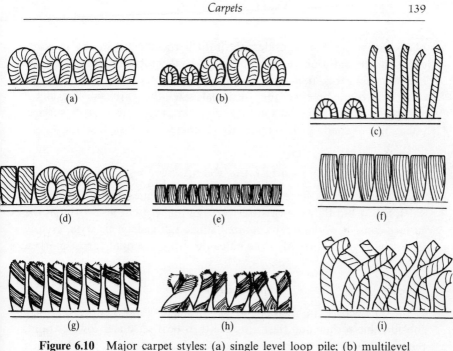

Figure 6.10 Major carpet styles: (a) single level loop pile; (b) multilevel loop pile; (c) multilevel cut and loop; (d) tip sheared; (e) velour; (f) plush; (g) saxony; (h) friezé; (i) shag.

The effect may be combined with the use of multicoloured yarn or of precoloured yarns to create a wide range of effects.

6.2.9 Cut pile velour

This style has a uniform tuft length and little or no integrity of the tuft so that the surface appears to consist of the ends of fibres arranged in parallel. The pile heights used are usually short, commonly 6 mm, and the density of packing is relatively high in order to provide a carpet with reasonable resistance to crushing. Such styles may be produced by weaving or tufting and bonding. Flocked or needle punched carpets normally have such a pile structure.

6.2.10 Cut pile plush

Carpets made from plied yarns with low twist levels have a surface in which the yarns from adjacent tufts tend to blend together. Such styles are called plush, velvet or velvet plush.

The pile heights are normally higher than those used for velour, often 12 to 19 mm. Low density versions with higher pile heights permit the pile yarns to lie over considerably from perpendicular to the backing, exposing the sides of the yarn bundles. Higher density and lower pile height versions maintain a more erect pile geometry. The optical appearance differences between the yarn ends and the yarn sides result in variations in the apparent lightness and darkness of the carpet surface, giving rise to shading effects that are characteristic of such styles.

6.2.11 Cut-pile saxony

These carpet styles have pile heights and densities like those of the plush family but they use twist-set yarns of higher twist levels so that the surface consists of readily recognized single tuft ends. This style, usually in a solid colour, currently dominates the US residential carpet market.

6.2.12 Cut-pile friezé

If the pile yarns are twisted so as to make a plied yarn that has considerable torque and the yarns are then heat set under low restraint, the resultant heat-set yarn has kinks distributed along its length. When placed in carpet, these yarns produce individual tufts that deflect at random locations along their length in random directions to make a surface in which the tufts are clearly seen and which has a nubby texture. A similar effect can be achieved with continuous filament yarns by imparting a mechanical crimp to the yarn during the continuous heat-setting operation. Considerable latitude in the final appearance is obtainable by varying the intensity and frequency of the kinking.

6.2.13 Cut-pile shag

When the pile height of the tufts is large and the density is relatively low, a carpet results in which the tufts are primarily oriented nearly parallel to the carpet backing and the sides of the yarns are clearly seen. These styles are called shags.

6.2.14 Cut and loop mixtures

Several methods such as tufting a mixture of cut and loop, shearing high-level loops after tufting, and selectively shearing portions of the tufted carpet are used to create styles containing both cut and uncut loops of pile yarns. Combining these with the tufting of various pile heights gives rise to a large number of carpet styles including those known as random shear, sculptured, and high–low sheared according to the final face appearance. These are primarily used in residential applications.

6.3 Carpet performance characteristics

Although it may at first appear that carpeting is used primarily as a means of creating a pleasing aesthetic character in the living and working environment, there are also effects on the physical well-being of those who use the carpeted area. The most important of these factors are

walking comfort
thermal comfort
acoustic comfort
absence of static generation
low flammability

Once a suitable result has been obtained that satisfies the aesthetic requirements and these personal requirements, it is obviously desirable that the carpet maintain these characteristics for as long a time as possible. This leads to the need to have high levels of the following factors:

resistance to material loss through wear
retention of the initial surface appearance
texture retention
colour retention
soil resistance
stain resistance

Finally, it is desirable that these characteristics be obtained at the lowest initial cost and the lowest total cost of ownership and maintenance.

Obviously, the relative importance of these factors is a complex function of the type of location in which the carpet is to be used and the level of wear to which it is to be exposed as well as of the aesthetic considerations of style and taste of importance to the users. Thus there is no single answer to the question 'What is the best carpet to purchase?' and probably no answer to the question 'Which is the best carpet fibre?'.

Much research has been done already on the development of means of measurement of these characteristic properties and the relating of the results to the carpet construction and fibre parameters. None the less, there are several of these properties for which the present state of our ability to test and to interpret the results is unsatisfactory and so appreciable future reseach will be required to clarify the situation satisfactorily.

6.3.1 Walking comfort

The cushioning effect of the floor covering on the person who walks upon it results in an impression of luxury and in a decrease in fatigue upon continued standing and walking about. These effects are due to the decrease in impact loading of the muscles caused by energy dissipation of the floor covering during the walking cycle. The human observer perceives this interaction as a sensation of 'cushioning' or 'softness' initially. A poorly energy absorbing surface will contribute to an increasing feeling of soreness and fatigue on continued exposure.

One of the early studies of these effects with the objective of quantitatively describing these sensations is due to W. Herzog.[2] His treatment analyses the physics of the walking actions in terms of the impact of the foot upon the floor. This was simulated in this work by studying the force–displacement curves obtained when a free-falling weight was allowed to impact on various floor covering materials. The most significant observation at this stage was that ranking of the maximum impact forces, which is a measure of the energy absorption by the floor covering, of a series of materials changes with the magnitude of the impacting weight. Thus, materials that might be very desirable for use as a covering for a running track would represent a poor choice for use in a residential living room or bedroom and would not be optimum for use in a business office or retail store environment.

Such research is still ongoing (see for example McFarlane and Watson[3]) and has shown that the subjective perceptions of the walker may be correlated to the absorption of impact loads by the surface. These properties include the entire floor covering system of pile fibre, backings and underlayments so that the properties of the carpet itself do not entirely determine the properties of the environment. Thus far there has not been any conclusive evidence that differences in mechanical properties between the pile fibres have a controlling influence on the energy absorption characteristics of the system.

In general, textile floor coverings used for sports surfaces in gymnasia and the like are dense and firm such as needlefelts and low, dense pile structures made with fibres that show good abrasion resistance and are used with little or no underlayment over hard surfaces such as concrete or hardwood. Heavy industrial areas are similarly carpeted.

Materials for use in commercial locations vary with the expected traffic loading and the pattern of use, i.e. long corridors where rapid walking is the norm versus enclosed offices where a more leisurely pace is expected. The carpet systems usually consist of low to medium pile height, moderate to high density structures, most frequently of loop pile construction and usually installed directly on the underlying surface.

The high pile materials are best used in the residential environment where the walking pace is slower, i.e. a 'strolling pace', and the feeling of luxury is enhanced by combining such materials with a compliant and relatively thick underlayment.

6.3.2 *Thermal properties*

Carpeted areas are commonly perceived as being 'foot warm' as opposed to areas of tile, ceramic or bare wood. The presence of a layer of fibre on the floor does have appreciable insulating effect and decreases the rate of heat transfer from the air to the colder floor surface. The magnitude of this effect of course depends upon the thermal gradient and is greater as the floor surface is less well insulated and is exposed to a lower temperature on the outer side.

The heat transfer of the flooring sytem may be characterized in terms of its R value, which measures the resistance to heat flow. In a study by the Carpet and Rug Institute,[4] it was shown that the overall R value of a flooring system was a linear combination of the R values of its components, i.e. the carpet pile, backing and underlayment materials. The carpet R value depends mainly upon the thickness and is relatively little influenced by the properties of the pile fibres. The insulating mechanism is largely the result of decreased convection and radiation and is increased by increasing the thickness of the structure. Since the conduction of heat is greater through the fibres than it is through still air, however, a structure with lower fibre content but of equal thickness and containing a larger proportion of entrapped air will have a somewhat greater insulating value than one of the same thickness with a higher density and greater fibre weight.

The magnitude of this effect has been reported in various studies to result in fuel savings of 5% to 13% from completely carpeting a school building in the United States and 9% to 13% in a parallel test of carpeted versus noncarpeted model homes in Japan. The importance of such changes is highly dependent upon the exact nature of the application; the carpet contribution may be estimated in the same manner as that used for other insulation methods.

6.3.3 *Acoustic comfort*

Carpeted areas are quieter than those with a hard surface, both because the movement of people and objects across the pliant flooring generates less sound and because the sound that originates within the area from other sources is partially absorbed by the floor covering. The absorption of airborne sound may be measured by standard tests[5] that measure the reverberation times of sounds in a reverberation chamber using various frequencies and present the result as a sound absorption coefficient indicating the completeness of the absorption. As with thermal transmission, the fibre composition is not usually a major determinant of the result; thickness and porosity have a larger effect and the underlayment plays a significant role.

A second sound absorption effect of carpeting is the absorption of impact sounds through the floor to the rooms below. This is also measured by standard tests[6] that measure the sound transmitted through the assembly of the carpet and a simulated standard floor layer by the action of a tapping machine. These effects are a function of the thickness and density of the carpet and of its underlayment but do not depend greatly on the nature of the pile materials.

6.3.4 Static generation

Static electricity is the frictionally induced accumulation of electric charges on a substance. This charge results from the separation of two dissimilar insulative materials. One material gains electrons and becomes negatively charged while the other material donates electrons and becomes positively charged. The polarity each material develops is dependent on the relative ranking of the materials in the triboelectric series. However, this series does not predict the magnitude of the charge, which is dependent on several factors including the nature of the materials, type of friction used and relative humidity.

In the case of carpets, static electricity is generated from repeated frictional contact between the shoe soles of a walking person and the carpet. This build-up of static electricity, for example with shoe soles of Neolite (a common rubber-like sole material) and nylon carpet fibre under low relative humidity conditions, can generate very high charges on the walking person. This can result in a disagreeable electrical shock when the person touches an electrical ground such as a doorknob. Relative humidity greatly impedes the static build-up. The lower the relative humidity the greater the static build-up. It is not uncommon for air-conditioned and heated offices and homes to reach low relative humidity conditions that yield annoying static shocks on carpets that are unprotected from generating static electricity.

Static electricity is measured by monitoring the voltage on a person, which is directly proportional to the charge build-up. Voltage rather than charge is measured because voltage equipment is more readily available.

Depending on the choice of shoe sole material and the carpet face fibre, static (maximum) levels of 15 000 V can be obtained. A static discharge at this level can be quite unpleasant. Most people will not experience a disagreeable shock sensation at or below voltage levels of approximately 3000 V. This human threshold level is the basis for specifying the acceptable 'comfort' level for many static-protected carpets.[7-9]

Various modifications to the carpet can reduce the static build-up on persons. These include the use of carpet fibre additives, topicals and antistatic fibres. Carpet fibre additives and topical materials, such as humectants and ionic conductors, change the surface chemistry of the carpet fibres. They tend to increase charge mobility, thereby reducing charge build-up. The drawbacks of these systems are that they may not work under prolonged periods of low relative humidity, are not durable and must be regularly reapplied, and may contribute toward accelerated soiling of the carpet.

The most common and permanent form of static protection for carpets utilizes antistatic fibres. These come in two major types, conductive and inductive. Both types contain a conductive phase most commonly made from either a carbon black material or a metal. They function by decreasing the level of charge build-up on a person by moving charge away from the person. The conductive fibres transfer the charge by conduction along the fibre surface. These materials include composite filaments that have exposed conductive material, commonly carbon, at some portion of the fibre surface or metallic filaments that are electrically conductive. The inductive fibres are composite materials in which an insulating polymer sheath surrounds a conductive core material, usually carbon. Transfer of the surface charge takes place by induction through the insulative sheath to the conductive core where it is then transferred by conduction. Although both types of fibres reduce static levels, only the surface conductive fibres can be measured using resistance techniques. Each antistatic fibre has some advantages and disadvantages compared to the others based on durability, colour and cost.

Carpet static protection for people is based on the static level threshold value for most people at or below approximately 3 kV. Accordingly, most commercial antistatic carpet installations have a warranty of approximately 3 kV. Since residential environments do not generally have areas in which maximum voltages can be reached (such as by walking down a long office hallway) and since residential carpet constructions may have a greater propensity to generate higher static levels than commercial carpet constructions, residential antistatic carpets typically have a warranty at or below approximately 5 kV.

However, since computers or other electronic equipment can potentially be damaged by static discharges at levels below the human perception threshold value, a more restrictive static level criterion is used for carpets intended for use in locations where such equipment is installed. Even though most, if not all, computers have some degree of shielding from static discharge from charged persons or other sources, carpets with lower static levels are available for protecting them. Typically, warranties at or below 2 kV are available. However, this assumes that the computer is shielded to a higher value in order to ensure protection from damage. This static level value will continue to decrease as improved antistatic carpets are developed. Resistance tests are often employed for this type of end-use. Typical values of 10^{-10} or 10^{-8} ohms or less are commonly specified. In order to meet this criterion, a surface conductive agent in the carpet pile and sometimes a conductive backing or adhesive must be used. The conductive backing or adhesive can also help to lower the static level.

6.3.5 Flammability

The combustion of organic materials in carpet form is dependent on a large number of carpet construction and fibre factors. Since filamentary materials have very high surface areas per unit weight, there is a large exposure to atmospheric oxygen. The normal course of the combustion process is one of heating from an external ignition source to reach the fibre decomposition temperature, followed by production of volatile pyrolysis products that form combustible mixtures with the air around the carpet. In order for the combustion process to be self-sustaining, there must be enough heat generated to maintain the material at a state of continuing pyrolysis; a sufficient surface area must be maintained to make an adequate supply of the volatile fuel; and there must be sufficient oxygen present to maintain a combustible mixture.

Fibres differ in their decomposition temperatures, the nature of the decomposition products, the percentage of oxygen required to maintain a combustible mixture, and the preservation of a large surface area. Selected combustion-related properties are listed in Table 6.1 for the fibres that are currently of most interest in floor coverings. The limiting oxygen index cited is the percentage by volume of oxygen required in the surrounding gas phase to support combustion of the fibre. In general, if this index is above the 21 per cent oxygen normally present in air the fibre will self-extinguish, while if it is below this value the fibre will continue to burn. However, the tendency of nylon and polyolefin materials to melt and pool, thereby decreasing their surface area, tends to decrease their apparent flammability. Also, the influence of blending of fibres of different chemical compositions produces materials whose properties are not simply the sum of those of their constituents. Thus, the flammability of acrylic carpet may be greatly reduced by blending a minor proportion of chlorine-containing modacrylic fibres since chlorine-containing pyrolysis products serve to act as free-radical terminators and quench the combustion. For these reasons, the flammability of carpets is determined for the entire carpet by use of tests that attempt to simulate and are intended to correlate with actual use conditions.

There is at present no international accepted standard for measuring carpet flammability. However, there are general similarities in the testing methods used. The levels for acceptance are a matter of local legislation

Table 6.1. Flammability properties of common carpet fibres

	Decomposition temperature		Ignition temperature		Limiting oxygen index/
Fibre	°C	°F	°C	°F	%
Wool	446	230	1094	590	25.2
Nylon	653	345	989	530	20.1
Polypropylene	–	–	570	349	18.6
Acrylic	549	287	986	530	18.2

and vary from country to country and even within countries by local ordinance.

The least demanding carpet test is the Methenamine Tablet Test which is described in US DOC FF-1-70 and 2-70, ASTM D 2859, Canadian Method 27.6, B.S. 6307:1982, and International Standard ISO6925-1982 as well as in many other similar test specifications.

The major fibres used in floor coverings—wool, nylon, polypropylene and acrylic—can all be constructed so as to pass the Methenamine Tablet Test provided that they have sufficiently close packing of the fibres. The acrylic carpets are most easily ignited in this test and usually require the addition of some proportion of chlorine-containing modacrylic fibres to act as flame quenching agents to avoid flame spreading.

The more severe Radiant Panel Test[10] measures the minimum radiant flux which must be supplied to a carpet in order to support combustion. High values of this critical radiant flux, CRF, indicate that the carpet is difficult to burn. For example, in the United States, a carpet installation intended for use on exit routes in health care facilities must have a CRF of at least 0.5 W cm^{-2}. To achieve this level it is usually necessary to maintain quite dense structures, to use as backing latex compositions that contain a high proportion of alumina trihydrate, which dehydrates and absorbs considerable energy on exposure to heat flux, and to use the carpeting in a direct glue-down form to permit dissipation of heat to the support rather than over a padding material that acts as a thermal insulator. By using appropriate combinations of these parameters, carpets which meet the requirement can be made of acrylic, polypropylene, nylon and wool.

Wool fibres form an insulating char on thermal exposure, while nylon and polypropylene fibres form a melt pool. The effect of the heat conduction properties of the padding is much more evident in the latter case since the char layer serves to insulate the remainder of the wool carpet from the radiant energy. Thus the CRF level of wool carpets decreases less when padding is used than does that of the polyamide or polypropylene carpets. In general it may be concluded that wool and wool blends with nylon or with polypropylene can pass the Radiant Panel Test over the broadest range of conditions, with nylon, acrylic and polypropylene following in that order.

6.3.6 Abrasion

There are several routes by which face fibres may be lost from carpets. These include possible catastrophic disruption of the carpet structure by mechanical means, the so-called shedding of carpets made from staple fibres, and the abrasive wear loss of segments of fibres broken off from longer initial lengths by the mechanical actions of traffic and maintenance.

The catastrophic damage event is not generally related to the nature of the fibre used in the carpet, and in any event is not related to abrasive action *per se.* In cases where the expected traffic is very severe, such as in industrial corridors exposed to heavily loaded wheeled devices and the like, the use of very short pile height, dense constructions and perhaps the use of cut rather than loop pile is preferred in order to present the minimum of snagging points on the surface of the carpet.

Shedding occurs by the working out of fibre lengths that were not anchored into the carpet backing during manufacture in response to the rubbing action of wear. The magnitude of the shedding depends upon construction factors such as the degree of parallelization of fibres during yarn preparation, the initial staple length distribution, the production of shorter fibres during the yarn and carpet preparation and the pile height in the carpet, modified by the degree to which the unanchored fibres have already been removed from the carpet sample during the finishing steps such as dyeing, brushing and shearing. The actual quantity of fibre lost in the shedding process is very small and normal maintenance such as vacuuming will remove it in the very early stages of use, so that the main consequence is the possible reaction of concern from the user on detecting quantities of fibre in the cleaning apparatus. For this reason, most manufacturers describe this phenomenon in their care and maintenance instructions, assuring the user that it is temporary and that negligible loss of fibre will result.

True abrasive wear involving fibre breakage is mainly found in the case of wool and of acrylic fibres and very rarely in the case of polypropylene or nylon. Work done by J. W. S. Hearle *et al.* at the Wool Research Organization of New Zealand (WRONZ) and others indicates that the mechanism of this fibre failure is primarily one of fatigue rather than of simple tensile failure. It is often preceded by filament splitting typical of such fatigue failure processes. Very recent work from the WRONZ group reported by G. A. Carnaby[11] has summarized the development of an equation useful for wool fibres that relates the fibre mass (M_{pn}) remaining after a number of treads (n) to the number of fibres per unit area (G_c), the fibre linear density (tex_f), an arbitrary length parameter (d), which is shorter than the pile length, and a term $[q(d, kn)]$ expressing the proportion of fibres of this length which will fail after this many tread cycles times an arbitrary factor (k) representing the probability that a particular fibre is loaded during a given tread event (Eqn 6.1):

$$M_{pn} = G_c \times \frac{tex_f}{10^5} \times \frac{d}{2} \times \frac{2 - q(d, kn)}{q(d, kn)} \qquad [6.1]$$

Ongoing work is attempting to define the applicability of the equation over a range of construction parameters.

Such work is hampered by the fact that currently available accelerated tests of carpet durability have very low correlations with the results of actual carpet wear trials, especially when comparisons are made between different fibres. The proper conduct of meaningful wear trials is in itself a matter requiring careful control. In order to obtain results in a reasonable time, such trials are often conducted under laboratory conditions using walkers who traverse the samples in a controlled and specified manner. Factors such as the presence or absence of turning and its degree if present, the nature of the shoe soles and the effect of wear on them, the severity of the walking gait, the effect of soil pickup on the carpet, the presence or absence of a recovery period during which there is no walking on the samples, the use of periodic maintenance such as vacuum, including its type and severity, the ambient conditions of temperature and humidity under which the test is conducted, and many more factors must be considered if the results are to have general validity. The alternative technique of installing test carpet samples in an area of very high traffic has the problem of making control of many of these factors impractical. It also usually introduces the need to rotate randomly the location of the items so as to equalize the wear intensity.

Normally, and especially in the case of nylon and polypropylene fibres, the useful life of a carpet is much shorter than the time required to wear away a significant quantity of the face fibre. Rather, it is customarily determined by objectionable changes in the appearance of the face material, especially when these cause patterns of wear to become evident, or by the appearance of soiled areas and stains that make the carpet unsightly. The tests used to measure these characteristics will be considered next.

Surface appearance retention

6.3.7 Texture retention

Texture change in this context is defined as the difference in the appearance of the carpet at two different stages of wear due to changes in such factors as the tuft configuration through flattening or separation of the yarn bundles, by matting of the fibres at the tuft surface or between tufts, and by fuzzing of fibres on the carpet surface. Other appearance changes such as colour change, soiling and staining are not intended to be included in the texture retention response.

There is no internationally accepted laboratory test method yet available for determining the degree of alteration of the surface texture of a floor covering in use. Also, despite general agreement that actual exposure to floor traffic followed by visual evaluation is the only method now available to obtain a realistic assessment of the texture retention, there is no universally accepted protocol for the conduct and rating scheme of such tests.

There is also considerable effort being placed on the objective rating of changes in the surface appearance of worn carpets. The bulk of this is involved in the application of image analysis techniques to the digitized local reflectance levels of the carpet surface.[12–15] Although promising progress has been made in deriving parameters from such images that correlate with visual assessment of the carpets within limited areas of fibre content and construction, the generalization of such techniques to the broad range of carpet styles to be evaluated remains to be accomplished. None the less, this is a most promising field of research that should be of considerable importance in providing detailed rather than overall information about the surface changes occurring during carpet wear processes.

The analysis of the response of the individual fibres in a carpet to the mechanical action of wear underfoot is extremely complex. There have been many attempts to deal with the mechanics of simple compression of the carpet structure.[16–24] Although none of these provides a really satisfactory model of the behaviour of carpets under compression–recovery cycles, it is generally accepted that the major factors that determine such behaviour are the stress–strain character of the individual fibres, the interfibre frictional restraints and the interyarn frictional restraints. On repeated cycling, the viscoelastic properties of the fibres, which can lead to stress decay and creep, will have an effect.

A useful qualitative model that is very similar to that discussed by Carnaby[11] considers that the initial compression of the carpet structure under a footstep results in a response in the tuft structure that is a combination of slippage between yarns, slippage of fibres within yarns and the bending and twisting of the individual fibres. This continues to compress the carpet pile structure until the recovery force of the compressed pile structure balances the compressive load. As the weight is removed from the footprint, the pile structure recovers until the recovery force is balanced by the frictional restraints. Since the fibres still have a residual stress, stress decay occurs. The pile geometry after this first cycle is more compressed than the initial untrodden geometry owing to the frictional restraints. A second compression cycle will repeat the mechanical deformations of the first cycle, but the fibres will in general have undergone some stress decay so that the degree of deformation required to generate a balancing recovery force will be somewhat greater that that experienced on the initial cycle. The result of continued exposure to foot traffic is a 'ratcheting down' of the pile structure to a limiting pile density.

The extent to which the carpet appearance is altered by the above process depends greatly upon the initial pile structure and appearance of the carpet, the mechanical properties of the fibres from which the carpet is made, and the detailed geometry of the tuft and yarn structure itself. For example, a carpet of the shag type, which is made from relatively long tufts of yarn spaced far apart in the carpet so that the individual tufts are not perpendicular to the backing but rather have a large degree of 'pile layover', will be compressed largely by loading the fibres on the side of the tufts. Such a structure will not have a high initial pile thickness, but will show relatively little loss in thickness during wear. The major factors degrading the surface appearance of these carpets will be the splaying out of fibres from the individual tufts and possibly their entanglement to form a matted surface. In such cases, the recovery properties of the fibres from torsion, extension and compression have a minor effect on the carpet performance. In the case of a velour or velvet style, on the other hand, the fibres are relatively closely packed and stand erect in the original carpet. The pile height is conventionally rather short and there is not much fibre-to-fibre frictional force involved. In this case, the recovery properties of the fibres have a large role to play in the retention of the carpet appearance, since a fibre with low recovery will soon take on a permanently bent configuration. The resulting contrast between the reflection from the fibre ends and that from the sides causes a distinct 'wear' pattern to be seen.

Loop pile carpets

Loop pile carpets are subject to the modes of distortion shown in Fig. 6.11. If the lateral packing is high and the pile height is low, then the flattening mode will predominate, while with less-dense constructions of higher pile height layover will be seen. At high tuft densities and

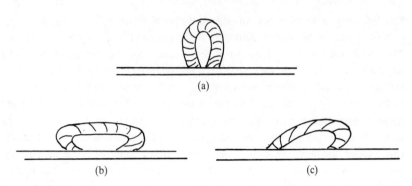

Figure 6.11 Loop pile distortion in wear: (a) original; (b) flattened; (c) layover.

especially with plied yarns, all of the fibres are capable of good texture retention. Such carpets are used for demanding areas of high traffic. As the pile density decreases, however, the fibres of lower recovery properties, polypropylene and polyester, suffer from extensive loop distortion with a pronounced change in appearance. The staple fibres, acrylic and wool, also have the problem of migration of fibres whose ends are not held into the carpet backing to the carpet surface, which creates fuzzing, and, if the fibres are relatively long, the problem of pilling, i.e., the formation of entangled balls of fibre on the carpet surface. Nylon with its high recovery used in continuous filament form can provide the broadest range of loop pile densities with good texture retention.

Velour carpets

Velour carpets, as mentioned above, require a fibre of higher recovery for good appearance retention, especially at lower densities and higher pile heights. Fibres that are relatively high in reflectance from the side will show a large difference in contrast from tip to side and will cause a large difference in appearance if the pile yarn leans over to expose the sides of the fibres to the observer. For this reason, wool and the delustred synthetic fibres offer the best appearance retention. In a recently published study[25] of the appearance retention of polyamide staple and of wool fibres in velour carpets, the effect of filament size of the polyamide was found to be the greatest factor with coarse fibres of 17–22 dtex giving far superior performances as compared to that with 4.4, 6.7 or 11 dtex fibres. For coarse fibres, the appearance retention increased with increasing pile weight. The use of constructions with high pile densities from yarn of coarse fibres was particularly favourable, as expected.

Saxony carpets

The performance requirements of the saxony carpets that are the most popular styles in the North American market and are very popular in the British market as well are more complex. These are cut-pile carpets made from yarns that have been twisted together and then set in the twisted form before being put into the carpet. The style owes its distinctive appearance to the presence of these compact cut ends of twisted yarns on the carpet surface. During wear, these carpet tufts may distort by becoming untwisted at the tips to show a brushy appearance, by buckling along their length, by leaning over to give a pile lay that exposes the side of the tuft, and by untwisting more extensively to permit splaying of the ends of the individual yarns composing the tufts. It is of major importance to the retention of carpet appearance during wear that the individual fibres in the twisted yarns be adequately set in that configuration. The process of twist-setting of yarns will be discussed in more detail later, but it should be stated here that the object of setting of the twisted yarn is to alter the state of the individual fibres from one in which there is considerable stored internal energy resulting from the deformation of twisting to a state in which the fibre has relaxed to a nearly zero energy state in the deformed geometry. This is achieved by heating, either dry or in a steam atmosphere, for the acrylic, polyamide and polypropylene fibres, and by wet treatment, often including a chemical step to break and reform disulphide linkages, in the case of wool fibres. By whatever means it is achieved, a high degree of twist-set is the most important factor in achieving good appearance retention in saxony style carpets. Carpet construction variables that improve the retention of surface appearance are the use of high ply twist levels, small to moderate size single yarns and yarns of lower bulk. High tuft densities also help, but there is an opposing requirement to maintain a minimum amount of twist along the individual tufts; this implies that a minimum pile height is needed so that the individual tuft has sufficient twist to hold the singles yarns in position. Increasing the pile density at constant overall weight by shortening the pile height may cause a worsening of the texture retention if the individual yarns have low enough twist to allow them to separate during compression. It is usually necessary to have at least one full turn of ply twist along the length of each tuft in order to get acceptable appearance retention, with higher values providing improved performance.

Of the major fibres, polypropylene provides the poorest appearance retention in saxony constructions. In this case, the heat setting is relatively easily overcome by mechanical working and significant matting of the pile and untwisting of the yarns is seen. Polyester saxonies perform relatively well if high tuft densities are used, but in more open constructions serious matting and tuft distortion are seen. Saxonies of wool, of acrylic and of wool blends with acrylic, polypropylene or polyamide of moderate to high tuft density perform very well. Saxonies of polyamide yarns, especially those from continuous filament yarns, have shown the ability to give good appearance retention performance over a very large range of pile fibre weights and tuft densities. Overall, the high recovery, heat settable polyamides are the most versatile of the major fibre classes.

Friezé carpets

Classical friezé yarns are made by producing a two-ply yarn with unbalanced twist so that the yarn is very twist lively and produces buckled sections and pigtail twisted segments when not held under tension. Such yarns are skeined after twisting and are then heat-set in steam autoclaves while in the relaxed and thus kinked form. Frequently these skeins are tumbled during setting to develop their kinky character. The tufts of carpets made from such yarns show a random pile direction with the pile yarns bent to varying degrees at random along their length. Because of the relatively high cost of this multistep process the popularity of this style has decreased recently. Wool was the dominant fibre and the friezé carpets were of relatively high price.

Recently, carpets of a similar appearance have been prepared from either staple or continuous filament polyamide yarns by a continuous heat-setting route that is much cheaper. The balanced-twist yarn produced by conventional cable twisting is mechanically crimped in a stuffer box immediately before the conveying belt of a Superba heat-setter. This crimped or 'textured' yarn is then heat-set relaxed as usual on the Superba. The carpet produced by tufting and dyeing such yarns has a random pile lay character with kinks and bends distributed along the tufts. The severity of the crimping process may be varied to produce effects that range from a subtle texture caused by the presence of a few gently curved tufts to an intensely irregular surface character caused by use of a highly kinked yarn. The overall appearance of the carpets from intermediately crimped yarns approximates the look of the classical friezé.

The appearance retention of the classical friezé carpet styles has been very good. The random surface hides changes in pile lay and the pile yarns tend to be compressed from the side rather than from the ends, which lessens the tendency to untwist or to bloom out at the tuft ends.

The 'textured' friezé styles made of polyamide yarns at the lowest levels of texture have floor performance similar to the corresponding saxony style carpets. Those that have appreciable random pile lay character show very good floor performance since they share the beneficial hiding tendency of the initially random surface. There has not yet been enough experience to establish the detailed effects of the fibre and yarn construction variables on appearance retention for these styles.

6.3.8 Colour retention

The requirement that the appearance of the carpet should not alter during use includes the proviso that the colorants used must be resistant to removal by abrasion (crocking), to change on exposure to light (light-fastness) or to atmospheric gases (fastness to ozone and oxides of nitrogen), and to removal by treatment with water or cleaning solvents.

Polyester fibres are available in homopolymer types that require pressure dyeing and/or the use of chemical 'carriers' to achieve practically useful rates of dye exhaust and in copolymer forms that may be dyed by these methods without use of carrier. Although both stock and yarn dyeing methods are used, the largest proportion of polyester carpets are dyed in carpet form after tufting and before application of latex and secondary backings. Dyes used are of the disperse dyeing class.

Polyolefin fibres such as the polypropylene fibres used in carpets are not exhaust dyed but are coloured by incorporating the dye or pigment into the bulk polymer before it is formed into filaments. With proper selection of the colorants and by use of suitable stabilizers to prevent degradation of the polyolefin by ultraviolet light, it is possible to reach sufficiently high levels of colour stability under exposure to moisture and to sunlight that these yarns may be used in outdoor locations. Because of the high abrasion resistance of large-denier filament polyolefin yarns, carpet-like 'artificial turf' for sports stadia may be made.

Acrylic fibres used for carpet manufacture contain sulphonated monomers in the polymer chains and may be dyed with dyes of the 'cationic' or 'basic dye' class. These fibres give dyeings that are quite wet-fast and which have good light-fastness. In addition, they are capable of providing dyeings with high colour purity or 'brightness' and can make carpets with very vibrantly vivid colorations when the style demands.

Polyamide fibres are usually exhaust dyed using dyes of the acid dye class. In staple form they may be stock dyed and in spun yarns or continuous filament yarns they may be skein or package dyed. However, the majority of carpets made from polyamide fibres are dyed in carpet form either using batch methods (beck or winch dyeing) or continuously. The continuous dye methods involve the application of the dye in concentrated aqueous form to the carpet surface, followed by steaming to effect penetration of the dye molecules into the filaments, with subsequent washing to remove excess dye and dyeing auxiliaries. The dyes may be applied to the entire carpet surface to obtain a solid colour dyeing by a variety of commercial methods such as foulard padding, roll and doctor-knife application, spray or foam application using equipment such as that made by Otting, Kuester and others, or by colour pickup in a nip such as Kuester Fluidye or Flexnip processes. Multicolour effects may be made by application of different colour solutions in a specific pattern by screen printing, roller printing or various computer-controlled jet applications such as the Millitron, Chromotronic and Titan dye machines. Most suppliers of polyamide carpet yarns supply them in several chemically different polymer compositions, which include those in which a sulphonated monomer is incorporated into the polymer chain so that the resulting fibre may be dyed with basic dyes similar to those used on acrylic fibres, plus products that have several different levels of amine-end concentrations so that they have different degrees of affinity for pickup of acid dyestuffs. By using these yarns in arranged patterns when making the carpet, and dyeing with mixtures of the acid and basic dye classes, multicolour patterned carpets may be made in a single dye application process. For the most demanding end-uses that require the highest levels of light-fastness, such as in the transportation industry, polyamides may be dyed with acid dyes that are metallized complexes. In addition, colorants may be added to the polyamide polymers before they are extruded into fibres in a manner similar to that used to colour the polyolefin fibres. Such materials may have extremely high resistance to water and light exposures and they also have been used in the preparation of carpet-like structures for outdoor use as 'artificial turf'.

Cleanliness

In addition to the effects on the appearance of the carpet during use of distortion due to abrasion, alteration of reflectance due to distortions of the fibre and yarn geometry at the carpet surface, and possible changes in the light absorption of the colorants used on the fibres, there is also the effect of soiling and staining. These last two factors alone, if their effects are severe, may render the installation so unsightly that it may be thought to have become unacceptably worn when there has been negligible abrasive fibre loss, pile distortion or fading of colour.

6.3.9 Soil resistance

Solid particulate soil materials may be held within a carpet structure by several mechanisms. The major ones are:

—Electrostatic attraction between fibre and soil
—Penetration of a soft fibre by a hard soil particle

—Adhesion of the soil particle to the fibre by an oily film on the fibre surface

—Mechanical entrapment of the soil between yarns and fibres in the carpet

Of these, the entrapment mechanism appears to dominate the dry soiling performance of carpets, with the oil bonding playing a minor role and the electrostatic attraction and penetration mechanisms usually negligible.

There are of course many specific situations in which a particular soiling mechanism is seen to an exaggerated degree. For example, oil attraction where a carpet is used next to a mechanical repair service area in an automotive garage, or in the kitchen area of a restaurant where oily aerosols are encountered. Another type of unusual result may be encountered when there is actual transfer of a substantive coloured agent from an adjacent area of poor crocking resistance to the carpet area. In general, however, the major factors that need to be addressed are the control of particulate soil accumulation and the facilitation of its removal by conventional maintenance vacuuming and occasional shampooing as well as the minimization of the severity of the visual change in appearance of the carpet surface while the soil particles are present there. The principal factors governing this will be discussed below.

Studies of the kinetics of soil pickup of carpets[26–31] have shown that the process is one of surface transfer of soil wherein the soil deposited from shoe soles is initially deposited on the top of the pile and then migrates through the pile with further traffic until a dynamic equilibrium is established. This equilibrium may be described by Eqn 6.2:

$$D = a(1 - e^{-KN}) \qquad [6.2]$$

where D is the grams of soil per 100 g of carpet after N cycles, a is the grams of soil per 100 g of carpet when N becomes infinite, N is the number of traffic cycles and K is an equilibrium constant.

The extent of soil deposition depends upon the fibre size, denier and cross-sectional shape, and the fibre surface energy. Thus coarse filaments of round cross-section will show the least soil deposition. Modification of the surface energy by addition of fluorocarbon surface coatings will decrease the soil deposition still further. Tests conducted on fibres of nylon 6,6 gave soil deposition values that fit the empirical equation

$$D = D_0 \times \frac{A}{A_0} \times \left(1 + 0.2 \frac{A_0 - A}{A_0}\right) \times \frac{E}{E_0} \times \left(1 + 0.5 \frac{E_0 - E}{E_0}\right) \qquad [6.3]$$

where D is grams of soil per 100 g of carpet, A is the carpet surface area and E is the fibre specific surface energy of the test fibre and D_0, A_0 and E_0 are the corresponding values for a reference sample. For uncoated nylon 6,6 the specific surface energy is 43×10^{-3} J m^{-2} while for a fibre coated with a fluorochemical finish it may be as low as 15×10^{-3} J m^{-2}. The equation predicts that with all other parameters unchanged this reduction in specific surface energy reduces the soil deposition of the treated sample to about half that of the untreated one.

The above treatment suggests that the extent of soil deposition may be minimized by using fibres with the lowest surface area and the lowest surface energy. Decreasing the surface area per unit weight of fibre requires the use of large filament size and round or nearly round cross-sections. The need to make carpets of a reasonably soft handle, rather than wiry or coarse to the touch, limits the filament sizes that are acceptable. Typical filament sizes for nylon are 17–22 dtex per filament with some samples as low as 6.7 dtex per filament for residential uses. Commercial applications use fibres of 17–22 dtex per filament with larger fibres 22–39 dtex per filament also used since a harsher handle is acceptable in this application. Corresponding sizes for polypropylene are 11–22 dtex per filament and for polyester 11–17 dtex per filament. Although round cross-section is preferred for decreased soil pickup, nonround cross-sections have advantages in decreasing fibre in yarn packing density and in providing light-scattering to hide the soil that is present. These factors have sufficient importance to the overall desirability of the carpet that they justify the use of nonround fibres despite their greater surface areas. The surface energy of the fibres may be reduced by the application of surface coatings such as fluorocarbon polymers. Although care must be taken to avoid on the one hand coating the filament with a soft material to which soil particles may adhere and on the other hand the creation of a nonadherent, brittle coating that soon wears off, there are many fluorocarbon-based antisoil treatments available for carpet fibres that decrease the soil pickup to a very desirable degree.

Fibres' optical properties are very important in hiding the effect of soil that is picked up by the carpet. Wool fibres are naturally delustred and have an irregular cross-section with a scaly surface that scatters light. Acrylic fibres have either a dogbone or a bean shape depending on spinning method—dry or wet. It is relatively difficult to alter this shape and so the only variable in this case is the filament linear density. The other synthetic fibres, however, may be prepared in a variety of shapes and sizes that can materially alter the effect of soil pickup on the apparent change in appearance. If one side of a clear, round rod is coated with a dark substance to simulate a soil particle, it can readily be shown that the apparent size of this particle is magnified when viewed through the fibre, since there is a lens effect due to the surface curvature. Thus the effect of the soil is magnified. When the filament cross-section is changed to trilobal, or in general to multilobal, then the magnifying effect is eliminated and the soil particle is hidden from view for many viewing angles. A similar soil-hiding effect results from the use of filaments that contain continuous or discontinuous longitudinal channels or voids along the fibre axis. A third means of reducing the visibility of the soil is to add a light-scattering agent inside the filament. These cases are shown in Fig. 6.12.

A major factor in determining the degree to which an observer will perceive the presence of a given quantity of soil on a carpet is the original colour of the carpet itself. Studies[31,32] have shown that the typical soil found inside residences and in commercial buildings causes the least apparent change in perceived colour on carpets on dark shades and on those of earth tone (grey and brown) hues (see Fig. 6.13). This plot of the appearance change expressed as ΔE calculated from tristimulus ratings of the carpet surface before and after soiling versus the original shade

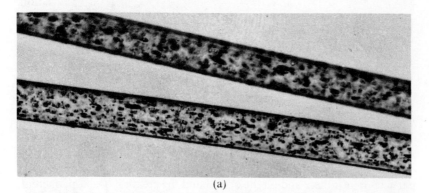

(a)

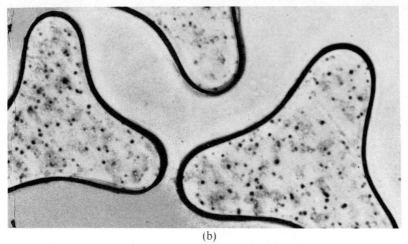

(b)

Figure 6.12 Filament shapes and lustres: (a) round delustred fibres (longitudinal); (b) trilobal highly delustred fibres. (*continued*)

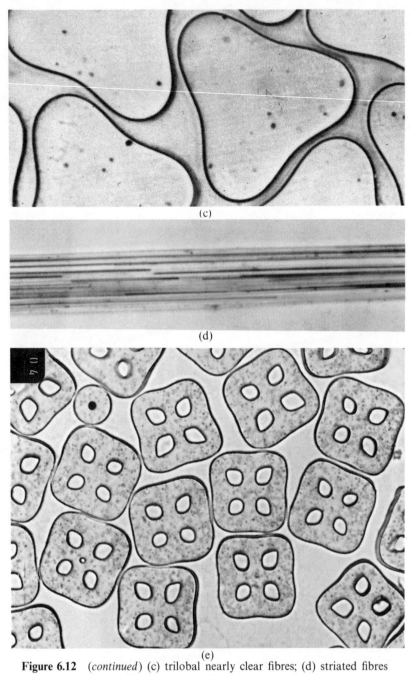

(c)

(d)

(e)

Figure 6.12   (*continued*) (c) trilobal nearly clear fibres; (d) striated fibres
(longitudinal); (e) hollow fibres (with one carbon core antistatic
filament). (Courtesy of DuPont Co. Fibers Photographic
Laboratory)

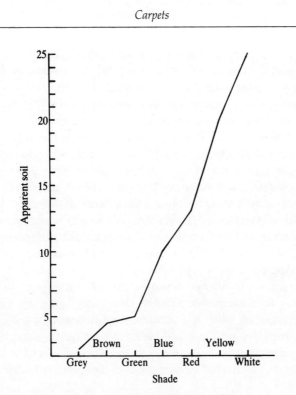

Figure 6.13 Effect of carpet shade on apparent soiling.

of the carpet illustrates the high degree of change for the white and yellow shades and the low degree for the grey, brown and green shades. Such ΔE values have been found to agree well with subjective ratings of the degree of apparent soiling. The demands of style do not permit the restriction of carpet colours to these highly hiding soil shades alone; the attempt to limit the severity of appearance change on soiling during use by other less restrictive means must be made.

The interaction of the soil deposition and soil hiding effects has been empirically related for fibres of nylon 6,6:

$$R = \frac{2.3}{0.3D_0 + 0.2} D \left(1 - 0.4C + 0.1h \frac{1 - C}{0.5} + 0.05L \right) + 1 \quad [6.4]$$

Equation 6.4 predicts the soil rating, R, on a 1 to 5 scale, where 1 represents no appearance change due to soiling. It assumes that a reference sample and a test sample of the same colour have been exposed to the same soiling conditions. In the equation D and D_0 are the calculated soil levels for the test fibre and the reference fibre, respectively, based on their surface area and surface energy values; C is the mass% of delustrant in the test fibre; h is the percentage of the test fibre cross-sectional area occupied by any voids; and L is a cross-section shape factor for the test fibre that varies from 0 for round to 1 for an essentially triangular cross-section to 2 for a deeply scooped trilobal shape.

The above rather crude model has been shown to give rather good agreement to actual carpet performance tests for a limited set of carpets of nylon 6,6 fibres.[26] Unfortunately, it has not been tested for polyester or for polypropylene fibres. None the less, it does serve to indicate the rough relative importance of the factors in controlling the soiled appearance of carpet fibres. By examining Eqns 6.4 and 6.3, one can see that the decrease in fibre surface energy always has a beneficial effect by decreasing soil deposition. Within the limitations of tactility and manufacturing difficulty, increasing the filament size to decrease the surface area always helps decrease soil pickup. The incorporation of delustrant, either of the light-scattering particulate type such as titanium dioxide or by internal light-scattering voids, always improves the soil rating and is the most powerful way to achieve this. Altering the external filament shape from round to multilobal, while less effective than delustring by the methods mentioned, provides sufficient soil hiding advantage to justify the use of multilobal rather than round cross-sections. These predictions have been confirmed in part by the independent tests of the German Textile Research Institute.[31]

6.3.10 Stain resistance

Another factor leading to undesirable appearance change of a floor covering in use is discoloration caused by staining. While all presently used fibre materials can be stained by some coloured materials, leaving a discoloured area that is difficult or impossible to restore to the original appearance, fibres of polypropylene, polyester and acrylic polymers, which are relatively difficult to dye, are more resistant to such incidents than are the polyamide-based materials. The latter are dyeable by anionic coloured agents—and many of the naturally occurring colours in foods as well as most of the artificial food colouring agents are of this class. Customer surveys carried out by several manufacturers of polyamide carpet fibres during the 1980s indicated that such staining was an item of considerable concern to the carpet users.

In 1986 in the United States a process for significantly increasing the resistance of nylon carpet fibres to staining by anionic materials was introduced. The positive reaction of the consumers to this new product concept has been so great that today at least 80 per cent of all residential carpet made of nylon in the United States has been treated to reduce this acid dye staining.

The essential mechanism of this stain resistance is the creation of an electronegative region at the fibre surface that repels the negatively charged coloured anions of the stain. This is done by adsorbing a polymeric, negatively charged material onto the fibre surface. In general, these materials are of the class of sulphonated phenolic resins commonly called syntans because they were initially used as synthetic tanning agents for use on leather. Such materials are water soluble or dispersible and have a high affinity for polyamides. Because of their large molecular size, they adsorb readily onto the fibre but penetrate slowly, thus creating a surface zone with a high concentration of negatively charged groups that provide a barrier to adsorption of the acidic staining agents.

The resistance is primarily a kinetic effect due to a reduction in the rate at which the fibre absorbs the acidic stain. The equilibrium capacity for the staining agent is only slightly affected. For this reason, the difficulty of attaining stain resistance and the quantity of antistain agent required to attain a given degree of resistance is a function of the diffusion characteristics of the nylon substrate. Fibres that have not been heat-set or that have been heat-set in hot air as by the Suessen process are easier to make stain resistant than are those that have been heat-set in steam as by the Superba process. Fibres of nylon 6,6 are usually made stain resistant at lower concentrations of resist agent than are those of nylon 6. Staple fibres, especially those that have not been treated with steam, are similarly made stain resistant at lower treatment levels than are continuous filament yarns whose diffusion characteristics have been increased by the thermal crimping process they have undergone. In all of these cases, the fibre with the lower diffusion rate is the easier to treat.

The resistance conferred is ionic in nature. Noncharged coloured materials such as disperse class dyestuffs, and some natural stains like those in mustard, are not resisted and the resist effect is larger with stains that contain a larger number of acidic groups in the molecule. The resistance is also significantly reduced as the temperature of staining is increased. In fact, some commercial processes use fibres that are treated with the stain-resist agent before they are dyed. In this case, the fibres dye to essentially equivalent full shade with acid dyes at elevated temperature as do untreated fibres but show much less staining at room temperature.

The composition of the stain-resist agent and the process used for its application require careful optimization and control. The agent must be sufficiently water soluble or dispersible to be applied in an effective quantity to the substrate; it should exhaust relatively completely onto the fibre for good utilization and to avoid effluent discharge problems; it should be cheap and effective in small amounts; it must not discolour the carpet or yellow with age; it must not adversely affect the handle, fastness to light or soiling of the carpet and it needs to be compatible with antisoil treatments of the fluorocarbon type. None the less, there are a considerable number of effective antisoil materials in use on nylon carpet fibres today.

The nylon carpet most resistant to appearance change due to soiling and staining by present technology contains large fibres that are delustred, preferable by internal voids, and that have a minimum surface area treated with a fluorocarbon and an antistain agent.

6.4 Twist setting of carpet yarns

Whenever carpet yarns are used in plied form in cut-pile constructions, they must be twist set to achieve the stability towards wear of the twisted yarns configuration. Since these carpet types—saxony, friezé and some velvets—represent a very large proportion of the carpet styles now used, it follows that a large volume of twist setting is practised in the carpet industry. The retention in wear of the carpet appearance of these styles is dependent upon achieving a very high degree of set in this process. Although heat is involved in the process, the details vary with the nature of the fibre being processed. A good description of the molecular differences between fibres and their influence on the setting process has been given by Hearle and Miles.[33]

The concept of setting in semicrystalline materials that show viscoelastic properties is difficult to define precisely. In a macro sense for the present case, one might define a setting process as one that converts the originally twist-lively and thus unstable plied yarn structure to a state in which there is little or no tendency to untwist when the end is cut and the yarn is set free to do so. Unfortunately, this simple concept does not describe the situation adequately. A plied polyamide yarn that has been stored on the package under winding tension for a period of time will have undergone appreciable stress decay and will not have a very large tendency to untwist. It may appear in this regard to be equivalent to another sample of the same yarn that has been heated under pressure in a steam autoclave. However, if the two yarns are steamed under no restraint or even if they are mechanically exercised by gentle cyclic tensioning, it is apparent that the yarn removed from the package becomes very twist-lively and untwists extensively, while the autoclaved yarn retains the twisted configuration almost completely. In the first case we speak of a temporary set that is not stable to steaming, while in the second case we say that the set is permanent.

We can distinguish these two cases by using the concept of a reference state. This state is defined as the state of minimum energy of the system. It is the state to which the system will return if the frictional restraints are removed and a degree of mobility is given, as by mechanically agitating the system, steaming it or immersing it in water. The yarn removed from the package has as its reference state the untwisted yarn form, but the yarn that has been autoclaved has had its reference state shifted to the twisted form. A permanent set is said to have been achieved when the reference state has been altered; mild relaxation treatments will not return the system to the original state again, since a new position of minimum energy has been established in a different geometric form. The temporary set case has not established a new position of minimum energy and thus the twisted yarn is returned to the original state on mild relaxation treatments. It retains its twisted form when initially released from restraint only because the stress has decreased owing to creep to the point at which the frictional restraints of the yarns are sufficient to prevent untwisting.

In real semicrystalline oriented polymer systems there are usually multiple minimum energy states possible. These are not all at the same energy level but are separated by energy barriers. If the system is treated so as to take it over an energy barrier into a new energy minimum then a new set is reached. If the energy minimum of the new set is lower than that of the previous set, the set tends to be permanent; otherwise it is likely to be temporary. For example, if a polyamide yarn that has been drawn so as to orient the chains along the yarn axis is extended moderately, exposed to high humidity and elevated temperature in the extended state, returned to the original conditions and then released it will have been set in the extended form. However, relaxing the yarn under the conditions of high humidity and high temperature will cause it to shrink back towards its initial state because the unextended state has the lower energy.

Other difficulties arise when the system consists of a multifibre yarn and even more so when it contains several yarns. In these cases there are interfilament frictional forces that are quite large and the individual filaments have very complex geometries of bending and torsion within the system. If this structure, no matter how well set it may be, is drastically disarranged, as might be the case with a tuft in a cut-pile carpet that is snagged and unplied by a pointed object, then the original geometry will not be regained even though it may represent the minimum energy location of the system.

The practical requirements for twist setting of yarns for use in cut-pile carpets is to achieve a stability of twist form that is maintained through the steps of carpet making, dyeing and finishing and that resists distortion by the mechanical wear and the cleaning processes that are involved once it is put in place on the floor.

Polypropylene fibres have only van der Waals attractive forces between chains. It is relatively easy to rearrange these bonds in twisted yarns so as to obtain a twist-set condition. This is usually done by one of the continuous heat-setting processes such as the Superba or Suessen. Only mild heating is required to obtain a set. However, the polymer structure may be distorted by high mechanical loads and the set may be lost in such cases. Thus far the texture retention of saxony carpets of polypropylene has only been acceptable in relatively dense, short pile, high twist constructions.

Acrylic fibres have a structure of long clusters of closely packed helical polymer chains in a matrix of less regularly organized material. The major interchain bonding is due to dipole interactions of the nitrile groups along the chains. These fibres may be twist set by heating in the 80–100°C range. The resultant yarns can be piece dyed and made into useful saxony carpets, but the texture retention of these is not very good.

Polyesters contain highly crystalline domains within a more randomly ordered matrix. Heat-setting is a process of structural rearrangement by melting of the smaller crystals, which have a lower melting point because of their larger surface area, and growth of the larger crystals so as to relieve the internal stresses and reach a new reference state. The set can be done in hot air as easily as in steam since the polyesters are not very water sensitive. Texture retention difficulties with polyester require its limitation to relatively dense carpet constructions for good performance. It is believed, however, that this is not due to a deficiency of setting but is rather a consequence of the chain rigidity, which gives rise to a yield point that is passed at high extensions, leading to a permanent structural rearrangement.

Polyamide fibres have the structure of crystalline material in a less ordered matrix, like polyester fibres have, but the amide groups of the polyamides are also capable of extensive interchain hydrogen bonding. This gives rise to an appreciable moisture sensitivity of the yarns, which is sometimes used to achieve desirable temporary set during processing. These fibres may be set in autoclaves using steam and in the continuous Superba steam units as well as in the hot-air Suessen units. Setting temperatures are commonly 120–140°C for steam setting and 180–220°C for dry heat. In general, the continuous filament yarns are set on the Superba equipment, which provides yarns of better uniformity of crimp, leading to more uniform carpet appearance. Staple yarns, which develop much less crimp during heat-setting than do filament yarns, are more commonly set on the Suessen equipment.

6.5 Bulkiness of carpet yarns

Over the years, useful floor coverings have been made from a wide variety of pile fibres covering a large range of physical properties. Fibres of very low breaking strength, especially the coarser cellulosics and the regenerated cellulose fibres, make functional floor coverings provided that the fibres are tightly packed and are therefore subject only to abrasive wear at the tips. For the majority of carpet applications, however, looser structures capable of greater energy absorption on compression are preferred and the fibre strength and recovery characteristics become more important. Also, since the cost of a carpet is most influenced by the cost of the pile fibres, the ability to achieve acceptable aesthetics and performance with the minimum quantity of pile fibre is of major economic importance.

The minimum weight of a given fibre that can provide acceptable performance in a specific carpet construction of yarn size, tuft spacing and pile height depends on the polymer density, the fibre in yarn packing factor and the optics of the fibre for the ability to provide the optical cover so that the carpet backing material does not show through the pile; the filament and yarn rigidity so that the compressive 'handle' of the carpet is not too limp and lean; and the flex and abrasion resistance as well as the recovery properties of the fibres and yarns so that the carpet appearance retention will be acceptable. It is the interplay of these factors in the various carpet types and over the range of severity of wear applications that largely determines the relative use of the particular fibres in each market segment. There is a cost–benefit relationship for each fibre in each application which, although it may be modified by aesthetic preference and custom factors, will in the long run determine the dominant selection of a particular material.

The size of a particular filament will depend on the density of the polymer from which it is made. As Table 6.2 shows, the range of densities of the common fibres is considerable. So if a carpet of polypropylene were made to match the pile yarn geometry of one of nylon with the filament cross-section areas held constant, the pile yarn weight of the poly-propylene carpet would only be 80 per cent of that of the nylon carpet. It would only contain 70 per cent as much weight as a similarly prepared polyester carpet.

Other factors that influence the covering power of carpet yarns involve the closeness of filament packing within the yarn bundle. The filament

Table 6.2. Density and filament size for various carpet fibres

Fibre	Specific gravity (g cm^{-3})	Relative diameter (constant denier)	Relative area (constant denier)
Polypropylene	0.91	1.0	1.0
Polyamide	1.14	0.89	0.80
Acrylic	1.11–1.19	0.87–0.91	0.76–0.81
Wool	1.34	0.82	0.68
Polyester	1.22–1.38	0.81–0.86	0.66–0.75

cross-sectional shape and the degree and type of crimp in the filament
are the important properties here. Wool fibres have somewhat irregular
cross-sections and carpet wools are blends that contain a considerable
range of fibre diameters. This prevents regular close packing of the
filaments and makes the wool yarn density considerably lower than the
individual fibre density. Similarly, the acrylic dogbone and bean shapes
have relatively inefficient packing. In this case also it is common to
use fibres of differing shrinkage levels in spinning the acrylic yarns
and to develop a yarn containing loopy fibres by relaxing the spun yarn
before preparing the carpet, to lower the yarn density and increase the
covering power further. The melt-spun synthetic staple yarns of polyester,
polypropylene and polyamide may be spun in cross-sectional shapes that
are designed to decrease closeness of filament packing and thus to increase
yarn bulk and cover. There are restrictions on the kinds of shapes that
may be used in the case of polyester and of polypropylene, since both
polymers have a tendency to fibrillate along the fibre axis if the shape has
sections with high radius of curvature. Polyamide fibres may be made
with more extensively modified shapes with good durability. The more
extensively modified shapes will permit less close filament-to-filament
packing even when rather high twist levels are used, as in high performance
saxony yarns. There may be an increase in effective bulk of as much as
5–6 per cent due to use of the optimum shape factors alone in the case of
polyamide yarns.

The filament crimp also serves to prevent close yarn packing. In the
case of staple yarns, much of the crimp is removed during the carding
and spinning processes of yarn formation and the packing density is rather
high. The bulked continuous filament (BCF) yarns prepared by hot fluid
jet bulking processes have a more effective type of crimp that results in
production of a less closely packed yarn structure. The crimp in such
yarns may be characterized as having a three-dimensional form without
sharp crimp nodes but with a radius of curvature that varies rapidly and
at random along the length of the filaments. Such a crimp form does not
permit easy close packing of the filaments, resulting in the production of
a voluminous, bulky yarn with very good optical cover and resistance to
compression. Because of this difference in the nature of the crimp, BCF
nylon yarns are usually about 6–7 per cent bulkier than are staple
yarns spun from filaments of the same cross-sectional shape. This means
that a carpet of continuous filament nylon can be made with a pile weight
6 per cent less than one made from staple yarns and will be perceived
subjectively to have the same compressive resistance, so that the consumer
views them as having equivalent weights of pile fibre. This increased
'value' of the BCF yarns helps to compensate for their usually greater
cost compared with staple. The use of more highly modified cross-sections

in the BCF yarns, the so-called 'extra body' yarns, increases this replacement level versus staple yarns to 10–12 per cent, especially when the comparison is made at relatively low pile yarn weights. This makes the use of BCF yarns particularly desirable in these large volume market segments.

6.6 Usage of carpet fibres

Worldwide there are about 2 billion kilograms of material consumed annually as carpet face fibre. More than half of this is nylon in either staple or bulked continuous form. Wool accounts for about 10 per cent, polyester and polypropylene fibres 10–15 per cent each and acrylic fibres about 5 per cent.

There are, however, very large differences in the relative consumption of the various fibres in different parts of the world. In the North American market, which accounts for about 55–60 per cent of the world face fibre market, nylon is dominant, representing over two-thirds of the usage, followed by polypropylene at 10–15 per cent, polyester at 5–10 per cent and wool at 3–5 per cent. In Western Europe, which represents 25–30 per cent of the total world market, relative use of nylon is about half of the total with wool and polypropylene about 15–25 per cent, and acrylic about 5 per cent. The United Kingdom, which consumes about 5 per cent of the world total, uses about half nylon and 20–25 per cent wool in addition to 15–25 per cent polypropylene and 5 per cent or less of acrylic or polyester fibre.

There are also important differences in the carpet styles used in the different market segments. The North American carpets are on the average relatively heavy at about 1050 g m^{-2} (30 oz yd^2) as compared to those in Western Europe and in the United Kingdom, where the average face fibre weight is only about 620 g m^{-2} (19 oz yd^2). Pile heights are correspondingly shorter and the tufts more densely packed in the latter cases also. These differences tend to make the importance of the cost of the face fibre represent a larger fraction of the total cost per unit area in North America than in the other two markets.

Construction methods also vary considerably. Tufting is predominant in the North American segment for about 90 per cent of the styles and solid colour carpets represent over half of the residential market, with half of these being continuously dyed. The majority of these styles contain twisted and continuously heat-set yarns. In contrast, the Western European market (including the United Kingdom in this case) is only about two-thirds tufted and about two-thirds of this is non-heat-set cut-pile velour.

Because of this diversity of manufacturing routes, coupled with the diversity of weights and styles, it is very difficult to generalize about the fibre preferences. Each of the major fibres can and does convert to carpets meeting the expectations of the consumers in many styles. Certain characteristics are expected from them that tend to differentiate their areas of use.

Wool, once the dominant carpet fibre, retains the image of producing a luxurious carpet with good performance characteristics in texture retention and soiling. It is very frequently the fibre of choice in the most expensive carpet styles. Nylon in its staple and continuous filament forms probably has the broadest spectrum of weights and constructions in which good performance characteristics, aesthetics and low cost are combined. Its excellent abrasion resistance, high recovery properties, good heat-settability, high dye stability, and in the newer modifications increased bulkiness, good soil and stain resistance coupled with ready cleanability are its strong points. Polypropylene has high abrasion resistance, high stain resistance, excellent colour stability in the pigmented yarn and high bulkiness as well as very low cost. It shares with nylon the areas where traffic demands are high, as in the commercial uses, and is expanding into the residential styles. In this case, however, there seems to be a limitation in using low density constructions because of a loss in carpet texture in use. Polyester carpets have high stain resistance and good abrasion but require the use of relatively densely packed carpet constructions to achieve good appearance retention in use. Acrylic fibres have good colour stability with use of cationic dyes and very brilliant colours. They are reasonably high in abrasion resistance and have a particularly soft and pleasant tactility. Because of this they are often preferred in the Far East where shoes are removed before the carpet is walked upon and where the carpet frequently serves as a mat to sit upon. Some blending with modacrylic fibres is normally required to achieve acceptable flammability characteristics.

New variants of the synthetic fibres are continually appearing from the fibre producers and new conversion techniques that can alter the balance of economics and/or aesthetics available from a particular fibre are being developed by the carpet machinery manufacturers and the carpet producers. These can appreciably shift the balance of fibre usage, as was the case when tufting, low-cost twisting via the 'two for one' or 'cable' twist routes, continuous yarn heat-setting, the production of nylons with increased bulkiness and the stain-resist nylon products were introduced.

There is a continual evolutionary progression in the production of carpeting with the objective of providing materials that meet the consumers' desire for aesthetics and performance at the lowest cost.

References

1. G Robinson, *Carpets*, The Textile Book Service, 1972, Ch. 1.
2. W Herzog, *Textile Inst. Ind.*, **9**, 126 & 153 (1971).
3. I D McFarlane, J M Watson, *WRONZ Communication No. C94*, 1984.
4. *Advantages of Carpet and Rugs in Energy Conservation*, The Carpet and Rug Institute, Dalton, GA, 1976.
5. ASTM C-423-66; BS 3638: 1963; ISO/R 354; DIN 52212.
6. BS 275: 1956; ISO/DIN 52210/DR 477.
7. M C Geoghegan, E Rivet, C D Malone, *Textile Res. J.*, **46**(5), 367 (1976).
8. G J Pontrelli, K Chakravarti, Allied Chemical device measures carpet static, *Textile Res. Inst. Abstracts*, 43rd Annual Research and Technology Conference, 1973, p. 23.
9. N Wilson, *Textile Inst. Ind.*, **10**, 235, 1972.
10. L G Hartzell, Development of a radiant panel test for flooring materials, NBSIR 74-496.
11. G A Carnaby, E J Wood, *J. Textile Inst.*, **80**(1), 74 (1989).
12. G A Carnaby, E J Wood, *J. Textile Inst.*, **80**(1), 78 (1989).
13. E J Wood, R M Hodgson, H S Lee, *WRONZ Communication No. C105*, 1987.
14. E J Wood in *The Application of Mathematics and Physics in the Wool Industry*, ed. G A Carnaby, E J Wood, L F Story, WRONZ/Textile Institute (N.Z. Section), Christchurch, 1988, p. 104.
15. R M Hodgson, E J Wood, H S Lee, *ibid.*, p. 237.
16. P Larose, *Textile Res. J.*, **23**, 730 (1953).
17. A El-Shiekh, S P Hersh in *Studies in Modern Fabrics*, ed. P W Harrison, The Textile Institute, Manchester, 1970, p. 159.
18. O H Culcuoglu, PhD thesis, University of Leeds, 1981.
19. S P Hersh, A El-Shiekh, *J. Textile Inst.*, **63**, 660 (1972).
20. T Horino, S Shimonishi, *J. Textile Machinery Soc. Japan*, **18**, 21 (1972).
21. K Kimura, S Kawabata, *J. Textile Machinery Soc. Japan*, **18**, 141 (1972).
22. B L Thomas, PhD thesis, University of Leeds, 1973.
23. D W Hadley, D Preston, *J. Textile Inst.*, **58**, 194 (1967).
24. J L Barach, *Textile Res. J.*, **19**, 194 (1949).
25. R Luning, L Tenzer, A Lehnen, *Melliand Textilberichte*, **69**, 319 (1988).
26. D Jose, P Hauck, P Singh, *Textile Progress*, **19**(3), 60 (1983).
27. E Rivet, R M Shellenbarger, *53rd. Annual Research and Technology Conference*, Textile Research Institute, 1983.
28. E Rivet, Book of papers, *18th Canadian Textile Seminar*, May, 1982.
29. E Brown, *Fundamentals of Carpet Maintenance*, P. A. Brown Associates, Leeds, 1982.
30. E Rivet, *Tufting Yearbook*, Textile Business Press, Manchester, 1985.
31. H Klingenberger, *Chemiefasern/Textilindustrie*, **33/85**, 53 (1983).
32. E Kratzsch, H Ninow, *Textil-Praxis*, **38**, 1298 (1983).
33. J W S Hearle, L W C Miles, *The Setting of Fibres and Fabrics*, Merrow Publishing Co., Watford, 1971.

CHAPTER 7

Nonwovens

P A SMITH

7.1 Introduction

It is an unfortunate fact that there is no internationally agreed definition of nonwovens, in spite of the fact that the International Standards Organization (ISO) published a definition in 1988 (ISO 9092: 1988). The problem arises partly from the very name 'nonwovens', which appears to include all types of fabric except woven fabrics. In an attempt to solve this problem the name 'fibre-formed fabrics' has been suggested, but has never become popular. The present situation is that many different countries have their own interpretation of the precise meaning of nonwoven.

The main areas of contention are the interface between paper and nonwovens and whether or not nonwovens may include spun yarns or continuous filaments either to bind the fibres together or to act as a reinforcement. For the purposes of this chapter the author has chosen to use the American Society for Testing Materials definition (ASTM D 1117-80). This has the advantage of being very short and simple to understand, but it must also be admitted that it is not very precise and that it avoids the contentious issues. The definition is as follows:

A nonwoven is a textile structure produced by the bonding or interlocking of fibres, or both, accomplished by mechanical, chemical, thermal or solvent means and combinations thereof.

It is assumed for the purposes of this chapter that 'mechanical bonding' includes bonding fibres with yarns and that nonwovens may also be reinforced by yarns.

It can be gathered from this definition that nonwovens are mainly composed of fibres forming a web or batt and that this web or batt is transformed into a nonwoven by some method of bonding. There are several methods by which the fibres may be laid down and also several different methods of bonding; moreover, most of the methods of bonding could be used in conjunction with most of the methods of batt formation, so that there are very many possible manufacturing routes, which could not be discussed individually. Instead the main methods of batt formation will first be described, followed by a description of the main methods of bonding, together with an explanation of which batt formation methods can be used with each method of bonding. Only after explaining how the nonwovens are made is it possible to discuss their properties and the roles of synthetic fibres in nonwovens.

7.2 Raw materials

In theory nonwovens can be made from any type of fibre, natural or synthetic, regenerated cellulose or reprocessed waste, or of course a blend of any of these. In fact the two main natural fibres, wool and cotton, are very rarely used. This is due in part to the presence of impurities in both fibres that are difficult to remove due to the shortness of the nonwoven process. It is also partly due to the cost of the natural fibres, particularly of wool, and to the fact that the attractive properties of the natural fibres do not show up in the finished nonwovens. An important exception to this rule is bleached cotton, which is being used in increasing quantities in nonwovens. A major reason for its use is that it is a very absorbent fibre; it is not too expensive because it is made from reclaimed waste and the bleaching process removes all impurities, leaving pure white fibres. Of the regenerated cellulose fibres, cellulose triacetate, cellulose diacetate and viscose, only viscose is used in large quantities and this is due to the relatively low price and the good water-absorbing properties of that fibre. The solvent-regenerated fibre Tencell has been recommended for nonwoven use but at the present time it is too soon to tell whether or not it will be used in large volumes.

Reprocessed textile wastes may be divided into two or three separate categories for convenience, though there are in fact very many types. The so-called 'producer waste' is one category, being waste made during fibre extrusion and consisting of tangled continuous filament of a known synthetic type. Fibre derived from producer waste is widely used in the nonwoven industry because it is always cheaper than the similar virgin fibre yet from the point of view of a nonwoven manufacturer the fibre properties are not significantly worse than those of the first-grade fibre. Reprocessed wastes containing exclusively wool are usually expensive and are not usually used in nonwovens because the benefits derived from using the fibre do not generally justify the cost. Reprocessed wastes from fabrics containing synthetic fibres, regenerated fibres and natural fibres mixed together in a form that makes it uneconomical to separate out the constituent fibres are usually cheap. These fibres are used in large quantities to make products such as carpet underfelts and mattress pads. Although the processes involved in making these are the same as those used in nonwovens, there is a certain resistance in nonwoven circles to including these products as nonwovens, though it is difficult to devise a suitable definition that would exclude the carpet underfelts and at the same time would include similar materials that are made from virgin fibre and that are normally thought of as nonwovens. Whatever the rights and wrongs, the situation is that underfelts and mattress pads are usually excluded from statistics on nonwovens; since the weights of fibre involved in their manufacture are enormous, it is important to check whether they have been included or excluded from any particular statistic.

Synthetic fibres are by far the most important component in nonwovens and may be used as a single component, as a blend of two synthetic fibres or as a blend of synthetic fibre with a cellulosic, either viscose or bleached cotton. Synthetic fibres are so important to the nonwoven industry that it can be argued that the dramatic upsurge of nonwovens since about 1950 has been largely due to the availability of synthetic fibres and, furthermore, that if the supply of synthetic fibres were to cease then most of the nonwoven industry would have to stop too. Although this topic will be discussed at greater length later, it seems appropriate to mention at this point the main reasons why synthetic fibres are preferred. The desirable properties are good tenacity, good resistance to abrasion, good recovery from extension, creasing and crushing, together with ease of processing and an attractive price. In some circumstances thermal plasticity is of enormous importance and in other circumstances resistance to mould and biological degradation are necessary. In general, synthetic fibres can produce every desirable property except high moisture absorption, and for that viscose or bleached cotton is used, most frequently in a blend with synthetic fibres.

Having briefly established the importance of synthetic fibres in the nonwoven industry, it is useful to look, again briefly, at which synthetic fibres are used. It is worthwhile to point out here that raw material costs form a major part of the cost of the finished fabric so that the cost-effectiveness of the fibre becomes more important than elsewhere in the textile industry. The two fibres that stand out in this respect are polyester and polypropylene, both giving excellent properties at a competitive price. In the past polyester was used much more widely than polypropylene, but more recently the supply (and number of suppliers) of polypropylene fibres has increased, so that polypropylene is now the leading fibre in the nonwoven market. To some extent this position is due to the large weights of polypropylene used in needlefelt carpets; in some statistics needlefelts are not classed as nonwovens, so that the position alters. Of the other major fibres, nylon used to hold a strong position but is now of relatively minor importance. The reason is not that any of the fibre properties are wanting, but simply due to price. In the competitive nonwoven market the use of nylon can only be justified if it can perform some function that cannot be achieved by another fibre at the same price; this is becoming increasingly rare, so that the use of nylon is continuing to decrease. On the other hand, acrylic fibre has never been used much in nonwovens, again due to its lack of cost-effectiveness in most situations, so it continues to be used only in small quantities.

It is well known that all these fibres are available on the market in branded form at a premium price and unbranded at a discounted price, perhaps also second-quality at a lower price still. In some sections of the textile trade the tendency will be to buy the branded fibre. The apparel market is one example, because here the cost of the raw material is a relatively small part of the cost of the final garment, because of the guarantee which is explicit or implicit in the brand, because the cost of a serious problem in processing would be high and because the retailer and the consumer value the fibre brand name on the garment. In the nonwoven industry in general none of these four conditions applies. It has already been pointed out that raw material costs are important because the cost of processing is relatively low. Although a serious problem in processing would be equally disastrous, it is much less likely to occur because the fibre only has to go through one critical process (carding) compared with the many critical stages in fibre-to-fabric by weaving or knitting. As a consequence, the fibre manufacturer's guarantee is not required. Finally, in most cases the nonwoven fabric is sold by its fibre type rather than with a brand name. It follows that as a general rule the nonwoven manufacturer will prefer to buy unbranded or second-quality fibre in order to improve the cost-effectiveness of the raw material.

Taking into account the stress given to cost-effectiveness, one could be forgiven for thinking that the expensive high performance synthetic fibres would not be used at all. In fact quite significant quantities are used in specialized products where these fibres are the only ones with the required properties. The properties usually required are resistance to high temperature and/or resistance to chemical degradation. These fibres will be discussed in more detail in later sections.

The current position is summarized in Table 7.1. It should be noted that two facts appear to have been distorted by the way the data have been collected. Bicomponent fibres and wood pulp were not collected as independent data in 1980, so that the table appears to show a big rise in the quantities used. In fact there would be little change in either material over the period 1980–1990.

Table 7.1. The percentage use of synthetic fibres in nonwovens in Western Europe

	1980	1990
Polypropylene	19	42
Polyester	24	22
Viscose	36	11
Bicomponent	–	4
Polyamide	7	3
Natural	10	3
Wood pulp	–	8
Other	4	7

From D Ward in *The Nonwovens Industry in Western Europe*, E.I.U., 1992, with permission of Textiles Intelligence Ltd

7.3 Fibre opening and blending

Synthetic fibre tow is produced in the desired linear density and will almost invariably have crimp inserted and heat-set into the fibres. It is then cut to the required staple length and packed under very high pressure into bales ready for transporting to the fibre user. The high pressure is necessary in order to keep down the cost of transport but it does make the fibre rather more difficult to process at the next stage. The first operation by the fibre user must always be an opening process that loosens the hard-packed fibres and returns them to a state very similar to the state they were in prior to packing. Fortunately the opening process can be done at high production rates and with very little labour, so that it is not an expensive process.

It is usual to blend the fibres at the same time as the opening process is taking place. This is obviously necessary if a blend of different fibres is to be processed, for example high and low melting point polyester fibres may be blended together or a blend of polyester with viscose may be required. Blending is also necessary if only a single fibre is used, in order to compensate for any variation in manufacture. Since variation in fibre properties may be present between different deliveries, between different bales and also within bales, it is best to blend from as many bales and deliveries as possible, but in practice this is difficult to achieve. Some manufacturers argue that blending makes it more difficult to identify a below-standard fibre; following this view blending should still be done but all the bales should be taken from a single delivery.

Using the short staple process blending may be done using automatic bale pluckers and automatic stack mixers with a final sawtoothed opener to carry out the final opening. In the long staple process the bales are opened and spread automatically in a large rectangular bin. When the first bin is filled with horizontal layers an automatic bin-emptying machine takes vertical slices from each layer and the material is spread out again in a second bin. After similar removal from the second bin the material is given a final opening before passing to the next stage.

7.4　Methods of batt formation

7.4.1　Carding

The textile card is a machine that has gradually developed over hundreds of years; its main purpose is to disentangle the fibres from the tuft form obtained after the opening process and to spread the disentangled fibres into the form of a uniform web. A typical card web may be from 5 to $10 \mathrm{~g~m}^{-2}$ in weight and will be roughly 1–2 mm thick. Its width will be effectively the working width of the card. The card used for short staple carding consists of a single swift and doffer, the carding being done between the swift and a series of narrow flats that move slowly above the swift. Owing to problems with the stability of these flats and the fact that the carding surfaces could touch if the flats started to vibrate excessively, short staple cards are usually limited to 1 m working width. This is a disadvantage for manufacturing nonwovens but short staple cards have a very high opening power in a small amount of space and are cheap in relation to their production, so they are sometimes used. More commonly long staple cards are used, often consisting of two swifts and two doffers. In these cards the carding action takes place between the swift and a number of rollers placed above the swift. The rollers are 300–400 mm in diameter so that they are much more rigid than the flats on the short staple card. For this reason long staple cards are usually 2.5–3 m wide, and can be made up to 4 m wide. This width is far more satisfactory for manufacturing nonwovens, and this type of card is more frequently used.

The technique of carding is a very involved one; it is not intended to go into any detail in this chapter, but any reader requiring more information is advised to go to the standard textile textbooks. Although all cards will tend to smooth out any irregularity fed to them, it is clearly desirable to feed the card as evenly as possible. Two basic methods have evolved, one using a weighing hopper and one using a volumetric feed. In a weighing hopper the fibre is fed into a weighpan until the weighpan just over-balances, and then the supply of fibre to the weighpan is cut off as quickly as possible. It has been found that the traditional mechanical weighpan is rather slow in operation and that it is also inaccurate. In comparison, the modern electronic weighpans are much more rapid and they are also more accurate because the weighing system is controlled by a small computer that compensates for any long-term changes in the hopper.

Traditionally it is always thought that weighing hoppers are not suitable for high production rates, so, in spite of the fact that electronic weighing hoppers can weigh accurately at high production rates, volumetric fibre feeds are used for higher production rates. This type of fibre feed operates by supplying fibre at a constant head, i.e., treating it virtually as though it were a liquid. Provided that the fibre is of uniform packing density this would result in a uniform feed to the card, but in practice the packing density always varies, particularly in the medium and long term, so that the feed from a simple volumetric feed is never uniform. For this reason, in the nonwoven industry volumetric feeds are generally used in conjunction with an automatic control that results in a levelness of feed approximately the same as achieved by a weighing hopper.

The actual production rate of a card depends on a number of factors, of which the fibre fineness and the age of the card are probably the most important. A modern card running on fine fibre (1.6 dtex) may produce about 35 kg m^{-1} h^{-1} while a similar card running on coarse fibre (15 dtex) could process about 150 kg m^{-1} h^{-1}. Thus, allowing also for the card width a short staple card may produce between 35 and 80 kg h^{-1} depending on the fibre fineness, and a 4-m long staple card could process up to 600 kg h^{-1}. Depending on the age, older cards would give lower productions.

The actual linear speed of the card web depends mainly on the card production rate and the web weight required. It will normally be in the range 40–120 m min^{-1}, though it may be found that other sections of the nonwoven line may require a lower maximum speed. The web itself consists of fibres that are more or less entangled, giving the web a low strength that helps in manipulating it. Many fibres are bent double (termed hooked fibres) so that it is impossible to say that they are parallel to each other in any real sense, but in spite of this it is true to say that all cards produce a marked fibre orientation. If any section of a card web is analysed into fibre segments it will be found that the great majority of fibre segments lie in, or close to, the direction of carding. This fact has an important bearing on the directional properties of nonwovens made from carded webs.

7.4.2 Parallel-laying

As mentioned above, card webs are usually too light in weight to be incorporated directly into a nonwoven. Several layers are generally used, which has the additional advantage that the multiple layers are more regular than the single layers themselves. The simplest way of producing multiple layers is by parallel-laying. This can be done in a number of ways but in each case the final batt ready for bonding consists of several layers of card web laid on top of each other so that each layer is parallel to the adjacent layers. It follows from this that the width of the batt must be equal to the working width of the card, which is a serious limitation, particularly if short staple cards are used. Another serious limitation is that, because all the card webs are parallel to each other, the fibre orientation in the final batt is the same as in the component card webs, i.e., far more fibres lie in the carding direction than across the carding direction. The terminology now changes to nonwoven terms; the direction in which the nonwoven is produced is called the machine direction and the other direction is called the cross-direction. When the batt is bonded, the strength in the machine direction will be due to the bonds reinforced by the fibres (or vice versa), but the strength in the cross-direction is mainly due to the bonds, since there are very few fibres in the cross-direction. The strength in the machine direction is substantially greater than in the cross-direction; the strength ratio is normally given as between 10:1 and 20:1, but the precise value depends on the amount of bonding. The pronounced weakness in the cross-direction combined with the associated low tear strength along the machine direction make parallel-laid nonwovens suitable only for situations where strength is required in one direction only (e.g. tapes) or for fabrics not requiring strength (e.g. wiping cloths and disposables).

More recently, however, special cards have been designed for the nonwoven industry that give much less orientation in the carding direction. Two devices may be used either separately or together to achieve this. One device is the randomizing doffer, which runs in the opposite direction to a normal doffer, so creating more air turbulence and hence disorienting the fibres. The fibres are then further disoriented in transferring from the randomizing doffer to the final normal doffer. The second method involves the use of two scrambler rollers, which are small diameter rollers covered with card clothing. The first scrambler roller runs slower than the fibres as they are fed in, and the second scrambler roller runs slower still. The result is that the tail ends of all fibres are running faster than the leading ends so that the fibres become buckled, so increasing the number of fibre segments lying in the cross-direction. Of course the weight per unit area of the card web is increased by the scrambler rollers, perhaps almost doubled. It is also clear that the card web is in an unstable state after leaving the scrambler rollers, so that great care must be taken not to stretch it back to its original form. As mentioned above, the randomizing doffer and the scrambler rollers may be used alone or together. Used alone either can produce a strength ratio in the region 4:1 to 6:1 and when they are used together a strength ratio between 1.5:1 and 2:1 is possible. As will be seen later, this is almost as isotropic as can be produced by any nonwoven process, so that one of the limitations associated with parallel-laying has been removed, but the width limitation still remains.

7.4.3 Cross-laying

Cross-laying is probably the most common method of producing a batt because it is a very versatile method, particularly in terms of the ranges of weight per unit area and batt width that can be achieved. In cross-laying, the card runs at right angles to a moving conveyor and the card web is traversed forward and back across the conveyor to form the batt. The mechanism, called a cross-layer, to perform the traversing is somewhat complicated and expensive, but cross-layers can be manufactured to lay any width from 1 m up to 15 m. Any cross-layer can easily be adjusted to vary the batt weight; reducing the speed of the conveyor automatically increases the weight of the batt but there is some limit on the exact choice of speed. These facilities mainly account for the popularity of cross-laying, but there is another important advantage in the fibre orientation. After cross-laying, the fibres lie mainly in the cross-direction and the machine direction tends to be weak; however it is standard practice to draw the cross-laid batt in the machine direction, which tends to increase the machine-direction orientation and at the same time reduces the cross-direction, leaving a strength ratio of about 2:1.

7.4.4 Air-laying

The air-laying process involves feeding preopened fibres to a rapidly rotating roller covered normally in metallic card clothing. This roller further opens the fibres and they are then thrown by centrifugal force into an air stream, which carries them to a perforated conveyor on which the air-laid batt is formed. Normally suction is required underneath the conveyor to draw the air stream through the conveyor, since if there is any tendency for the air stream to be reflected back by the conveyor this would disturb the deposition of the fibres. Air-laying can be divided into two broad categories, for heavy batts and for light batts. For heavy batts (above 200 g m^{-2}) a very simple preopening is sufficient because, although the rotating roller in the air-lay machine is unable to remove the tuft formation of the fibres, in a heavy batt this does not show. In this case air-laying is an economical method of batt formation. For lightweight batts below about 100 g m^{-2} the residual tuft formation would show as irregularity in the batt, and until quite recently it was standard practice to card the material before feeding to the air-lay machine, which made the process relatively expensive. More recently an air-lay process has been developed that produces a lightweight batt without precarding, provided that the fibre has been adequately preopened.

The characteristics of air-laid webs have been much misunderstood. This is probably because the first air-lay machine had the term 'rando-' included in its name, which has resulted in many people using the term 'random-laid'. Random-laid implies that the batt should be isotropic and many people apparently believe this is so, but in fact the movement of the perforated conveyor tends to orient the fibres in the machine direction and strength ratios up to 2.5:1 can be found. The fibres or tufts of fibres are thrown down onto the conveyor individually and there is therefore no interfibre entanglement as in the case of carded batts. This makes air-laid batts extremely weak, which presents difficulties for the manufacturer, but the weakness does not affect the final fabrics. Another important characteristic of air-laid batts is that the fibres can be made to lie at a substantial angle to the plane of the fabric, whereas in parallel-laid and cross-laid batts the fibres lie entirely in the plane of the fabric. This property of the air-laid process makes the batts more bulky and more resilient to compression, both very useful characteristics in high bulk nonwovens.

The three processes discussed so far, parallel-laid, cross-laid and air-laid, are collectively known as dry-laid and currently represent almost exactly half of all nonwoven manufacture, cross-laying being much the most important of the three. The importance of the dry-laid processes is gradually diminishing, but this is entirely because some of the newer processes, in particular spun-laid, are expanding at a rapid rate.

7.4.5 Wet-laid

The wet-laid process is based on the papermaking process and machines are modifications of papermaking machines. Modifications are necessary because wood pulp fibres are very short, whilst the fibres used in wet-laying, although short by textile standards, are very long compared with wood pulp (6–20 mm). One of the advantages of the wet-laid process is that it is easy to blend together textile fibres and wood pulp. Since wood pulp costs roughly one-third of the price of textile fibres and the definitions allow up to 50 per cent of wood pulp to be incorporated, it is clear that significant cost savings can be made. It is also clear that such nonwovens must have paper-like properties; they are used mainly in the areas of coverstock, disposables and wiping cloths. Although the process lends itself to making lightweight products ($15–30 \text{ g m}^{-2}$) with a dense, paper-like character, it is also possible to make thicker, heavier products from 100 per cent coarse synthetic fibre, but this is not usual. Although paper machines have a number of different designs at the 'wet end' and run at very high speeds, the machines used in the nonwoven industry all operate on the same principle (inclined wire) and run at very much lower speeds. However, although the output speeds are low in comparison with the paper industry, they are very high in comparison with textile cards, and this is another advantage of the wet-laid process.

Although it seems reasonable to assume that the fibres in the water dispersion are completely random, when the fibres are deposited onto the moving inclined wire they are pulled into the machine direction so that wet-laid fabrics tend to have greater strength in the machine direction. At present wet-laid fabrics represent a little over 10 per cent of the nonwoven total; this percentage is gradually falling as the production of spun-laid is increasing while the production of wet-laid remains practically constant.

7.4.6 Spun-laid

Spun-laying combines extrusion, drawing and laying down of the filaments into the batt. Although in theory either dry or wet extrusion could be used, in practice only melt extrusion is utilized, which means that spun-laid fabrics are entirely made from synthetic fibres. The obvious advantage of the spun-laid process is that it starts with the polymer chips and ends with a completed fabric ready for sale, the whole operation being completed in one continuous production line. However, this is a simplistic view; the production equipment is massive and very expensive. If it were not possible to match the production rates of each individual section of the process then everything would have to run at the speed of the slowest unit and the whole process could be uneconomic. It is therefore clear that one of the most important features of a spun-laid plant is the matching of production rates. Another major advantage is that the filament is introduced into the batt without cutting or breaking, so that the filaments in the batt are endless. Since one of the greatest difficulties in nonwoven manufacture is bonding the fibre ends, a batt without any fibre ends is a great advantage, allowing the nonwoven to be bonded with less bonding agent. Although the spun-laid process eliminates the low production processes such as carding and cross-laying, the capital cost of a spun-laid plant is very high and the total processing cost seems to be roughly the same as for the dry-laid process.

In practice spun-laid plants are 4 m wide since there is a demand for fabric of this width, for example as carpet backing. The chips are melted, extruded and cooled in the normal way except that the spinnerettes are packed closely together in order to get adequate production per unit width of plant. After cooling, the drawing either may be done by godet rollers in the usual way or alternatively it can be done aerodynamically. In this method the filaments are taken into a long tube and the force needed to cause drawing is created by strong air currents blowing down the tube. After the drawing stage the filaments have to be separated from each other, since a group of filaments lying together would cause a faulty appearance in the nonwoven. It is suggested in the patents that this can be done by turbulent air, by bouncing the filaments from a stationary plate, or by charging all the filaments to produce mutual repulsion. Finally, the filaments are laid on a moving conveyor; as in other cases already considered, the movement of the conveyor will cause orientation of the filaments in the machine direction. To counteract this the filaments have to be traversed in the cross-direction, which can be achieved easily in the case of aerodynamic drawing by simply swinging the tubes backwards and forwards. By increasing or decreasing the velocity of the swinging movement relative to the movement of the conveyor, batts can be made with either a machine-direction or a cross-direction bias or alternatively practically isotropic. As mentioned above, spun-laid fabrics are showing the highest level of growth, owing mainly to the range of desirable properties that can be produced rather than any ability to undercut the price of other nonwovens.

7.4.7 Flash-spun

This is really a method of extrusion to yield fibres of low linear density without the problems of using a very fine spinnerette, but it is used in producing certain types of spun-laid batts, so it is included here. The polymer, usually polyethylene but polypropylene could probably also be used, is dissolved in a solvent and is extruded under conditions of temperature and pressure so that the solvent boils suddenly as the pressure is released on leaving the extrusion holes. The result is that the polymer is blown into a series of bubbles, giving it a large surface area and consequently a very thin wall thickness. The material is then drawn and fibrillated to give a network of very fine fibres; of course these are very irregular in shape and linear density but they are on average very fine and serve their particular purpose excellently. After laying and bonding, the fine fibres yield a fabric with a very small pore size that is not only waterproof but also proof to many other liquids with lower surface tensions. The product, Tyvek, is used for making protective clothing principally for use in the chemical, nuclear and oil industries, but also in many other locations. The fact that it is permeable to water vapour means that it is comfortable in wear, and because it is fairly cheap the garments are normally regarded as disposable.

7.4.8 Melt-blown

Although this method of production has been known for about 40 years, it is only recently that it has been widely exploited. Like the flash-spun method it is a way of producing very fine fibre without using normal extrusion. The polymer is melted and extruded in the normal way but before it is allowed to cool it is subjected to a very strong blast of hot air that blows the molten material into a mass of fine filaments. A separate blast of cold air quenches the filaments and the two airstreams combined are strong enough to draw and break the filaments into fibres. The fibres are collected from the air stream onto the surface of a perforated roller, using suction to hold the fibres against being dislodged by the air current. The principle of collection is exactly the same as in air-laying and for the same reason the movement of the perforated roller will cause some bias towards fibre orientation in the machine direction. In a similar way there is also no fibre entanglement to give cohesion to the fibre batt, but in this case the very fine fibres form such a multitude of interfibre contacts that the batt is strong enough to be manipulated without difficulty.

7.5 Methods of bonding

7.5.1 Chemical bonding

Chemical bonding involves sticking the fibres together with some form of adhesive. In the past, natural rubber latex and cellulose xanthate solution have been used as bonding agents, but currently virtually all adhesives are synthetic latices. In principle, chemical bonding could be applied to every type of batt; in the case of the three types of dry-laid batts it is a very standard method of bonding. With wet-laid batts it is equally normal, the adhesive often being applied on-line with batt preparation, though off-line application is also used. It is possible to chemically bond spun-laid products, but it is not very common and other methods are preferred. Flash-spun batt is always thermally bonded (see section 7.5.2). Melt-blown batt can be chemically bonded, but is frequently used in an unbonded form (in which case it is not strictly a nonwoven).

The synthetic latex is supplied at about 50 per cent solids content, i.e., 50 per cent polymer–50 per cent water, and must be diluted to about 25 per cent solids content before use. The most common method of application is saturation, i.e., the batt is carried into the diluted latex to saturate it and then is led to a squeeze roller that removes excess latex. The material then passes through a long hot-air dryer that first drives off the water and finally cures the polymer. This method of saturation suffers from one major problem, which is that the quantity of liquid picked up is rather high. This makes the drying process too long and expensive and also leads to binder migration, which is a movement of the binder latex during the drying process. For this reason, saturation is now more frequently carried out by foam impregnation, where the latex is first beaten into a foam before being applied to the batt. Using this method the liquid pickup can be reduced. Saturation in latex followed by a heavy squeeze normally results in a thin, dense fabric, but it is possible to produce a more open, lofty fabric if coarse hydrophobic fibres are used that are able to recover after the compression of the squeeze rollers.

Impregnated fabrics tend to be inextensible and inflexible owing to the fact that all fibres are bonded at every point where they touch or come into near contact with other fibres. All fibres are also completely covered in binder, which drastically alters the handle of the fabric, though many different types of binder are available that give hard, strong fabrics ranging to softer, weaker ones.

In order to obtain softer nonwovens with a more textile feel, the latex may be applied by print-bonding. Many different patterns are used, from isolated dots through parallel lines to intersecting networks, but all have the same basic aim of binding the fibres in some sections and leaving them free in others. Clearly the bonded section produces strength while the unbonded sections yield handle and flexibility. To a large extent the fabric properties are determined by the two relative areas, but the binder and the pattern are also important.

Latex may also be applied in spray-bonding, followed again by hot-air drying. In this case it is usual to apply only a low weight of binder that does not significantly reduce the thickness of the original fibre batt. The final nonwoven is weak and will not withstand much direct abrasion but is very useful in spite of that, especially in insulation.

7.5.2 Thermal bonding

In recent years chemical bonding has gradually dropped out of favour, being replaced by thermal bonding. However, the rate of growth of nonwovens as a whole has been so great during the same period that the actual usage of latex binder has not significantly declined; it simply has not increased as it would otherwise have done. The problems with chemical bonding are the time and energy needed to evaporate the water and the slight fear that the binder may cause irritation if worn in contact with the skin. In contrast, thermal bonding uses only those chemicals already present in the fibres; the bonding can take place in a fraction of a second and the energy used is small, being theoretically only the heat needed to melt those polymers taking part in the bonding process. There will inevitably be a lot of additional heat lost in practice, but in spite of this the energy costs of thermal bonding are much lower.

The melting component can take a number of forms. Frequently, thermoplastic fibres are used but it is unusual to use a single component in this case, because when the melting point is reached all the fibres disintegrate at the same time and all the textile structure is lost. Instead a blend is made of thermoplastic and nonthermoplastic fibres, or alternatively of two thermoplastic fibres, perhaps of the same chemical type, but one with a low melting point and the other with the normal melting point. A variant of this is to use bicomponent fibres, usually of the sheath–core type, with the polymer of low melting point as the sheath. These fibres still retain their textile form at temperatures several degrees above the lower melting point, which is a big advantage to the nonwoven structure during thermal bonding. Again the bicomponent fibres do not need to be used as the sole component but can be blended with conventional fibres to reduce raw material costs. A final alternative is to use thermoplastic powder that can be dropped evenly onto the fibres as the batt is being formed.

Just as there are many forms of thermoplastic elements, there are many ways in which the heat can be applied, leading to a range of quite different products. Heat may be used without any pressure at all, usually by passing the batt through a hot-air oven. Bonding takes place as the fibres cool; provided that a good blend of fibres has been chosen, the batt thickness is not reduced much by the bonding process and the product is a high loft fabric similar in many ways to the spray-bond nonwovens discussed in section 7.5.1. A blend containing too much single component thermoplastic fibre would not perform well in this application because the whole structure would collapse on the application of heat.

A second method is to apply heat with a limited amount of pressure to produce a fabric with medium density, the actual range depending on the amount of pressure applied. There are two ways of applying this method; in one the fibre batt is preheated before being compressed by a lightweight calender. The other system uses a belt calender, in which the batt passes between a belt and the heated calender roller. The pressure on the batt can be adjusted by varying the tension in the belt, and to obtain greater penetration of heat into heavier weight batts it is possible to heat the belt as well.

The most common method is to use heavily weighted calender rollers to apply heat at great pressure. There are several variants of calenders but they can be simplified into two types: area-bonding and point-bonding calenders. Area-bonding calenders, as the name implies, have a smooth surface and contact the whole area of the batt. Extremely stiff inextensible materials that scarcely seem to be textile in nature can be made in this way if a high proportion of melting fibres is used with area-bonding, but if a low percentage of bonding fibres or bonding powder is employed there are fewer interfibre bonds and weaker but more flexible nonwovens are obtained. In many cases, and particularly in spun-laid batts, it is not easy to produce the correct mixture of bonding and nonbonding fibres. In order to avoid the stiff, inextensible fabrics that would otherwise be produced, thermal bonding is carried out with a heated calender in which one of the rollers is smooth and the other roller is deeply engraved with a pattern, so that the fibres are only melted and compressed together where the two rollers actually touch. At points very close to this (of the order of 50–100 µm) the fibres may melt but there is effectively no pressure, so that very little bonding takes place. This form of bonding is termed point-bonding and is very common for spun-laids. The situation is very similar to print-bonding described in section 7.5.1; fabric strength may be increased by increasing the total area of bonding, but only at the expense of reducing the flexibility and textile handle of the nonwoven. As in the case of print-bonded nonwovens, the bond size, the bond spacing and the bonding pattern are very important. Most frequently bonds are about 1 mm square with about 2.5 mm between bonds, the bonds taking up about 10 per cent of the total area. The bonds are usually in either a simple square or a simple diamond pattern, but continuous line patterns are also used.

7.5.3 Needlefelting

Strangely, the needling process was in use long before the modern nonwoven industry started. Needlefelting was used in 1870 to make two main products, carpet underfelts and horse bankets, both from cheap reprocessed waste. With the introduction of synthetic fibres the needleloom was seen as a way of manufacturing felts without the need to use expensive wool fibres. At the same time the introduction of synthetic fibres into felt improved the abrasion resistance and resistance to chemical and biological degradation enormously in comparison with wool felts. It is also interesting to note that since the synthetic felts could be used for about ten times longer than wool felts, a very big shake-up was forced on the traditional wool felt industry by the introduction of synthetic fibres.

The action of a needleloom is relatively simple; the batt to be needled is drawn between two plates, the lower one called the bed plate and the upper one the stripper plate. Both plates are drilled with a series of holes to correspond with the needles in the needleboard. The needleboard is reciprocated up and down rapidly; on the down stroke the needles are pushed through the batt and the bed plate and on the up stroke the stripper plate holds the batt down as the needles are withdrawn from it. In the brief time that the needles are lifted clear of the batt, it is moved forward a short distance which is termed the advance. In order to increase production, modern needlelooms often have two or more needleboards; needle densities of 5000 to 20 000 needles per metre width are typical.

The needling process works because each needle has a number of barbs that point down towards the needle tip. As the needles go down the barbs engage with fibres from the batt and push them downwards through the batt; when the needles rise again the fibres disengage from the barb and are left in loops perpendicular to the surface of the nonwoven. These loops are responsible for the entanglement, which is the basic structure of a needlefelt. The other change that takes place during needling is that the density of the fabric increases greatly; this is caused mainly by the innumerable downward forces caused by the downward movement of the needles, but the entanglement is caused by the fibre loops.

The tenacity of needlefelt is surprisingly high considering the random way in which it is made, but the breaking extension is very high, often from 30 to 60 per cent. If the nonwoven is released before breaking, it is found that there is very little recovery from extension. In some applications, such as needled carpets, the load–extension properties are of little importance, but for many other uses the poor recovery from extension is a serious disadvantage. In order to improve the recovery it is conventional to include either yarns or woven fabric (scrim or base fabric) into the structure. Since most needled fabrics have been cross-laid, they have better mechanical properties in the cross-direction than in the machine direction. In the less stringent situations, such as needled blankets, it is often sufficient to include yarns running in the machine direction to obtain adequate stability. (It is a happy coincidence that this is also the easiest direction for insertion.) In other situations, where good dimensional stability is required in both directions, a woven fabric is incorporated into the structure. As was noted earlier, some definitions of nonwoven exclude these fabrics owing to the presence of yarn and/or woven fabric, though it will normally only represent 10–15 per cent by weight.

A needlefelt is a flat, featureless structure, which is very desirable for an industrial fabric, but where it is used as a floorcovering some additional features are desirable. One standard method of improving the appearance is to blend two or more colours together to form the batt, and the inevitable slight variations in blend proportions give the carpet an attractive patchy appearance. An alternative method is to structure the carpet; the batt is first needled in the usual way and is then taken to a structuring needleloom. Here special needles known as fork needles are used. The fork is at the end of the needle and is designed to hold far more fibres than do the barbs on normal needles. For this reason structure needling is a much more drastic process. Each fork produces a large loop of fibres at every stroke, but if a drilled bed plate were used the previous loop would be damaged by the next descent of the needle. To prevent this, structure needlelooms use lamellar bed plates formed from a series of rectangular bars separated by spaces. The loops are formed in the spaces and simply slide along in the space as the fabric is moved forward.

Needlefelting can be applied to any dry fibre batt provided that it is heavier than about 150 g m^{-2}. Below this weight it is found that needling is not effective; the explanation is thought to be that there are not sufficient normal fibres for the loop fibres to be pushed into. Since wet-laid batts are normally lightweight, and it would be inconvenient to dry them prior to needlefelting, they are not needlefelted. Similarly parallel-laid batts are normally too light, so the materials used for needlefelting are cross-laid, air-laid and spun-laid.

7.5.4 *Hydroentangling*

This method of bonding used to be known as spun-lacing, but for various reasons the more modern name of hydroentangling is now preferred. It has been mentioned above that needlefelting is not successful with lightweight nonwovens; hydroentangling was invented in the attempt to find an alternative process similar to needling that could be used on the lighter weight fabrics. The needling process is replaced by a number of water jets, which operate through very fine nozzles but at a very high pressure. The nozzle has to be designed very carefully so that the water jet remains as a straight column and does not break up into a series of drops, because although these would have the same inertia their energy would be spread over a wider area of the batt, so that the effect on the fabric would not be the same. Although the needling process is a reciprocating one, the action of the jets is continuous; the necessary variation may be caused by more and less dense sections of the batt passing under the jet or it may be due to the conveyor under the batt, which will sometimes reflect the jet back again and sometimes allow it to pass through. It is clear from this that the conveyor belt may play quite an important part in the entangling process and this is in fact the case. In a typical process between 4 and 12 sets of jets may be needed to achieve the required degree of entanglement.

To be suitable for the hydroentangling process, fibre should be flexible— i.e., in the range 0.9 dtex to 3 dtex—and perhaps with a cross-section that increases flexibility, e.g. oval or flat. It should also be remembered that some fibres change modulus and tenacity quite markedly on wetting. The fibre length is also quite critical; if fibres are too long there are two few ends to entangle, but fibres can also be too short to form a coherent fabric. Lengths between 20 mm and 80 mm are suitable, depending mainly on the linear density. One very interesting development is fibres designed especially for hydroentangling. These are extruded at 1.7 dtex and can be carded without any problem. However, they are designed to split into eight segments under the action of the water jets, becoming fine micro-fibres of 0.2 dtex.

7.5.5 Stitch-bonding

This process involves passing the fibrous batt through a warp knitting machine modified in such a way that the needles can penetrate the batt without damage. In normal warp knitting the warp yarns are fed to the needles by guides, the needles draw the new loops through the old loops, and the fabric moves forward ready to start the next cycle. In stitch-bonding exactly the same thing happens with the needles passing through the batt so that the loops of yarn are on one face of the nonwoven and the yarns connecting the loops are on the other. The batt is both held together by the forces between the yarns and at the same time reinforced by the yarns. As in the case of the needled fabrics discussed above, these particular stitch-bonded nonwovens are not classified as nonwovens by some countries owing to the use of yarns in the con-struction. Another nonwoven, about which there is no dispute, can be made on the same machine without the use of yarn. In this case when the needles move forward they entrap a few fibres from the batt and pull them through to the other face of the nonwoven. On the next cycle another group of fibres is pulled through the loop made by the previous ones. The final fabric has rows of loops on one side and small holes from which the fibres have been drawn on the other. With suitable long fibres in the batt this can be quite a stable fabric, but it cannot achieve a tenacity approaching that of the structure containing yarns.

It is possible to make several other structures on the same machine with only minor modifications. In one case either unconnected 'weft' threads or unconnected 'warp' and 'weft' threads may be held together by the 'knitting-through' technique described above. In other cases loops of pile yarn may be held down to an existing fabric (woven or nonwoven) to form a single-sided pile fabric. Both of these fabrics contain only yarn and/or preformed fabric and no fibre batt at all, so they are likely to be described as nonwoven only by the most generous definitions. A final fabric may be made that is roughly on a par with the needlefelt containing a woven supporting fabric. In this case the needle penetrates the fabric (again it can be woven or nonwoven) and the fibre batt is fed to the hooks of the needles. As the needles are withdrawn they pull substantial tufts of fibre through the fabric to form loops; as before, each succeeding loop is pulled through the previous one to form an endless chain of loops on one side of the fabric. The other side of the fabric is covered by fibre but it is tightly locked into the fabric. The final fabric is normally raised and may be used for artificial fur or as the pile lining for winter outerwear.

7.6 Specific nonwoven products

It is intended in this section to discuss the major nonwoven products and some of the interesting minor uses from the points of view of methods of production, fibres used and potential for future growth.

7.6.1 Coverstock

Coverstock is at present the largest single outlet for nonwovens in the three major production areas, Western Europe, USA and Japan. In Western Europe coverstock formed 26 per cent of the total by weight in 1990 and the figure was 36 per cent in Japan in 1991. Coverstock is used in disposable nappies (diapers), incontinence pads and feminine hygiene products to form a separating layer with adequate abrasion resistance between the body and the absorbent layer. In the early years the design was based on permeability to ensure that the coverstock would distribute urine to the absorbent rapidly enough to prevent flooding. At that time the most common fibre used was viscose and the coverstock was just as wet as the absorbent, resulting in damp or even wet skin in contact with the product. More recently the concept of the dry liner was developed, which is a hydrophobic coverstock still able to allow the flow of urine into the absorbent but preventing flow in the reverse direction (wet-back) provided that the absorbent itself is not too wet. More recently still, superabsorbents have been used to reduce the physical size of the pad while still retaining sufficient absorbency. The two fibres most suitable for modern coverstock are polyester and polypropylene, since they are naturally hydrophobic and can be thermally bonded. Viscose can be used together with a hydrophobic binder but is at a disadvantage compared with fibres that are naturally hydrophobic. The other desirable properties in the fibres are fineness (about 1.5 dtex) to obtain a smooth soft handle, and complete freedom from any chemical irritants. Since the product is a disposable one, there will be demand in the future for biodegradable fibres, but at present the coverstock is quite minor in comparison with the waterproof cover, which is not currently biodegradable, but development has already started to alter this.

Coverstock may be produced by a wide variety of routes. It is a lightweight fabric in the range $18-25 \text{ g m}^{-2}$ and the batt may be made by parallel-laying with randomizing, air-laying, wet-laying and spun-laying. Bonding is preferably by thermal bonding, but can also be done by foam-bonding or print-bonding. Since each method of batt preparation could theoretically be used with each bonding method, there are 12 different possible routes, but they are not all used. The preferred routes are parallel-laying or spun-laying with thermal point-bonding to give a soft textile feel.

The use of coverstock has been increasing in Western Europe at the rate of 10 per cent per annum. For this rate to continue the use of disposable nappies would have to spread to other countries, because some countries already have almost 100 per cent usage. The position in North America and Japan is less certain, but it nevertheless seems reasonable to predict a continuing growth of 10 per cent per annum, which is fairly low compared with other sections of the nonwoven industry.

7.6.2 Nonwoven floorcoverings

The use of nonwoven floorcoverings is very widespread, both geo-graphically and on a microscale. They are not only used in all the major manufacturing countries but are also produced and used in the Middle East and India. The actual location of use can be in offices, showrooms, shops and public buildings; in private houses, where they are found more commonly in kitchens and bathrooms; in cars, where a lighter weight version is used not only on the floors but also on the lining of the boot (trunk), on parcel trays and sometimes as the complete covering of the doors and walls of the car and the headliner. They are also used frequently in hotels and other buildings as a wall covering in corridors and in the lifts (elevators). Unfortunately, it is virtually impossible to collect the data for worldwide production or consumption because in some countries needled carpets are classified as nonwovens but in other countries they are not classified at all. There is no doubt that the total is very large and that it is probably increasing.

The method of production is by either air-laying or cross-laying followed by needling. In many cases a scrim fabric may be included to improve the dimensional stability, which is a critical property in any floorcovering. If the carpet is to be structured, a scrim is not used, since it would interfere with the structuring. Structured carpets are back-coated with latex to improve dimensional stability and to hold the fibres together. The flat needlefelts may be either back-coated to leave a face un-contaminated by the latex or in some cases impregnated with resin to give the maximum wear life.

The major fibre used for this product is polypropylene, usually in high linear density. The main reasons for this are the very good abrasion resistance and the good resistance to staining. The other big advantage is that the many fibre suppliers will also supply spun-dyed fibre either in a range of standard colours or to match any desired sample. In comparison, polyester, which is also quite widely used, has similar or rather better physical properties, but is supplied écru and has to be dyed. The large manufacturers cannot arrange to keep a similar range of spun-dyed polyester to compete with the polypropylene colour service, though spun-dyed black polyester is available.

7.6.3 Geotextiles

The four major characteristics of a geotextile are reinforcement, separation, filtration and drainage, but it is unlikely that any one material would be expected to perform in all four ways. The most common nonwoven products are a spun-laid thermally bonded fabric designed to give high strength and good tear strength used for reinforcement and separation, and a spun-laid or cross-laid batt bonded by needlepunching into a thick fabric to give good filtration and drainage properties. Again the two major fibres are polyester and polypropylene, with nylon being used occasionally. The main property requirements are high tenacity and good resistance to biodegradation. The use of geotextiles has increased enormously since the first introduction in about 1970. It is currently increasing at about 12 per cent per annum, but it seems likely that this rate may increase partly owing to a new use for geotextiles in environmental protection and partly because the present transport and service infrastructure is wearing out and will need repair or replacement.

7.6.4 House furnishings

The most common use of nonwovens actually seen in the home are stitch-bonded fabrics used for curtains, and mattress and divan covers. These are most frequently made from viscose with polyester stitching threads, which makes the printing process both cheaper and easier. The demand for nonwoven curtains seems to have fallen but this has been compensated by the heavy penetration of nonwovens into the bedding market. (It should be noted that these stitch-bonded fabrics are frequently not classified as nonwovens.)

In addition to this there are at least as many nonwovens that are not seen in the home. These include needled carpet underfelt made from waste fibre, which is still used to a limited extent. A very similar material is also used in mattress manufacture and in sprung furniture as a pad over the springs. The padding in upholstered furniture is frequently high loft nonwoven made by either air-laying or cross-laying followed by spray bonding or thermal bonding. The fibre will almost always be polyester, the main characteristic required being good recovery from compression. The filling material for duvets, pillows and cushions will usually be similar material made in the same way but often using polyester hollow fibre, which gives the same insulation properties while cutting down the total weight of the wadding in proportion to the volume of space within the fibre. The space inside the fibre does not have any bigger effect than this because the volume of space inside the fibre is negligibly small compared with the volume of air trapped in the wadding itself. The covers underneath upholstered chairs, armchairs, divan bases and settees will normally be spun-laid thermally bonded nonwoven exactly similar in type to that used in geotextiles. A very similar material is also used as the primary backing for tufted carpets, though woven polypropylene tape is also frequently used. The main property required is good tenacity and excellent dimensional stability, but in the case of carpet backing there is the further requirement that the nonwoven should not be damaged by the needles in the tufting process, or change dimensionally in the heat of the carpet finishing process. It has been found that rather weaker bonding than normal allows the filaments to move out of the way of the needles, so that the final tufted carpet is stronger. Again polyester and poly-propylene are the main fibres, in particular because they are not affected by water absorption; hygral expansion is very important in a carpet that is 4 m wide. The growth in this area has been about 4 per cent per annum in recent years; since the market and the competition have not changed at all, it seems likely that future growth will be at about the same rate.

7.6.5 Wipes

The area of wipes is a very wide one, including disposable wiping fabrics for use in the home and in industry, wet wipes mainly for personal use in the home and semidurable fabrics for dishcloths and dusters. Most of these are made by the parallel-lay or wet-lay route and are either impregnated with a soft resin or print-bonded to give a soft, flexible handle. Alternatively, these fabrics can now be hydroentangled, giving the advantages of freedom from lint and chemical binder. Since the fabrics have to be water-absorbent and neither strength nor resistance to abrasion are important, viscose is the most important fibre in this area and synthetic fibres are not much used.

An exception to this is when the material is required to absorb oil, in which case polyethylene and polypropylene are very good natural absorbents. Very large quantities of melt-blown material are used in cleaning up oil spills, the advantage being that the fibres are in a very fine form so that they can absorb many times their own weight of oil. Polyethylene and polypropylene have the additional advantage that when used on water they float naturally and do not need to be held up artificially by floats. In spite of the big upsurge in the use of wipes that was expected with the introduction of wet wipes, this area has shown a growth of only 9 per cent per annum, significantly below the nonwoven average.

7.6.6 Apparel

With very few exceptions, nonwovens are not used as the outer surface in apparel. This section can therefore be discussed under three sub-headings.

Interlinings

Interlinings were one of the major nonwoven products in the early days and are still an important sector, but, as a very mature product, there is relatively little growth potential. An interlining is required to be inextensible without being rigid in bending, a combination that is difficult to achieve. It must also have excellent recovery from extension and bending. In most cases interlining is made by cross-laying followed by foam-bonding using a high quality latex. However, more recently an alternative route has become available—cross-laying followed by hydroentanglement. Polyester is the major fibre used owing to its high initial modulus and its excellent elastic recovery.

Quilting

With the exception of a few specialist garments that are quilted with duck down, all quilted garments are filled with nonwoven materials. The general method of production is cross-laying or air-laying followed by spray-bonding or thermal bonding. In some cases the manufacturers find that, since the filling is protected on both sides, it can be adequately bonded by the quilting sewing alone, which gives a significant saving in bonding costs. It is very rare to find the latter materials mentioned at all and it is not clear whether they are strictly nonwovens, though they are clearly of that general class. Virtually all filling materials are of polyester, owing to the excellent recovery properties and to the availability of the hollow fibre alternative.

Protective clothing

Protective clothing may be intended either to protect the wearer from dirt, dust, etc., or to protect from dangerous substances, radioactive chemicals, corrosive chemicals, etc. For most situations a disposable nonwoven material is available, but not for every one. The most common material is polyethylene spun-laid into a very fine-fibred material that is finally thermally point-bonded. The fine fibres give a high degree of protection against most liquids and polyethylene is inert to most but not all chemicals and solvents. Other forms of material can be made by wet-laying, parallel-laying or cross-laying, followed by foam-bonding or print-bonding; the problem is to obtain adequate strength, especially tear strength and seam strengths, without making the material too stiff to be comfortable in wear. More recently the alternative of hydroentanglement has offered a bonding route that gives a more textile drape; material of this type is now being made up into protective clothing. Either polyester or polypropylene can be used in these applications.

7.6.7 Footwear

Synthetic leather base. The production of the nonwoven base for synthetic leather is very complicated, involving the blending of normal polyester, heat-shrinkable polyester and viscose. The viscose produces the required absorbency and the polyester gives strength and dimensional stability. The blended fibres are cross-laid and very heavily needled into a dense felt. The density is then further increased by heating the felt to cause the shrinkable polyester fibres to shrink. The nonwoven is then impregnated with polyurethane resin, not to bind the fibres together but to fill in the voids between the fibres. All these processes are carried out on fabric three or four times the desired weight per unit area, because needling is more effective on heavier weights and it is more economic to do it this way. Consequently, the fabric has to be slit into layers and finally the surfaces are ground to make them smooth and free from wrinkles. The leather-like surface layer is put on in a final coating operation.

Shoe linings. The linings for shoes are made in a similar way, only the weight per unit area of the substrate and the coating process being different.

Shoe stiffeners. The many nonwoven products required to give a shoe its rigidity and its dimensional stability are made from thermally bonded fabric that can be shaped and moulded by the further application of heat. Polypropylene or polyester fibres are suitable. Since all the three products discussed in this section can only be used in the shoe industry, it is not surprising to find that the rate of growth is very low, only about 2 per cent per annum.

7.6.8 Filtration

As there are many different forms of filtration and the filter media differ very widely in each case, it is only possible to discuss each type separately.

Coarse air filtration

Air filtration refers to the filtration of relatively clean air at or near normal room temperatures. It is used principally in the air intakes of air conditioning plants in offices, factories and public buildings. No attempt is made to remove all dust and impurities; hence the use of the term 'coarse'. (Some filters on the air intakes of car engines have similar media, though others use paper filters.) The filter media are made usually by cross-laying and spray-bonding or thermal bonding, giving a thick, lofty material. In many cases the filter may be stabilized by a small amount of woven fabric, thereby excluding it from being a nonwoven by many definitions. The filter works by the dust particles being adsorbed onto the surface of the fibres whenever an impact takes place. It is clear therefore that finer fibres, having a greater surface area, will remove more dust, but finer fibres also produce a bigger pressure drop and become more quickly blocked by the accumulation of dirt. Consequently, modern filters have multiple layers in which the air first meets coarse fibres and moves on to the finer fibres after being partially cleaned in the earlier layers.

Since the fibres in these filters act simply as rigid rods to collect the dust, no special properties are required and it could be inferred that any fibre would serve. This is not entirely true because the possibility of a fire in the filter must be considered. Since the combustion products of acrylic fibres are poisonous, these fibres would not be suitable. In practice polypropylene and polyester are cheap and are available in the correct linear density, so they are generally used.

Fine air filtration

Fine air filters are used mainly in two situations, filtering air for a 'clean room' and in personal respirators. In both cases it is necessary to remove a high proportion of sub-micrometre particles. This can be done using very fine fibres, for example melt-blown glass or melt-blown polyolefin. (It is doubtful whether either of these is a nonwoven because the fibres are not further bonded after collection.) However, filters of this type have to be quite heavy per unit area in order to remove sufficient impurity; consequently the pressure drop across them is high. This means in a static installation that additional power is needed, but in a personal respirator either a larger area of filter is required or breathing is harder, neither of which is desirable. Improved filtration efficiency at a much lower pressure difference can be obtained using an electrostatic filter in which the fibres themselves are charged. The first filter of this type was made with wool and a selected resin in powder form; the fibres and powder are blended together prior to carding. During carding the interactions between the fibres and the resin cause electrostatic charging that is stable even after months of storage. Electrostatic filters can now be made from synthetic fibres in a number of ways. The earliest medium is known as electret, which is formed by fibrillating a fine sheet of thermoplastic film that has been charged by a corona. Surprisingly it is found that the charge on opposite faces of the fibres is stable to the fibrillating process and to storage. The melt-blown process (section 7.4.8) may also be modified by holding the extrusion nozzle at a very high potential relative to the collecting surface. For some reason some of the fibres are charged by this process while others are not, so that the final filters contain fibres with a high potential difference. Finally, it has been found that if polypropylene fibres are blended with modacrylic fibres prior to carding, the fibres will charge each other triboelectrically in the same way as in the wool–resin filter. The mixed fibre filter is very useful becaue it gives a good filtration efficiency with a very low pressure difference. It should be mentioned that in some of the cases mentioned the filters are sufficiently dense without any bonding process. For this reason they may not be classified as nonwovens.

The most recent development in electrostatic filters is to apply an external electrostatic field to a filter without any intrinsic charge. This approach is also found to work well; the fibres of the filter have to be sufficiently conducting to carry the charge from the electrode into the body of the filter, but must not conduct so well as to form a conducting path between the electrodes. For obvious reasons filters needing an external field are suitable only for static filters, not for personal respirators.

Gas filtration

Although the gas filtered is most usually air, gas filtration conventionally differs from air filtration either because the contaminating load is much greater or because the air is very hot or chemically noxious, or possibly all three conditions apply at the same time. Since the contaminant quickly builds up into a dust cake on the filter surface, it is necessary to clean the filter periodically, in many cases very frequently. Since this mechanical action tends to distort the fabric, all gas filtration media are very substantially needlefelted and contain a woven support. For this reason they may not be classified as nonwovens by some definitions. The fibres used for gas filtration are extremely wide-ranging. Where conditions permit, polypropylene and polyester are used, but for the higher temperatures and special chemical conditions, very special (and very expensive) fibres are used. These include homopolymer acrylic fibres, the aramid fibres, PEEK fibres, PTFE fibres and stainless steel. It is also worth mentioning that when filtering inflammable materials, such as coal dust or sawdust, there is a possibility of either fire or explosion caused by the filter becoming charged during the filtering process and eventually discharging with a spark. In a situation like this it is necessary to make the filter conducting and this is done by blending in a low percentage of conducting fibre such as epitropic fibre or stainless steel.

Wet filtration

Nonwovens have not been so successful at penetrating the durable wet filtration market, where plate and frame presses are used. One major problem is that the wet cake tends to stick to the nonwoven media; developments in smooth surface finishes will probably improve this aspect in the future. However, many disposable filters are used for filtering the coolants used in machining metals. These are frequently supplied in a roll form and are wound slowly past the filtering point, and the roll is disposed of after one use. The fabrics most frequently used in this application are wet-laid, probably cellulosic, but a significant amount of spun-laid material is also used, being of either polypropylene or polyester.

According to the statistics, the whole of the nonwoven filtration industry has shown a growth of about 10 per cent per annum, slightly below the nonwoven average. This seems to be strange because there has been a strong focus on ecological issues in recent years and there is no doubt that one of the major ways of cleaning the environment is to filter the liquid and gaseous effluents from factories. However, it has been noted above that certain filter materials may not qualify as nonwovens under some definitions, which would of course exclude them from the statistical data. It is therefore thought that nonwovens as defined in this chapter are in fact growing at a higher rate than 10 per cent per annum.

7.6.9 Medical and surgical

Nonwovens are used in a vast number of different medical end-uses, which makes it difficult to generalize about them, but they have one major characteristic in common; they are all disposable. The dressings and swabs have to be absorbent, so these will be made mainly from viscose rayon or bleached cotton. The use of disposable sheets, pillow cases, pillow covers, examination gowns, etc., depends very much on the standard of living in the country concerned. In the United States and Sweden, disposable materials have been used for some time, but their use in other Western Europe countries has been delayed. It would appear that a change is gradually taking place because the rate of growth in this area in Western Europe has been 20 per cent per annum, whereas it has only been 7 per cent in the developed US market.

The development of an acceptable nonwoven surgeon's gown has been a lengthy process. Nonwovens have always shown a clear advantage of being disposable and therefore not requiring sterilization. However, most nonwovens in the past have lacked a textile drape and handle and there has been a strong resistance from surgeons for this reason. More recently, the hydroentangled nonwovens have offered greatly improved drape and handle. In the United States at the present time hydroentangled products account for almost half of the total medical market. Although in the past the disposable market tended to be dominated by viscose rayon, at present the use of polypropylene and polyester is tending to increase in this area at the expense of viscose.

7.7 Summary

It has been shown that the nonwoven industry is a very wide one, covering many different forms of manufacturing technique. Equally, both the properties and end-uses of nonwovens are very wide-ranging. As a direct result of this versatility the nonwoven sector continues to show substantial real growth, partly by natural growth in an established market but also by sudden breakthroughs into new areas where there was previously no established demand.

Throughout all these developments the nonwoven industry has depended almost exclusively on man-made fibres, i.e., the synthetic fibres but including also viscose rayon. During the decade 1980–1990 the use of viscose rayon has fallen very rapidly, mainly because it no longer shows any cost advantage. As a consequence, roughly three-quarters of the fibres used in nonwovens now are synthetic, and more than half of the synthetic fibres are polypropylene. It might be inferred from this that the two industries depend on each other, but that is not really so. It is clear that the nonwoven industry depends heavily on the synthetic fibre industry, but the reverse is not true. The nonwoven industry is only a small part of the textile industry and uses roughly 1.1 million tonnes per year of synthetic fibre. This is large enough to be important to the synthetic fibre industry, large enough to make it worthwhile to develop specialized fibres, but not large enough to claim any form of dependence. However, the nonwoven industry is still expanding rapidly.

Further reading

Chemiefasern/Textilindustrie (English edition), E355 (1981).

GE Cusick, *Nonwoven Conference Papers*, UMIST, Manchester, 1983.

Journal of the Textile Institute, **62**, 1 (1971).

P Lennox-Kerr (ed.), *Needle Felted Fabrics*, Textile Trade Press, Manchester, 1972.

A T Purdy, *Developments in Nonwoven Fabrics*, Textile Institute, Manchester, 1983.

Textil Praxis International, 1206, 1325 (1980).

Textile Institute and Industry, 260 (1976); 216 (1980); 178 (1981).

Textile Progress, Vol 5 No. 3, Vol 7 No. 2, Vol 12 No. 4, Textile Institute, Manchester.

CHAPTER 8

Rubber composites

S K CHAWLA

8.1 Introduction

Composite materials, defined as materials containing two or more distinct phases, have become a way of life in almost all engineering applications. High performance or rigid composites made from glass, graphite, Kevlar, boron or silicon carbide fibre in epoxy matrix, because of their application in aerospace and space vehicle technology, have been vastly glamorized. A great deal of analytical work towards the understanding of such composites is available. On the other hand, composites based on fibre and rubber materials, in spite of having the widest industrial application ranging from pneumatic tyres, conveyor belts, hoses, power transmission belts, marine applications, to skirts for air cushion vehicles, have been paid very little attention in terms of the understanding of the basic mechanics of such composites — mainly owing to the complexity of their constituent material properties. These fibre-reinforced rubber (FRR) composites are characterized by the extremely low stiffness of the rubber matrix compared to that of the reinforcing cords. This large difference between elastic moduli of commonly used fibres and matrix rubber makes the analytical formulation much more difficult. For example, the pneumatic tyre consists of a large number of components designed to serve a specific function with maximum effectiveness and at the same time provide synergism to produce desired performance. Because of the complex geometry, heterogeneous and anisotropic material characteristics, and complex load distribution of a tyre, determination of composite properties of such a structure is essential for engineering analysis and its performance in the face of ever-increasing demands from the consumer for smooth ride, good handling, and longevity. Figures 8.1 and 8.2 show the schematic and design complexity of all-season high performance radial passenger and typical truck tyres, respectively.

The characteristic features of cord–rubber composites have produced various theories such as netting theory, cord inextensibility theory, classical lamination theory, modified lamination theory, and three-dimensional

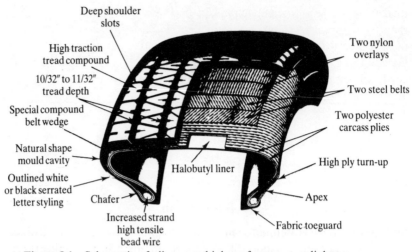

Deep shoulder slots

High traction tread compound

10/32″ to 11/32″ tread depth

Special compound belt wedge

Natural shape mould cavity

Outlined white or black serrated letter styling

Chafer

Increased strand high tensile bead wire

Halobutyl liner

Two nylon overlays

Two steel belts

Two polyester carcass plies

High ply turn-up

Apex

Fabric toeguard

Figure 8.1   Schematic of all-season high performance radial tyre.

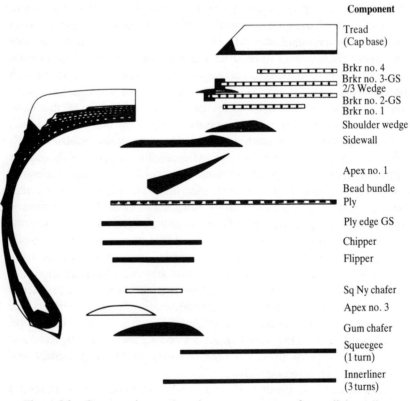

Component

Tread (Cap base)

Brkr no. 4
Brkr no. 3-GS
2/3 Wedge
Brkr no. 2-GS
Brkr no. 1

Shoulder wedge

Sidewall

Apex no. 1

Bead bundle
Ply

Ply edge GS

Chipper

Flipper

Sq Ny chafer

Apex no. 3

Gum chafer

Squeegee (1 turn)

Innerliner (3 turns)

Figure 8.2   Cross-section and various components of a radial medium truck tyre.

theory. This chapter briefly explains some of the basic concepts of these theories with useful results and applications, by reviewing the evolution of synthetic fibres as the reinforcement for rubber products; the manufacture and property development of nylon, polyester, and aramid fibres; cord–rubber adhesion development and mechanism; and constituent cord and rubber properties.

8.2 Synthetic reinforcement for tyres

The first pneumatic tyre patent was issued by R. W. Thompson, a British civil engineer, in 1845. However, it was not until 1888 that an acceptable textile-reinforced pneumatic tyre was built by J. Dunlop, an Irish veterinarian, that forms the basis of the modern day tyre. Dunlop used Irish linen as the reinforcing material, which owing to its cost was gradually replaced by cotton around 1900. The continuously increasing demands on the tyre to provide load carrying capability, and severe performance requirements such as traction, abrasion resistance, low rolling resistance, durability, and safety led to the development of various reinforcing materials. The evolution of fibres from cotton to Kevlar as tyre reinforcement is depicted in Table 8.1. Fibre producers have been constantly modifying the processes to enhance existing fibre properties or developing new fibres with expanded ranges of properties that might lead to the development of new, innovative tyre designs. The stress–strain curves and properties of most commonly used tyre yarns are shown in Fig. 8.3 and Table 8.2, respectively.

Nylon, polyester and aramid — the prominent synthetic tyre cords — offer very different properties and accordingly are used in entirely different applications. For example, nylon offers an attractive combination of high strength and excellent fatigue resistance, and is generally the preferred reinforcement for bias truck, off-the-road and aircraft uses and as an overlay material for high performance radial passenger tyres. Nylon cord

Table 8.1. The evolution of tyre reinforcement fibres

Cord	Year introduced
Cotton	1900
Wire (Europe)	1936
Rayon	1938
Nylon	1947
Wire (USA)	1955
Polyester	1962
Glass fibre	1966
Aramid	1974

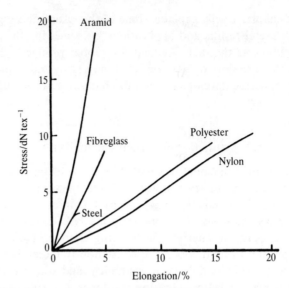

Figure 8.3 Stress–strain curves of industrial yarns.

Table 8.2. Comparative properties of tyre yarn

	PET[a]	Nylon 6,6	Rayon	Kevlar
Typical size/dtex	1110	1400	1830	1670
Relative density	1.39	1.14	1.52	1.44
Tenacity/N dtex^{-1}	6.6	8.4	4.4	20.3
Initial modulus/N dtex^{-1}	97	44	106	530
Elongation at break/%	12	19	15	2.2
Moisture regain/% (at 25°C, 55% RH)	3–4	5–7	10–12	7–8
Thermal shrinkage/% (at 177°C in air)	7–9	9–11	5–7	<0.1
Flex fatigue/% retained	95+	95+	95+	75
T_g/°C	70	50	NA	NA

[a] Dimensionally stable polyester.
NA = not applicable.

tyres provide excellent impact resistance and fatigue performance. However, owing to poor dimensional stability, lower glass transition temperature and lower modulus, the main deterrents to application of nylon fibres as carcass reinforcement for radial passenger tyres have been flat spotting[1,2] and performance at high speeds. Polyester fibre, on the other hand, owing to its high modulus, high glass transition temperature and superior dimensional stability, is the preferred carcass reinforcement for radial passenger and light truck tyres in the United States and is rapidly replacing rayon or nylon in other parts of the world. The polyester cord tyre provides excellent high speed performance, and improvements in tyre deformation, flat spotting, tread wear, handling, tyre uniformity and durability. It also offers better moisture resistance and maintains its properties in the wet state. Aramid, which offers exceptionally higher strength, higher modulus and outstanding dimensional stability but poor compressive fatigue strength, has found limited use in some tyres as belt reinforcement material. The concept of hybrid or merged cords,[3] designed by combining different yarns together during the cabling step, can expand the range extensively and provide balanced tyre cord properties. For example, the hybrid cord containing aramid and nylon provides improved fatigue resistance over all aramid cord and maintains excellent strength, modulus, controlled elongation and shrinkage. The known commercial application of hybrid cord, containing two aramid strands twisted with one strand of nylon, is General Tire's trademark 'Aralon', used as a breaker fabric in off-the-road tyres. Steel cords remain the predominant reinforcement for belts in radial passenger and light truck tyres, and for carcass and belt for radial medium truck tyres.

8.2.1 Nylon

This topic is covered in references 4–16.

Nylon is truly the first synthetic fibre derived from petroleum products. It represents a family of polymers, of which nylon 6,6 and nylon 6 fibres are most commonly used in the tyre industry. Nylon 6,6 is made from adipic acid and hexamethylene diamine. The adipic acid is produced by oxidation of cyclohexanol with concentrated nitric acid. Hexamethylene diamine is made from adipic acid by first feeding molten adipic acid and preheated ammonia over boron phosphate catalyst at 360°C, converting it to adiponitrile, which in turn is reduced to diamine by hydrogenation. The hexamethylene diamine and adipic acid are then dissolved in methanol and, upon mixing, nylon salt is precipitated, which after purification is polymerized in an inert atmosphere in the presence of acetic acid as stabilizer at 280°C for 4 hours. The water from the nylon salt is allowed to escape. The molten polymer is extruded, quenched in cold water, and reduced to nylon chips.

Nylon 6 polymer is made from caprolactam, which in turn is synthesized from cyclohexanol. The synthesis process consists of the following steps. (1) Cyclohexanol is dehydrogenated into cyclohexanone using copper as a catalyst. (2) Cyclohexanone is reacted with hydroxylamine to produce cyclohexanone oxime. (3) The oxime is then subjected to Beckmann transformation by treating with sulphuric acid, converting it to caprolactam. The polymerization of caprolactam is conducted at high temperature with water at 10 per cent by weight. The water reacts with caprolactam to form aminocaproic acid, which in turn polymerizes to form nylon 6.

The nylon fibre is produced via a melt spinning process. Nylon chips (nylon 6,6 or nylon 6) are fed through a hopper into a heating grid for melting. The molten polymer is pumped through a sand or screen pack and then through spinnerette holes where fibres are formed and extruded. The fibres are then quenched, combined into yarn and taken up on a bobbin for further drawing and heat treatment for development of properties. In some cases, the process is continuous in that the fibres are simultaneously spun and drawn.

The screen pack, beside performing the function of filtration, exerts a strong influence on the fibre properties because of a change in the rheological behaviour owing to screen pack porosity and pack pressures. The flow of polymer melt through spinnerette capillaries leads to some molecular orientation. During quenching or solidification, the fibre experiences more shear and additional orientation along with development of fine crystalline structure. The solidified fibres are subjected to heat and stresses during the drawing process, further enhancing molecular orientation and crystallinity and setting the fibre properties.

8.2.2 Polyester

This topic is covered in references 17–20.

The starting raw materials for polyester fibre are ethylene glycol and terephthalic acid. The glycol and terephthalic acid are polymerized through polycondensation in vacuum at high temperature into polyethylene terephthalate. The polymer is extruded, solidified, dried and cut into chips for fibre forming. The intrinsic viscosity of polymer used for industrial fibre applications such as tyres is about 1.0 dl g^{-1} ($M_r = 40\,000$) in contrast to polymer IV of 0.6 dl g^{-1} ($M_r = 20\,000$) used for textile purposes.

The polyester fibres are extruded via melt spinning, quenched, have finish applied and are drawn to achieve high molecular orientation and crystallinity, and therefore desirable fibre properties. The drawability of fibre has a strong effect on its ultimate properties. In general, the drawing process of polyester fibres consists of a two-stage draw followed by a relaxation. The first-stage draw is approximately 500 per cent at temperatures slightly higher than the glass transition temperature, T_g, of the polymer, followed by about 125 per cent elongation at a temperature ranging between 200°C and 230°C. The application of finish decreases the static friction of the yarn and improves the yarn to cord efficiency. The finish can also help in promoting the adhesion between the cord and rubber. The rubber composition is also important for the chemical degradation of polyester cord and its adhesion to the rubber. Rubber stocks that minimize the presence of amines provide good performance of polyester cords.[21,22]

8.2.3 Aramid

The introduction of aromatic polyamide fibres as a rubber-reinforcing material has been the most significant development in synthetic fibre technology since the introduction of polyester.[23,24] The presence of aromatic and amide groups along a polymer chain results in extremely high strength and high modulus fibres. The most commonly used polyamide (poly(*p*-phenylene terephthalamide)) is made by dissolving *p*-phenylenediamine in a mixture of hexamethylphosphoramide and *N*-methylpyrrolidone, cooling and then adding powdered terephthaloyl chloride with constant stirring. The crumb-like acidic product is ground with water in a blender or mill and then filtered. The filtered polymer is washed several times with soft water to remove solvent and HCl, followed by a drying step at 120–140°C.

The resultant polymer is dissolved in concentrated sulphuric acid at 20–24 per cent polymer concentration. The aramid fibres are extruded from this optically anisotropic solution using a dry-jet wet spinning process.[24] The aqueous system is used as the coagulation medium, and the solidification of fibre takes place owing to evaporation and diffusion between polymer filament and the bath. The extruded fibre is washed with aqueous alkaline solution to remove acid and then dried with air at 10°C under a tension of about 0.3 g dtex^{-1}. The fibres thus produced have very high molecular orientation and crystallinity, and exhibit excellent tensile properties. The properties of fibres can be altered by a heat treatment. Heating fibre under tension, in general, increases its modulus and reduces yarn elongation.

Aramid fibres have been engineered for the reinforcement of tyres and mechanical goods including hoses, conveyor and power transmission belts. The most successful application of aramid fibres in radial tyre has been in the form of a reinforcement material for belt. The poor compressive fatigue properties and marginal adhesion to rubber inhibit the application of aramid fibres as a carcass reinforcement in tyres. However, research is continuing to improve both these fibre properties and thereby expand the use of aramid fibres in rubber product applications.

8.2.4 Fabric preparation for tyres

The spun yarn on beams is twisted, corded and then woven into a fabric form that is suitable for tyre use. The first step of the process involves twisting of yarn plies themselves with the desired number of turns per unit length. Two or more spools of twisted yarn are combined and twisted in opposite directions with the same or different amounts of twist to form a cord. When the ply twist and cord twist are equal and in opposite directions, the cord is termed a balanced cord — otherwise it is known as unbalanced. Balanced cord is most commonly used in the tyre industry, though there is some discussion of the possible use of unbalanced cords in tyres. In any case, the twist in yarn and in cord provides durability and therefore fatigue resistance at the expense of strength and modulus. The optimum balance between strength and fatigue resistance depends on the requirements of the end-product. A thorough mathematical explanation of the effect of yarn and cord twist on properties, from a mechanics point of view, is given in reference 25.

The spools of twisted cord are supplied to conventional shuttle looms for weaving on 62-inch-wide drums at a predetermined cord spacing. The woven fabric contains fill threads to maintain constant spacing between the cords and to facilitate factory handling. The rolls of woven fabric are then transferred to the treating unit for adhesive application under controlled process conditions of time, temperature and tension to provide the desired properties for end-use. The adhesive provides a bond between the cord and rubber. Cord–rubber adhesive systems are discussed in detail in section 8.3. The predetermined temperatures and tensions provide optimum treated cord properties such as strength, modulus and dimensional stability.

The fabric is processed in a typical fabric processing unit. The process consists of a first dip application, drying and heat treatment, second dip application, drying and heat treatment, followed by cooling. In general, tension is applied during the first heat treatment zone followed by relaxation in the second heat treatment zone. The adhesive is applied by immersion. The amount of dip and dip penetration into a cord are controlled by immersion time and cord tension. The drying zone residence time and temperatures are specified to reach a proper state of cure for aqueous adhesive systems. The time, temperature and tension conditions and their response depend on the cord material. Nylon and polyester materials, which are thermoplastic in nature, have a tendency to return to their unoriented forms, producing cord shrinkage, which adversely affects the uniformity of tyres. However, this phenomenon of cord shrinkage can be controlled somewhat by stretching the fabric in the first heat treating zone and then relaxing the fabric in a controlled manner in the second zone. In general, higher relaxation in the second zone results in lower cord shrinkage, and higher net stretch (the difference between stretch in the first zone and relaxation in the second zone) and higher relaxation results in higher strength and low modulus cords. The temperature conditions are also used as variables to control cord properties: higher temperatures decrease cord strength and modulus, probably owing to degradation of material properties such as intrinsic viscosity.

There are several published reports on the heat treatment of nylon cords.[26–30] The heat treating conditions used in the tyre industry vary between 177°C and 218°C for temperature and between 7 per cent and 16 per cent for stretch. The residence time varies between 20 seconds and 60 seconds. All these conditions serve the purpose of meeting the product requirements of the individual company. The heat treatment conditions for polyester cord are different from those for nylon: the heat treatment temperatures are considerably higher, whereas the net stretch is considerably lower. Aitken *et al.*[20] have conducted a thorough study on the processing of polyester cords and reported that heat treatment conditions have a pronounced effect on the static and dynamic properties of polyester cords. The very important property of fatigue resistance has been found to be particularly sensitive to the treatment temperatures. The optimum fatigue rating was obtained at a treating temperature of 246°C, 4 per cent stretch in the first zone and 3 per cent relaxation in the second zone, with residence time of 90 seconds in both zones. High temperature treatment may be good for fatigue rating and dimensional stability but can result in lower strength cord. Also, they concluded that upon heat treatment, higher molecular mass polymer yarns provide higher yarn tenacity, lower strength loss and improved fatigue performance over lower molecular mass polymer yarns. It is therefore necessary that heat treatment conditions be optimized to obtain balanced cord properties, and different yarns should be compared at their respective optimum treatment conditions.

The heat treatment condition for aramid cords is much simpler than for nylon or polyester, owing to the fact that no high shrinkage forces are involved and there is no need to control the dimensional stability of the aramid fabric.[31] The treatment of aramid fabric is carried out by controlling tension rather than stretch, as very little stretch is involved. Because of the very high thermal stability of aramid fibres, the aramid fabric can be treated at much higher temperatures than for polyester or nylon cord fabrics; the process is sometimes limited by the maximum operating temperature of the heat treating unit. In general, the heat treating temperature of the aramid cord fabric is between 232°C and 260°C, coupled with tension of about 1 g dtex^{-1} with a residence time of 60 seconds. Higher tension will further increase the modulus but may cause loss in cord strength. Table 8.3 summarizes the range of time, temperature and tension used for nylon, polyester and aramid cords in the tyre industry.

Table 8.3. Heat treatment conditions for nylon, polyester and aramid used in the tyre industry

	Set temperature/°C	Net stretch/%	Tension/cN tex^{-1}
Nylon 6,6	177–246	3–7	–
Polyester	204–246	0–4	–
Aramid	232–260	<1	8.8

8.3 Cord–rubber adhesion

The most critical aspect of the performance of cord–rubber composites is the development of adhesion between relatively high modulus fibres and low modulus rubber. It is only through adequate adhesion that the tyre cord is able to reinforce the rubber and provide desirable composite mechanical properties for tyre performance. Tyres, conveyors and transmission belts require a very high level of adhesion because of their severe service conditions. There is a significant amount of published literature on the subject. Takeyama and Matsui[32] published the first comprehensive review of tyre cord adhesion to rubber in 1969, covering rayon, nylon and polyester. In 1985, Solomon[33] presented an updated review covering developments in polyester and aramid fibre adhesive technology as well as the effects of exposure to ozone, UV, humidity, heat and chemicals on adhesion. Iyengar,[34,35] Albrecht,[36] and Erickson[37] have described the effects of rubber compounding and physical properties on nylon and polyester cord adhesion. Iyengar,[38] in a 1987 article, discussed the effects of polymer–fibre interface, fibre surface, fibre finish, rubber properties and environmental exposure of cords on adhesion. The adhesion of steel cords to rubber is provided through well known zinc or brass plating technology. The adhesion of nylon, polyester and aramid cords to rubber is the subject of discussion in this chapter.

8.3.1 Adhesion mechanism

There are very significant differences between synthetic fibres and elastomer rubber from a chemical as well as a mechanical point of view. The polarity and the high modulus of fibres are very different from the nonpolar nature and low modulus of rubber. The development of adhesion between fibre and rubber can occur through mechanical, chemical and molecular interactions. Mechanical bonding is obtained by mechanical entanglement and is critical where high surface area is a factor in promoting adhesion, as is observed in the short fibre reinforcement of rubber. Chemical bonding is achieved through chemical interaction between the adhesive, the fabric and the rubber. It is therefore desirable that adhesives have two-faceted molecular polarity and reactivity. Molecular bonding is achieved through molecular interdiffusion between adhesive and substrate (fibre and rubber) and specific physicochemical activity such as hydrogen bonding. In general, a combination of all three mechanisms is responsible for development of adhesion.

The resorcinol–formaldehyde–latex (RFL) adhesive system developed in the early 1940s is still in use throughout the rubber industry. 2-Vinylpyridine–butadiene–styrene is the common latex used in the adhesive recipe. However, compatibility with different types of elastomers can be achieved by selecting the proper type of latex. For the best performance from RFL dip, special attention must be given to the recipe and the maturing and baking conditions, as they affect adhesion as well as the appearance and performance. The latex component makes the adhesive layer flexible and is mixed with a rubber layer through secondary bond and co-vulcanization. The RF components react with the hydrogen bonding groups (NH) in the nylon. The NH groups in aramid are hidden between bulky aromatic groups and are not easily accessible for hydrogen bonding. Therefore, the development of adhesion through hydrogen bonding in aramid is much more difficult than in nylon. The normal RFL treatment does not provide adequate adhesion for polyester because of lack of functional groups that promote hydrogen bonding. However, the polar end groups and higher concentration of carboxyl groups in polyester are beneficial to development of adhesion owing to better wettability and chemical activity. The finish additives, consisting of lubricants, emulsifiers, antioxidants and antistatic ingredients applied to the yarn to facilitate the yarn spinning and cord twisting processes, can aid in promoting adhesion by improving the wettability of the fibre by aqueous adhesive systems. The most significant design requirement for the adhesive systems is the selection of adhesive components specific to the fibre–rubber system. For example, RFL adhesive used for nylon was found to be inadequate for polyester substrate. Good wettability to the fibre is another important requirement for continuous rather than spotty coating of cord. Balanced and two-faceted reactivity as well as mechanical compatibility of adhesive for bonding high modulus fibre to low modulus rubber are also essential for adhesion development. The adhesive can provide a transitional modulus to accommodate stresses in the system. The adhesive or dip pickup and cure conditions can affect the adhesion and must be optimized. Overcuring of the dip or inadequate dip pickup can lower the adhesion. Flexible, heat- and fatigue-resistant adhesives are necessary to meet the requirement of cord–rubber composites for industrial applications.

Table 8.4. RFL adhesive recipe for nylon

RF resin solution		RFL formulation	
Water	238.4 g	Gentax latex, 41%	428.0 g
Resorcinol	11.0 g	RF solution, 6.5%	465.0 g
Formalin 37%	16.2 g	Water	107.0 g
Sodium hydroxide	0.3 g		

From T. Takeyama and J. Matsui[32] with permission of *Rubber Chemistry and Technology*.

8.3.2 *Adhesive for nylon*

The resorcinol–formaldehyde–latex (RFL) system is the most commonly used adhesive for nylon. RFL adhesives are prepared first by forming an RF resin solution prior to adding latex.[32] The RF resin solution is prepared by dissolving resorcinol, formaldehyde and sodium hydroxide successively in water. A reaction takes place that is allowed to continue for 6 hours at 25°C. The RFL is made by the addition of RF resin to rubber latex and is matured for 6 hours at 25°C. The mixture is aged for 18–20 hours before use. The recipe for commonly used RFL adhesive for nylon is given in Table 8.4.[32]

Adhesion is affected by the composition of RFL, maturing conditions, latex used, resorcinol-to-formaldehyde ratio and ratio of latex to the RF resin, and conditions of dipping and heat treatment. The amount of dip penetration and dip pickup depend on the type of fibre as well as on the lubricant or finish applied to the yarn. It is important to consider the effective dip pickup and not the total pickup, as fully penetrated RFL does not contribute to the adhesion. It has been observed that dip penetration up to the second layer of filaments is most effective in obtaining maximum adhesion. Dip pickup of 6–8 per cent of cord weight is considered most suitable. An effective heat treatment is required to form highly cross-linked structure in RFL and strong interaction between adhesive and fibre. Both over- and under-treatments are undesirable for adhesion. Commonly acceptable time and temperature conditions for nylon 6,6 are 220–230°C and 30–60 seconds, respectively.

8.3.3 *Adhesive for polyester*

The conventional nylon adhesive system based on the resorcinol–formaldehyde–latex formulation was found to be unsatisfactory for use with polyester owing to the hydrophobic nature of its surface. As a result, a number of adhesives for bonding polyester fibres to rubber have been developed. Canadian Industries Ltd[39,40] suggested an aqueous two-dip system based on a poly(vinyl chloride), polyamide–polyamine combination. The second solution was standard RFL used for nylon. This adhesive system was used to a limited extent in some commercial polyester tyre cords but was not widely accepted owing to the specialized technique and equipment required for preparing stable emulsion. Polyisocyanate solution in organic solvent was developed[41,42] to treat polyester followed by the standard RFL treatment. The isocyanate-treated cords provided excellent adhesion and good shelf-life but the highly toxic nature and sensitivity to water of isocyanate limits its wide acceptance in the tyre industry. Subsequently an aqueous solution of blocked isocyanate, using phenol as the blocking agent, was developed. Phenol-blocked isocyanate does not decompose in water and is marketed by Dupont as Hylene MP. Shoaf[43] used an aqueous solution of blocked isocyanate and epoxide as a subcoat. The polyester cord was heat treated for 1 minute at 240°C before applying the standard top coat RFL, resulting in an excellent adhesion to rubber. Several other subcoat dips based on chemical activation of polyester surfaces have been patented across the industry.[33] A mixture of Hylene MP and RFL has also been found to be a good adhesive for polyester cord.[44]

The systems so far described are two-dip systems, requiring the application of a subcoat followed by a second RFL dip. However, a single-dip system is desirable from economic considerations. Canadian Industries Ltd[20,45] developed a single-dip aqueous adhesive system called N-3 that consisted of a reaction product produced by mixing triallyl cyanurate, resorcinol and formaldehyde at the preferred composition shown in Table 8.5.[20] The ratio of resorcinol to formaldehyde in the RFL formulation used with N-3 additive plays an important role in the level of adhesion obtained. The N-3/RFL-dipped polyester cord processed by heat treatment conditions of drying at 107°C for 120 seconds followed by heat treatment at 254°C for 45 seconds, preferred dip pickup being in the range of 5–7 per cent, provides excellent adhesion to rubber.

ICI developed a new single-dip adhesive system (Belgian Patent 688424), utilizing a mixture of resorcinol–formaldehyde–latex and an

Table 8.5. N-3 additive composition (relative)

Resorcinol	100
Triallyl cyanurate	24
Red lead catalyst	0.25
37% Aqueous formaldehyde	28
Water	400
28% Aqueous ammonium hydroxide	28

From R. G. Aitken *et al.*[20] with permission of Lippincott and Peto, Inc.

Table 8.6. Pexul single-dip recipe (relative)

Penacolite R-2200 RF (70%)	3.4
Sodium hydroxide	0.2
Water	18.8
Pexul (20%)	30.0
VP latex (38%)	31.8
Water	12.6
Formaldehyde	1.0
Water	1.0
Total	98.8

From T. S. Solomon[33] with permission of *Rubber Chemistry and Technology.*

aromatic additive called Pexul (or H-T), or the polyester cord adhesive. The additive is a condensation product of chlorophenol and resorcinol with methylene. The bonding mechanism is attributed to the sorption or penetration of Pexul into the polyester fibre surface. The final Pexul single dip is made according to the recipe given in Table 8.6.[33] Several other single-dip adhesive systems for polyester cord have been patented and reported over the years.[33] These systems are in some use in the tyre industry. More recently, Hisaki *et al.*[46] suggested the use of carboxylated VP latex in conjunction with conventional VP latex to provide sufficient adhesion as well as protection from hydrolysis of polyester by amine compounds, especially in large-size tyres that generate a significant amount of heat with increased mileage.

Another important development in polyester adhesive technology occurred when Fibre Industries Inc.[47] demonstrated that the surface of polyester filament can be activated through the use of a finish solution during the fibre spinning process so that it requires only one top coat of standard RFL adhesive for optimum adhesion of polyester fibre to rubber. The finish solution composition consisted of 0.1 per cent sodium carbonate, 5.0 per cent glycidyl ether, 5 per cent spin finish containing 60 per cent dimethylsiloxane, 5 per cent spin finish containing 75 per cent ethoxylated sorbitan monooleate and 25 per cent ethoxylated octylphenol, and 84.9 per cent water. The finish is applied to the spun yarn prior to drawing at elevated temperatures.

Subsequently, several patents describing polyester fibre surface activation processes using epoxides in combination with other activating ingredients have been issued in the United States, Europe and Japan.[33] For example, Maeta[48] used dimethylaminopropylamine along with Eponite 100 as part of the spin finish, Kigane[49] used blocked isocyanate α,ω-polybutadiene glycol to activate the surface of polyester, Bhakuni[50] added tris(hydroxy-ethyl)cyanurate or triallyl cyanurate to a standard tyre yarn spin finish.

Also, several attempts have been made to modify the polyester fibre surface by grafting or plasma-treating in order to improve adhesion.[33] However, the most widely used methods for adhesion development use single dips that utilize a mixture of RFL and aromatic additives such as Pexul or double-dip systems in which the fibre surface is pretreated with an epoxy compound followed by RFL treatment or a modification.

8.3.4 Adhesive for aramid

Aramid fibres, like polyester, do not provide satisfactory adhesion to rubber when coated with a conventional RFL type adhesive system. Iyengar[51] discussed in detail the adhesion mechanism of aramid fibres to rubber. The development of adhesion by hydrogen bonding is difficult with aramids, as amide groups are hidden between aromatic groups. Therefore, for an aramid type substrate entropic, i.e. diffusion type, bonding becomes important. Thermodynamic compatibility between the substrate and the adhesive is the key requirement for promoting adhesion. This compatibility is achieved when the solubility parameters of substrate and adhesive are matched. The most commonly used adhesive system for aramid fibres is a two-dip system developed by Dupont.[31] The formulations for the subcoat (coded as IPD 38) and the top coat (coded as IPD 39), and preferred treatment conditions, are listed in Tables 8.7 and 8.8, respectively. The addition of protective waxes to RFL minimizes adhesion

Table 8.7. Dip formulation for aramid tyre cord

Subcoat (IPD 38)			
Diglycidyl ether of glycerol (epoxide)	2.22		
10% Sodium carbonate, anhydrous (catalyst)	0.37		
5% 'Aerosol' OT, 75% solids (wetting agent)	0.56		
Water at room temperature	96.85		
	100.00		
Topcoat (IPD 39)	*Wet*		*Dry*
Water	141.0		–
Ammonium hydroxide, 28% aqueous	6.1		1.7
Preformed resorcinol–formaldehyde (RF) resin, 75% aqueous	22.0		16.5
Mix and add without ageing to			
VP latex, 41% aqueous	244.0		100.0
Mix and add:			
Formaldehyde, 37% aqueous	11.0		4.1
Water	51.4		–
To use, add:			
Carbon black dispersion, 25% aqueous HAF black	60.3		15.1
Wax dispersion, 4% active solids based on	12.1		5.4
RF resin + latex + formaldehyde			
Total	547.9		142.8
Solid %		26.1	

From J. R. Willis, The Tire Technology Conference, Clemson University, Oct. 1984.

Table 8.8. Preferred adhesive treatment conditions for aramid tyre cord

	First oven	Second oven
Oven exit tension/cN tex^{-1} (gf denier^{-1})	8.8 (1.0)	2.6 (0.3)
Exposure temperature/°C	220–240	215–230
Exposure time/s	60	60
Dip	IPD 38 (subcoat)	IPD 39 (topcoat)

From J. R. Willis, The Tire Technology Conference, Clemson University, Oct. 1984.

loss due to ozone.[52] Several other two-dip adhesive systems for adhesion of aramid fibres to rubber have been proposed.[33]

8.3.5 *Evaluation of adhesion*

Evaluation of adhesion between the tyre cord and rubber is conducted in the laboratory by utilizing both static and dynamic methods. Static adhesion tests are either shear or peel type and are performed at room temperature, at elevated temperature, after ageing, and under extreme humidity conditions. In shear adhesion tests, also known as H-pull, U-pull and TCAT, the adhesion is measured as the amount of force required to pull out a single cord embedded in a block of rubber. The TCAT (tyre cord adhesion test) procedure[53] involves the pulling of a cut cord versus a continuous single cord in H or U tests. In a peel test (ASTM D-2630 2-ply strip peel test), two layers of cords separated by a layer of rubber are pulled apart by peeling at an angle of 180°. The peeling force is the measure of adhesion and is influenced by rubber thickness, cord spacing, cord angle and peel rate.

The dynamic evaluation of adhesion is measured by the number of cycles required to reach a certain limiting value. A number of test methods, based on compression, shear, or flex deformation applied to the specimen, have been developed. The Goodrich disc fatigue tester,[54] utilizing compression–extension type of deformation, the Scott tester in which a two-ply specimen is subjected to vertical vibration under certain weight, and the dynamic flex adhesion tester,[55] applying compression fatigue by flexing over a spindle to a two-ply fabric specimen, are the most commonly used methods for dynamic adhesion evaluation.

8.4 Cord mechanics

Two or more plies of yarn are twisted together to form a cord. Yarn consists of a large number of continuous, oriented filaments. The intrinsic yarn properties depend on the morphology or the microstructure, which are functions of yarn material and the processing conditions as discussed earlier. Cord as the reinforcing element of cord–rubber composite must meet certain strength, modulus, dimensional stability, thermal stability and chemical stability criteria. Cords may be characterized by a linear relationship between stress and strain tensors, expressed as

$$\sigma_i = Q_{ij}\varepsilon_j \quad \text{or} \quad \varepsilon_i = S_{ij}\sigma_j \quad i, j = 1, \ldots, 6 \qquad [8.1]$$

where σ, ε, $[Q]$ and $[S]$ represent stress component, strain component, stiffness matrix and compliance matrix, respectively. The stress–strain relation for cords, which are considered as transversely isotropic, i.e. in one plane mechanical properties are equal in all directions, can be expressed as

$$\begin{Bmatrix} \varepsilon_1 \\ \varepsilon_2 \\ \varepsilon_3 \\ \gamma_{23} \\ \gamma_{31} \\ \gamma_{12} \end{Bmatrix} = \begin{bmatrix} S_{11} & S_{12} & S_{13} & 0 & 0 & 0 \\ S_{12} & S_{11} & S_{13} & 0 & 0 & 0 \\ S_{13} & S_{13} & S_{33} & 0 & 0 & 0 \\ 0 & 0 & 0 & S_{44} & 0 & 0 \\ 0 & 0 & 0 & 0 & S_{44} & 0 \\ 0 & 0 & 0 & 0 & 0 & 2(S_{11} - S_{12}) \end{bmatrix} \begin{Bmatrix} \sigma_1 \\ \sigma_2 \\ \sigma_3 \\ \tau_{23} \\ \tau_{31} \\ \tau_{12} \end{Bmatrix} \qquad [8.2]$$

which has only five independent constants, namely, extensional Young's modulus, extensional Poisson's ratio, transverse Young's modulus, transverse Poisson's ratio and shear modulus. In reality, the twisted cord exhibits a nonlinear stress–strain relationship. Most applications of cord–rubber composites require only extensional Young's modulus, extensional Poisson's ratio and shear modulus. The influence of transverse Young's modulus and transverse Poisson's ratio on cord–rubber characteristics has been found to be negligible.

Organic cords are often viscoelastic in nature, i.e. some energy is lost during cyclic loading. This dissipated energy results in heat generation and therefore higher operating temperature, affecting material properties. In addition, cord properties (strength, elongation, fatigue) are significantly affected by cord twist.[56,57]

8.5 Rubber elasticity

The physical properties of rubber compounds are affected by their constituents and the processing parameters.[58–60] Various hypotheses have been developed to formulate rubber elasticity, with different degrees of success.[61–65] However, the most important property of rubber is its ability to undergo large deformation and return to its original shape. Rivlin[66,67] presented a general stress–strain relation for rubber, assuming the material to be isotropic in the unstrained state and incompressible in bulk. It follows that the stored energy W is a symmetric function of extension ratios λ_1, λ_2 and λ_3. In terms of strain invariants, the extension ratios are expressed as

$$I_1 = \lambda_1^2 + \lambda_2^2 + \lambda_3^2$$

$$I_2 = \lambda_1^2 \lambda_2^2 + \lambda_2^2 \lambda_3^2 + \lambda_3^2 \lambda_1^2 \qquad [8.3]$$

$$I_3 = \lambda_1^2 \lambda_2^2 \lambda_3^2$$

Considering the incompressibility of rubber, $I_3 = 1$ and λ_3 can be eliminated. Therefore, strain energy density is a function of two independent variables only, and in more general form,[62] neglecting higher-order terms, can be expressed as

$$W = C_1(I_1 - 3) + C_2(I_2 - 3) \qquad [8.4]$$

where C_1 and C_2 are constants. This is the famous Mooney–Rivlin equation proposed by Mooney.[68] In simple tension or compression Eqn 8.4 leads to the stress–strain relation

$$\sigma = 2C_1\left(\lambda^2 - \frac{1}{\lambda}\right) + 2C_2\left(\lambda - \frac{1}{\lambda^2}\right) \qquad [8.5]$$

and, for simple shear,

$$\tau = 2(C_1 + C_2)\gamma \qquad [8.6]$$

where τ is the shear stress, γ is the shear strain and $2(C_1 + C_2)$ is the modulus of rigidity.

For the purposes of cord–rubber composite analysis, rubber is considered as a homogeneous, isotropic material with two independent elastic constants, Young's modulus E_r and Poisson's ratio v_r.

8.6 Undirectionally cord-reinforced rubber (UDCRR)

Consider a calendered undirectionally reinforced single-ply cord–rubber system as shown in Fig. 8.4. For such a thin lamina, plane stress conditions exist and, therefore, the elasticity law can be expressed as

$$\left\{ \begin{array}{c} \varepsilon_L \\ \varepsilon_T \\ \gamma_{LT} \end{array} \right\} = \left[\begin{array}{ccc} S_{11} & S_{12} & 0 \\ S_{12} & S_{22} & 0 \\ 0 & 0 & S_{66} \end{array} \right] \left\{ \begin{array}{c} \sigma_L \\ \sigma_T \\ \tau_{LT} \end{array} \right\} \qquad [8.7]$$

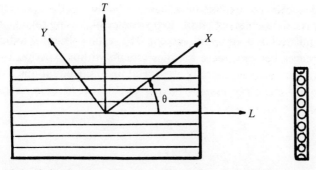

Figure 8.4 Calendered unidirectionally reinforced single-ply cord–rubber lamina.

or

$$\left\{ \begin{array}{c} \sigma_L \\ \sigma_T \\ \tau_{LT} \end{array} \right\} = \left[\begin{array}{ccc} Q_{11} & Q_{12} & 0 \\ Q_{12} & Q_{22} & 0 \\ 0 & 0 & Q_{66} \end{array} \right] \left\{ \begin{array}{c} \varepsilon_L \\ \varepsilon_T \\ \gamma_{LT} \end{array} \right\} \qquad [8.7]$$

where the elements of compliance matrix [S] are given by

$$S_{11} = \frac{1}{E_L} \qquad S_{12} = -\frac{v_L}{E_L} = -\frac{v_T}{E_T}$$

$$[8.8]$$

$$S_{22} = \frac{1}{E_T} \qquad S_{66} = \frac{1}{G_{LT}}$$

and the elements of stiffness matrix $[Q] = [S]^{-1}$ are expressed as

$$Q_{11} = \frac{E_L}{1 - v_L v_T} \qquad Q_{22} = \frac{E_T}{1 - v_L v_T}$$

$$Q_{12} = \frac{v_L E_T}{1 - v_L v_T} = \frac{v_T E_L}{1 - v_L v_T} \qquad [8.9]$$

$$Q_{66} = G_{LT}$$

where E_L, E_T are Young's modulus along directions L and T; v_L and v_T are primary and secondary Poisson's ratio; and G_{LT} is the shear modulus of the composite.

In order to characterize the stress–strain response of composite materials, it is essential to define the compliance or stiffness matrix, which in turn depends on the engineering properties. Several authors,[69–76] using the principles of micromechanics and considering the orthotropic elasticity law, have expressed the engineering constants of composites in terms of their constitutive material properties. Walter and Patel[69] compared predicted values of engineering constants based on these theories with experimental data obtained on single-ply composites using rayon, polyester, aramid and steel as cord reinforcement. The results obtained indicate that the differences between these theories remain within experimental error. However, the most commonly used Halpin–Tsai equations[71,72] for a cord–rubber single-ply lamina having circular cord cross-sections are expressed as

$$E_L = E_c V_c + E_r V_r$$

$$E_T = E_r \frac{1 + 2V_c}{1 - V_c}$$

$$G_{LT} = \frac{G_r[G_c + G_r + (G_c - G_r)V_c]}{G_c + G_r - (G_c - G_r)V_c} \qquad [8.10]$$

$$v_L = v_c V_c + v_r(1 - V_c)$$

$$v_T = v_L\left(\frac{E_T}{E_L}\right)$$

where V_c is the volume fraction of cord given by the relation $\pi R^2/t$ (epi) (R = radius of cord, t = thickness of ply, epi = cord ends per inch of ply); V_r is the volume fraction of rubber; and v_c and v_r are the Poisson's ratio of cord and rubber, respectively.

It should be noted that the Halpin–Tsai equations based on laminar theory are realistically valid only for small deformations. However, Odon Po'sfalvi[77–79] has attempted, with some degree of success, to predict effective elastic properties of orthotropic cord–rubber composites from the principle of virtual work, considering the composite to be compliant in the non-cord direction and therefore that it must undergo large deformations.

Industrial applications use more complex cord–rubber composite. In general, the direction of the applied load is not coincident with the geometrically natural or principal axis of the system. An excellent example of such a 'generally orthotropic' system is helically wound cord-reinforced hose. The response of such a system is complicated by shear strains developed due to off-axis loading. Referring to Fig. 8.4, if X, Y represent the natural or load axes and L, T the principal lateral and transverse cord material directions, then the elasticity law for UDCRR sheet with respect to the X, Y coordinate system can be written as

$$\begin{Bmatrix} \sigma_x \\ \sigma_y \\ \tau_{xy} \end{Bmatrix} = \begin{bmatrix} \bar{Q}_{11} & \bar{Q}_{12} & \bar{Q}_{16} \\ \bar{Q}_{12} & \bar{Q}_{22} & \bar{Q}_{26} \\ \bar{Q}_{16} & \bar{Q}_{26} & \bar{Q}_{66} \end{bmatrix} \begin{Bmatrix} \varepsilon_x \\ \varepsilon_y \\ \gamma_{xy} \end{Bmatrix} \qquad [8.11]$$

and

$$
\left\{ \begin{array}{c} \varepsilon_x \\ \varepsilon_y \\ \gamma_{xy} \end{array} \right\} =
\left[\begin{array}{ccc} \bar{S}_{11} & \bar{S}_{12} & \bar{S}_{16} \\ \bar{S}_{12} & \bar{S}_{22} & \bar{S}_{26} \\ \bar{S}_{16} & \bar{S}_{26} & \bar{S}_{66} \end{array} \right]
\left\{ \begin{array}{c} \sigma_x \\ \sigma_y \\ \tau_{xy} \end{array} \right\}
$$

where terms of the transformed reduced stiffness matrix $[\bar{Q}]$ and transformed reduced compliance matrix $[\bar{S}]$ are expressed as

$$\bar{Q}_{11} = Q_{11} \cos^4 \theta + 2(Q_{12} + 2Q_{66}) \sin^2 \theta \cos^2 \theta + Q_{22} \sin^4 \theta$$

$$\bar{Q}_{12} = (Q_{11} + Q_{22} - 4Q_{66}) \sin^2 \theta \cos^2 \theta + Q_{12}(\sin^4 \theta + \cos^4 \theta)$$

$$\bar{Q}_{22} = \bar{Q}_{11} \sin^4 \theta + 2(Q_{12} + 2Q_{66}) \sin^2 \theta \cos^2 \theta + Q_{22} \cos^4 \theta \qquad [8.12]$$

$$\bar{Q}_{16} = (Q_{11} - Q_{12} - 2Q_{66}) \sin \theta \cos^3 \theta + (Q_{12} - Q_{22} + 2Q_{66}) \sin^3 \theta \cos \theta$$

$$\bar{Q}_{26} = (Q_{11} - Q_{12} - 2Q_{66}) \sin^3 \theta \cos \theta + (Q_{12} - Q_{22} + 2Q_{66}) \sin \theta \cos^3 \theta$$

$$\bar{Q}_{66} = (Q_{11} + Q_{22} - 2Q_{12} - 2Q_{66}) \sin^2 \theta \cos^2 \theta + Q_{66}(\sin^4 \theta + \cos^4 \theta)$$

and

$$\bar{S}_{11} = S_{11} \cos^4 \theta + (2S_{12} + S_{66}) \sin^2 \theta \cos^2 \theta + S_{22} \sin^4 \theta$$

$$\bar{S}_{12} = S_{12}(\sin^4 \theta + \cos^4 \theta) + (S_{11} + S_{22} - S_{66}) \sin^2 \theta \cos^2 \theta$$

$$\bar{S}_{22} = S_{11} \sin^4 \theta + (2S_{12} + S_{66}) \sin^2 \theta \cos^2 \theta + S_{22} \cos^4 \theta \qquad [8.13]$$

$$\bar{S}_{16} = (2S_{11} - 2S_{12} - S_{66}) \sin \theta \cos^3 \theta - (2S_{22} - 2S_{12} - S_{66}) \sin^3 \theta \cos \theta$$

$$\bar{S}_{26} = (2S_{11} - 2S_{12} - S_{66}) \sin^3 \theta \cos \theta - (2S_{22} - 2S_{12} - S_{66}) \sin \theta \cos^3 \theta$$

$$\bar{S}_{66} = 2(2S_{11} + 2S_{22} - 4S_{12} - S_{66}) \sin^2 \theta \cos^2 \theta + S_{66}(\sin^4 \theta + \cos^4 \theta)$$

Equations 8.11 through 8.13 indicate that generally orthotropic cord–rubber ply exhibits a complex response in terms of coupling between (i) shear strain and normal stress and (ii) normal strain and shear stress. Thus, in a natural coordinate system, generally orthotropic ply behaves as anisotropic laminae with stress–strain relations under plane stress conditions expressed as

$$
\left\{ \begin{array}{c} \varepsilon_x \\ \varepsilon_y \\ \gamma_{xy} \end{array} \right\} =
\left[\begin{array}{ccc} S_{11} & S_{12} & S_{16} \\ S_{12} & S_{22} & S_{26} \\ S_{16} & S_{26} & S_{66} \end{array} \right]
\left\{ \begin{array}{c} \sigma_x \\ \sigma_y \\ \tau_{xy} \end{array} \right\} \qquad [8.14]
$$

where

$$S_{11} = \frac{1}{E_x} \qquad S_{12} = -\frac{v_L}{E_x} = -\frac{v_T}{E_y}$$

$$S_{22} = \frac{1}{E_y} \qquad S_{66} = \frac{1}{G_{xy}} \qquad\qquad\qquad [8.15]$$

$$S_{16} = \frac{\eta_{LT,L}}{E_x} = \frac{\eta_{L,LT}}{G_{xy}} = \lambda_x \qquad S_{26} = \frac{\eta_{LT,T}}{E_y} = \frac{\eta_{T,LT}}{G_{xy}} = \lambda_y$$

The engineering constants in the natural (x, y) coordinate system for orthotropic cord–rubber laminae are expressed as

$$\frac{1}{E_x} = \frac{1}{E_L} \cos^4 \theta + \left(\frac{1}{2G_{LT}} - \frac{v_{LT}}{E_L} \right) \sin^2 2\theta + \frac{1}{E_T} \sin^4 \theta$$

$$\frac{1}{E_y} = \frac{1}{E_L} \sin^4 \theta + \left(\frac{1}{2G_{LT}} - \frac{v_{LT}}{E_L} \right) \sin^2 2\theta + \frac{1}{E_T} \cos^4 \theta$$

$$\frac{v_{xy}}{E_x} = \frac{v_{yx}}{E_y} = \frac{v_{LT}}{E_L} - \frac{1}{4} \left(\frac{1 + v_{LT}}{E_L} + \frac{1 + v_{TL}}{E_T} - \frac{1}{G_{LT}} \right) \sin^2 2\theta$$

$$\frac{1}{G_{xy}} = \frac{\cos^2 2\theta}{G_{12}} + \left(\frac{1 + v_{LT}}{E_L} + \frac{1 + v_{TL}}{E_T} \right) \sin^2 2\theta \qquad [8.16]$$

$$\lambda_x = \sin 2\theta \left[\frac{\cos^2 2\theta}{E_L} - \frac{\sin^2 2\theta}{E_T} + \left(\frac{v_{LT}}{E_L} - \frac{1}{2G_{LT}} \right) \cos 2\theta \right]$$

$$\lambda_y = \sin 2\theta \left[\frac{\sin^2 2\theta}{E_L} - \frac{\cos^2 2\theta}{E_T} + \left(\frac{1}{2G_{LT}} - \frac{v_{LT}}{E_L} \right) \cos 2\theta \right]$$

The coefficients λ_x and λ_y are known as coupling coefficients and couple normal strains to shear stresses and vice versa. Figures 8.5 and 8.6 show the variation of elastic constants with cord angle θ for nylon cord–rubber laminae.[82]

8.7 Multi-ply system: classical lamination theory

Laminated composites used in industrial applications are composed of numbers of orthotropic laminae. The material properties of each ply vary with material constituents along with cord end count, cord angle and rubber thickness. The relations connecting stress and strain in such a system are complicated by the fact that when subjected to tensile load, it experiences twisting and bending as well as stretching. The stretching and bending strains are related to the derivatives of in-plane and out-of-plane displacements, respectively. The elasticity law governing the behaviour of

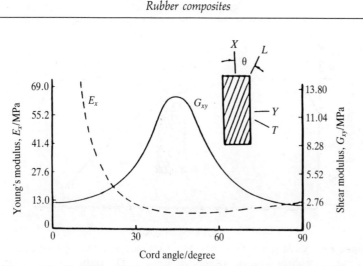

Figure 8.5 Variation of Young's modulus and shear modulus with cord angle for one-ply nylon–rubber system. (Redrawn from J. D. Walter *et al.*[82] with permission of the Tire Society, Inc.)

such laminates is expressed as

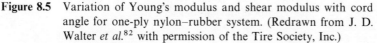

$$
\begin{Bmatrix} N_x \\ N_y \\ N_{xy} \\ M_x \\ M_y \\ M_{xy} \end{Bmatrix} = \begin{bmatrix} A_{11} & A_{12} & A_{16} & B_{11} & B_{12} & B_{16} \\ A_{12} & A_{22} & A_{26} & B_{12} & B_{22} & B_{26} \\ A_{16} & A_{26} & A_{66} & B_{16} & B_{26} & B_{66} \\ B_{11} & B_{12} & B_{16} & D_{11} & D_{12} & D_{16} \\ B_{12} & B_{22} & B_{26} & D_{12} & D_{22} & D_{26} \\ B_{16} & B_{26} & B_{66} & D_{16} & D_{26} & D_{66} \end{bmatrix} \begin{Bmatrix} \varepsilon_x^0 \\ \varepsilon_y^0 \\ \gamma_{xy}^0 \\ K_x^0 \\ K_y^0 \\ K_{xy}^0 \end{Bmatrix} \qquad [8.17]
$$

where (N_x, N_y, N_{xy}) and (M_x, M_y, M_{xy}) are stress and moment resultants and $(\varepsilon_x^0, \varepsilon_y^0, \gamma_{xy}^0)$ and (K_x^0, K_y^0, K_{xy}^0) are laminate mid-plane strain and curvatures, respectively. The elements of the stiffness matrix are provided by

$$
[A_{ij}, B_{ij}, D_{ij}] = \sum_{m=1}^{N} \int_{hm-1}^{hm} [1, -Z, Z^2] Q_{ij}^m \, dZ \qquad [8.18]
$$

where Q_{ij}^m is the stiffness matrix element of the mth layer shown in Fig. 8.7.

Akasaka[81] calculated stiffness constants obtained at the crown area of radial passenger tyres composed of three plies of cord–rubber laminae in the belt and carcass, two inter-ply rubber laminae and the tread rubber. The terms $A_{16} = A_{26} = 0$ are due to the presence of two breaker plies at $+\theta$ and $-\theta$, and terms $D_{16} = D_{26} = 0$ are due to the presence of one breaker at $+\theta$ above the mid-plane and a second breaker at $-\theta$ below the mid-plane. The engineering properties of a balanced, symmetric composite

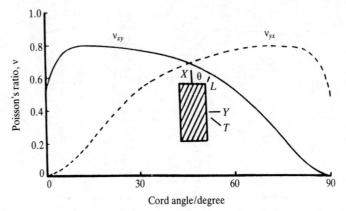

Figure 8.6 Variation of Poisson's ratio with cord angle for one-ply nylon–rubber system. (Redrawn from J. D. Walter *et al.*[82] with permission of the Tire Society, Inc.)

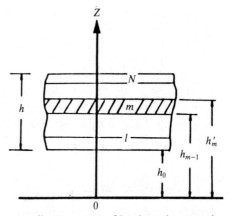

Figure 8.7 The coordinate system of laminated composite.

can then be expressed in terms of constants A_{ij} and thickness as[82]

$$E_\phi = E_x = (A_{11}A_{22} - A_{12}^2)/hA_{22}$$

$$E_\xi = E_y = (A_{11}A_{22} - A_{12}^2)/hA_{11}$$

$$v_{\phi\xi} = v_{xy} = A_{12}/A_{22} \qquad\qquad [8.19]$$

$$v_{\xi\phi} = v_{yx} = A_{12}/A_{11}$$

$$G_{\xi\phi} = G_{xy} = A_{66}/h$$

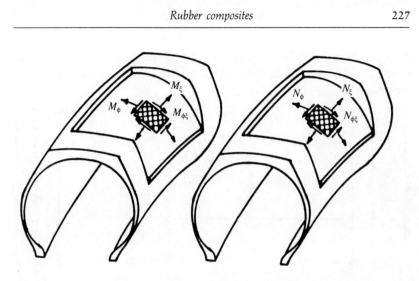

Figure 8.8 Moment and stress resultants acting on tyre structure (ϕ, ζ system).

where E_ϕ, E_ξ, $\nu_{\phi\xi}$, $\nu_{\xi\phi}$ and $G_{\phi\xi}$ are meridional Young's modulus, circumferential Young's modulus, major Poisson's ratio, minor Poisson's ratio and shear modulus, respectively, in the (ϕ, ζ) coordinate system used in tyres as shown in Fig. 8.8. It appears possible to express the general trend of behaviour of cord–rubber composites used in various industrial applications. However, owing to various simplifying assumptions made in the derivation of the equations, one must not rely completely on the calendered properties of elastic constants without some experimental verification whenever possible

8.8 Interlaminar deformation

A characteristic feature of cord–rubber composite laminates is the significant amount of inter-ply shear deformation. The plane stress assumption in the analysis of cord–rubber laminates ignores the effect of interlaminar shear stresses and deformations. The primary consequence of interlaminar shear stress is delamination-induced failure within the boundary region. Works of Walter,[83] Turner and Ford,[84] Lou and Walter,[85] Stalmaker, Kennedy and Ford,[86] Hirano and Akasaka,[87] and DeEskinazi and Cembrole,[88] utilizing experimental, analytical and finite-element tools to model interlaminar deformation phenomena, with specific applications to belt edges in tyres, are of great significance to compliant cord–rubber composites. From a theoretical point of view, Whitney[89] developed a mathematical procedure to calculate accurately the mechanical behaviour of thick laminated composites by extending the laminate plate theory to include effects of transverse shear deformations. The governing elastic law for laminate in terms of force, moment and shear resultants is expressed as

$$
\begin{Bmatrix} N_x \\ N_y \\ N_{xy} \\ M_x \\ M_y \\ M_{xy} \\ Q_y \\ Q_x \end{Bmatrix} = \begin{bmatrix} A_{11} & A_{12} & A_{16} & B_{11} & B_{12} & B_{16} & 0 & 0 \\ A_{12} & A_{22} & A_{26} & B_{12} & B_{22} & B_{26} & 0 & 0 \\ A_{16} & A_{26} & A_{66} & B_{16} & B_{26} & B_{66} & 0 & 0 \\ B_{11} & B_{12} & B_{16} & D_{11} & D_{12} & D_{16} & 0 & 0 \\ B_{12} & B_{22} & B_{26} & D_{12} & D_{22} & D_{26} & 0 & 0 \\ B_{16} & B_{26} & B_{66} & D_{16} & D_{26} & D_{66} & 0 & 0 \\ 0 & 0 & 0 & 0 & 0 & 0 & K_1^2 A_{44} & K_1 K_2 A_{45} \\ 0 & 0 & 0 & 0 & 0 & 0 & K_1 K_2 A_{45} & K_2^2 A_{55} \end{bmatrix} \begin{Bmatrix} \varepsilon_x^0 \\ \varepsilon_y^0 \\ \gamma_{xy}^0 \\ K_x^0 \\ K_y^0 \\ K_{xy}^0 \\ \gamma_{yz}^0 \\ \gamma_{zx}^0 \end{Bmatrix}
$$

[8.20]

where

$$
A_{ij} = \int_{-h/2}^{h/2} C_{ij} \, dZ \qquad i, j = 4, 5
$$

[8.21]

and K_1, K_2 are shear correction factors and Q_y, Q_x are shear resultants.

Equation 8.20 is fairly complex and requires serious computational capability to predict terms of the stiffness matrix. However, for a simple case of laminate subjected to uniaxial tension, Lou and Walter[85] discussed the inter-ply shear phenomenon using the mechanics of materials approach for angle-ply laminate. For two-ply ($+\theta$ and $-\theta$) cord–rubber laminate of width $2b$ and thickness h between plies, as shown in Fig. 8.9, the inter-ply shear strain γ_{xz}, distributed over the width, is expressed as

$$
\frac{\gamma_{xz}(y)}{\varepsilon_x} = \frac{2nAE_c(1 + v)\sin\theta\cos\theta(\cos^2\theta - v\sin^2\theta)}{E_r h + 2nAE_c(1 - v^2)\sin^4\theta} \frac{\mu t \sinh \mu y}{\cosh \mu b}
$$

[8.22]

where

$$
\mu = \frac{1}{h}\left[\frac{1}{12} + \frac{nAE_c(1 + v)\sin^2\theta\cos^2\theta}{E_r h + 2nAE_c(1 - v^2)\sin^4\theta}\right]
$$

As expected, Eqn 8.22 indicates that the inter-ply shear strain is maximum at the free edge ($y = \pm b$) and zero along the belt centre ($y = 0$), and is of opposite sign at each belt edge. This is verified experimentally in reference 85 for a polyester cord–rubber two-ply laminate, as shown in Fig. 8.10. The inter-ply shear phenomenon is not of just localized significance but is the combined effect of various parameters used in the design.

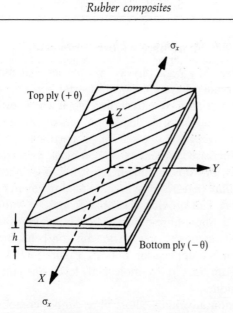

Figure 8.9 Two-ply cord–rubber laminate subjected to uniaxial tension.

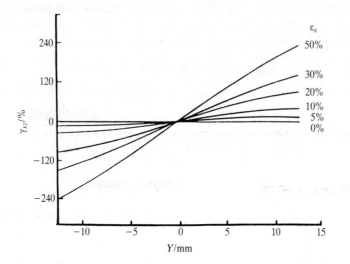

Figure 8.10 Interlaminar shear strain across width of a balanced two-ply polyester cord–rubber laminate at different extensions. (Redrawn from A. Y. C. Lou and J. D. Walter[85] with permission of *Rubber Chemistry and Technology*.)

8.9 Short fibre rubber composites

Rubber by itself has very poor physical properties that make it difficult to process it and render it suitable for industrial applications. Rubber is therefore commonly reinforced with carbon black which, owing to its very fine particle size and uniform dispersion, enhances tensile strength, modulus and tear strengths — essential properties for performance of rubber products. Increased loading of carbon black increases the rubber's physical properties, but this is limited by processing difficulties. On the other hand, short fibre (natural or synthetic) reinforcement of rubber, even at very low loading, can provide desirable properties without adversely affecting processing. Short-fibre-reinforced rubber composites have found use in product such as hoses, power transmission belts, brakes, clutches and gaskets and limited application in tyres. Figure 8.11 shows the range of short fibres from macro to molecular level and the diversity of preparation techniques.

The most common advantage short fibre rubber composites offer are the stiffness enhancement and improved fracture resistance and crack growth. The magnitude of property enhancement depends on the material type, the geometrical dimensions and the processing. Specifically, the stiffness of such composites is influenced by rubber modulus, fibre orientation, fibre concentration, fibre aspect ratio, dispersion, and inter-facial adhesion between fibre and rubber matrix. These parameters affect the reinforcement potential and anisotropic behaviour of the composites. Boustany and Coran[90] developed a relationship for composite modulus (E_{comp}) in terms of rubber modulus (E_r), fibre aspect ratio (l/d) and fibre volume fraction (V_f) for a cellulose fibre type, expressed as

$$E_{comp} = E_r\{1 + KfV_f[26 + 0.85(l/d)]\} \qquad [8.23]$$

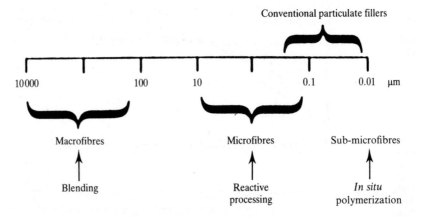

Figure 8.11 Range of sizes of short fibres from macroscopic to molecular level and diversity of preparation technique.

where K is constant and f is a function of fibre orientation. The effect of fibre orientation is expressed by a simple relation of the form[91]

$$\frac{1}{E_\theta} = \frac{\cos^2 \theta}{E_L} + \frac{\sin^2 \theta}{E_T} \qquad [8.24]$$

where E_θ is the modulus of the composite for fibre oriented at θ degrees with respect to the direction of testing; E_L is the longitudinal modulus and E_T is the transverse modulus of the composite. The experimental verifications of these equations are given in references 91, 92 and are applicable for any fibre type.

Foldi[93,94] and Watson and Frances[95] have given a rigorous explanation of processing and the use of milling procedures to obtain fibre alignment and orientation in elastomers, along with property characterization and applications. Foldi has shown that Kevlar fibres offer significant improvement in elastomer characteristics for high wear, high temperature applications even at loadings of less than 10 per cent. O'Connor[96] has studied, in detail, the properties of short-fibre-reinforced elastomers using cellulose, nylon, glass, carbon and aramid as fibres. Typical properties of these fibres are given in Table 8.9. The load transfer in fibre-reinforced composites takes place through interfacial shear stress between fibre and elastomer matrix. For maximum transfer of strength, it is essential to have strong interfacial adhesion between fibre and rubber surface. The treatment of fibres with adhesive agents significantly improves the composite yield strength and modulus in the longitudinal fibre direction, whereas the properties in the transverse direction, which are primarily dependent on the rubber matrix, are not affected as much by increased fibre–matrix adhesion.[96] Figures 8.12, 8.13 and 8.14 show the effect of fibre content, i.e. volume fraction, on Young's modulus, tensile strength and tear strength in longitudinal directions. Properties of rubbers reinforced with carbon and glass fibres are not as good, probably owing to damage of these fibres during processing because of their brittle nature.

Table 8.9. Typical properties of reinforcing short fibres

Fibre	Density/mg m^{-3}	Fibre diameter/μm	Tensile strength/GPa	Tensile modulus/GPa
Glass	2.52	13	2.41	72.4
Carbon	1.80	8	2.76	220.7
Cellulose	1.50	12	0.52	20.7
Aramid	1.44	12	2.76	62.1
Nylon	1.14	25	0.69	5.5

From J. E. O'Connor[96] with permission of *Rubber Chemistry and Technology*.

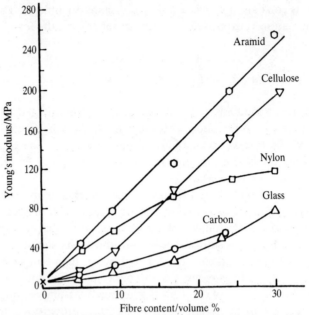

Figure 8.12   Effect of fibre content on Young's modulus of short-fibre-
reinforced natural rubber. (Redrawn from J. E. O'Connor,
Rubber Chem. Technol., **50**, 945 (1977) with permission of
Rubber Chemistry and Technology.)

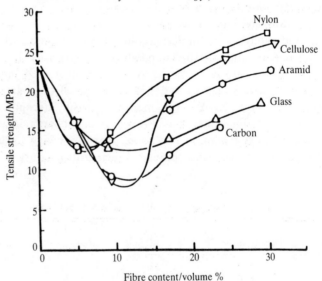

Figure 8.13   Effect of fibre content on tensile strength of short-fibre-
reinforced natural rubber. (Redrawn from J. E. O'Connor,
Rubber Chem. Technol., **50**, 945 (1977) with permission of
Rubber Chemistry and Technology.)

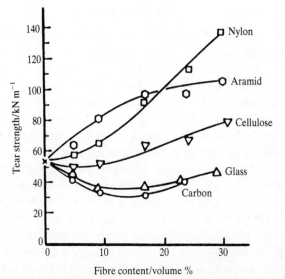

Figure 8.14   Effect of fibre content on tear strength of short-fibre-reinforced natural rubber. (Redrawn from J. E. O'Connor, *Rubber Chem. Technol.*, **50**, 945 (1977) with permission of *Rubber Chemistry and Technology*.)

The dynamic behaviour of short-fibre-reinforced composite is critical to the performance and industrial application of such materials. The stress–strain behaviour is influenced by shear deformation between interfacial regions and surface temperatures of composites undergoing complex dynamic fatigue. Ashida *et al.*[97] obtained the dynamic moduli *E'* and *E"* for nylon and PET-reinforced short fibre composites as a function of temperature. The characteristics of dynamic modulus curves depend on the fibre type. Mashimo *et al.*[98] studied the effect of dynamic fatigue for PET-fibre-reinforced composite systems with 0, 5, 10 and 15 per cent fibre volume fraction at 0° orientation. They observed a significant decrease in stress for all composites up to 5000 cycles and then the maintenance of an almost constant value. The rate of temperature rise and temperature build-up is much higher for composites with higher fibre loading. This can be critical when designing composites for applications that are subjected to continuous dynamic strain cycles.

The short fibre reinforcement of rubber can result in composites having a variety of properties. However, to enhance particular product performance, serious consideration must be given to the design of appropriate fibre–rubber composites, considering the effects of processing, orientation and volume fraction to meet and satisfy static, dynamic and thermal load requirements.

8.10 Summary

The basic concepts of cord-reinforced rubber (CRR) composites are described, with emphasis on application and usefulness in product structural analysis. The fundamental properties of cord–rubber composites play a significant role in product design, performance and finite-element analysis. Akasaka et al.[99,100] have used these basic concepts to explain the phenomenon of buckling of tyre belts under in-plane bending moments transmitted during vehicle cornering; and the load–deflection characteristics and contact pressure distribution between road and tyre. Walter et al.[82,83] applied these principles to obtain the mechanism of obstacle enveloping characteristics, tread wear, vibration characteristics, ply stress behaviour and stress analysis of tyres. However, these concepts are not developed to the extent that one can accurately predict performance characteristics of either tyres or other rubber products such as belts, hoses, air springs, diaphragms, etc., subjected to large deformation under static and dynamic loads. There is potential for future research work to expand the existing theories to include large deformation behaviour of cord–rubber composites, the effect of twisted cord on the composite properties, and prediction of product performance, both analytically and experimentally.

Acknowledgements

The author thanks all the scientists for their work referenced in this chapter; Mrs Pat Hagey for typing the manuscript and assistance in preparation of illustrations; and The Goodyear Tire & Rubber Company for permission to publish.

References

1. P V Papero, R C Wincklhofer, *Rubber Chem. Tech.*, **38**, 999 (1965).
2. W H Howard, M L Williams, *Rubber Chem. Tech.*, **40**, 1139 (1967).
3. E R Barron, presented at the 129th meeting of Rubber Division, Amer. Chem. Soc., New York, April 1986.
4. W H Carothers, J W Hill, *J. Am. Chem. Soc.*, **54**, 1559 (1932).
5. W H Carothers, J W Hill, *J. Am. Chem. Soc.*, **54**, 1266 (1932).
6. W H Carothers, US Patent 2130523 (1938).
7. W H Carothers, US Patent 2130947 (1938).
8. W H Carothers, US Patent 2130948 (1938).
9. W H Carothers, US Patent 2163636 (1939).
10. Dupont, British Patent 811349 (1959).
11. Dupont, British Patent 889144 (1962).
12. Monsanto, British Patent 955903 (1964).
13. Chemstrand, Belgian Patent 615991 (1962).
14. Chemstrand, French Patent 1323811 (1968).

15. Toyo Rayon, Japanese Patent JP 42-25502 (1967).
16. Toyo Rayon, Japanese Patent JP 42-27572 (1967).
17. J R Whinfield, *Textile Res. J.*, **23**, 289 (1953).
18. F J Kovac, T M Kersker, *Textile Res. J.*, **34**, 999 (1965).
19. F J Kovac, C R McMillaen, *Rubber World*, **152**(5), 83 (1965).
20. R G Aitken, R L Griffith, J S Little, J W McLellan, *Rubber World*, **151**(5), 58 (1965).
21. Y Iyengar, *J. Appl. Polymer Sci.*, **15**, 267 (1971).
22. Y Iyengar, *Rubber Chem. Tech.*, **46**, 422 (1973).
23. S L Kwolek, US Patent 3600350 (1971).
24. H Blades, US Patent 3767756 (1973).
25. J W S Hearle, P Grosberg, S Backer, *Structural Mechanics of Fibers, Yarns and Fabrics*, Wiley (Interscience), New York, 1969.
26. E Mukoyama, *J. Textile Machinery Soc. Japan*, Proc. 20, 706 (1967).
27. R G Patterson, H H McCrea, D E Howe, *Rubber World*, **138**, 409 (1958).
28. D E Howe, *Dupont Tire Yarn Tech. Rev.*, 7 (1960).
29. C A Litzler, *Rubber World*, Southern Rubber Group, Dallas, Texas, February 22 (1963).
30. C A Litzler, *Rubber Age*, **105**, 27 (1973).
31. J R Willis, The Tire Technology Conference, Clemson University, Oct. 30, 1984.
32. T Takeyama, J Matsui, *Rubber Chem. Tech.*, **42**, 159 (1969).
33. T S Solomon, *Rubber Chem. Tech.*, **58**, 561 (1985).
34. Y Iyengar, *Rubber World*, **148**(6), 39 (1963).
35. Y Iyengar, *J. Appl. Polymer Sci.*, **15**, 261 (1973).
36. K D Albrecht, *Rubber Chem. Tech.*, **46**, 981 (1973).
37. D E Erickson, *Rubber Chem. Tech.*, **47**, 213 (1974).
38. Y Iyengar, *Rubber World*, **197**(2), 24 (1987).
39. J S Little, *Can. Textile J.*, **78**(19), 57 (1961).
40. Canadian Industries Ltd, US Patent 3051594 (1962).
41. Burlington, US Patent 3240659 (1966).
42. Burlington, US Patent 3240649 (1966).
43. C J Shoaf, US Patent 3307966 (1967).
44. Goodyear, US Patent 3268467 (1968).
45. Canadian Industries Ltd, US Patent 3318750 (1967).
46. H Hisaki, Y Nakamo, S Suzuki, *Tire Sci. Tech.*, **19**(3), 163 (1991).
47. Fiber Industries Inc., British Patent 1323804 (1973).
48. K Maeta *et al.*, Japanese Patent JP 46/17300 (1971).
49. K Kigane *et al.*, Japanese Patent JP 46/27782 (1971).
50. R Bhakuni, US Patent 3718587 (1973).
51. Y Iyengar, *J. Appl. Polymer Sci.*, **22**, 801 (1978).
52. Y Iyengar, *J. Appl. Polymer Sci.*, **19**, 855 (1975).
53. D W Nicholson, D I Livingston, G S Fielding-Russell, *Tire Sci. Tech.*, **6**, 114 (1978).
54. K Kawai, Akron Rubber Group Symposium, Akron, Ohio, Jan., 1970.
55. ASTM Standards, Part 28, Rubber; Carbon Black; Gaskets, Method D430–59, Dynamic Testing for Ply Separation and Cracking of Rubber Products, pp. 233, American Society of Testing and Materials, Philadelphia, 1968.
56. S Backer in *Mechanical Behavior of Materials*, ed. F A McClintock, A S Argon, Addison-Wesley, Reading, MA, 1963, p. 675.

57. J O Wood, G B Redmond, *J. Textile Inst.*, **56**, T191 (1965).
58. J D Ferry, *Viscoelastic Properties of Polymers*, 2nd edn, Wiley, New York, 1970.
59. G Kraus, *Adv. Polymer Sci.*, **8**, 155 (1971).
60. A B Davey, A R Payne, *Rubber in Engineering Practice*, Maclaren & Sons, London, 1964.
61. P J Flory, *Statistical Mechanics of Chain Molecules*, Wiley (Interscience), New York, 1969.
62. L R G Trelor, *Physics of Rubber Elasticity*, 2nd edn, Oxford University Press (Clarenden), London and New York, 1958.
63. J P Cotton, B Farmonx, G Tannink, *J. Chem. Phys.*, **57**, 290 (1972).
64. R G Kirsti, W A Kruse, J Schelten, *J. Makromol. Chem.*, **162**, 299 (1972).
65. A N Gent, *Science and Technology of Rubber*, ed. F R Eirich, Academic Press, Orlando, 1978, Ch. 1.
66. R S Rivlin, *Phil. Trans. Roy. Soc. (London)*, **A241**, 379 (1948).
67. R S Rivlin, *Rheology Theory and Applications*, ed. F R Eirich, Academic Press, New York, 1956, Ch. 10.
68. M Mooney, *J. Appl. Phys.*, **11**, 582 (1940).
69. J D Walter, H P Patel, *Rubber Chem. Tech.*, **52**, 710 (1979).
70. S K Clark, *Rubber Chem. Tech.*, **56**, 372 (1982).
71. J C Halpin, S W Tsai, *AFML-TR*, **67** (June 1969).
72. J E Ashton, J C Halpin, P H Petit, *Primer on Composite Materials: Analysis*, Technomic, Stamford, CT, 1969.
73. V E Gough, *Rubber Chem. Tech.*, **41**, 988 (1968).
74. G Tangorra, Proceedings of International Rubber Conference, Moscow, Khimiya, Moscow, 1971.
75. T Akasaka, M Hirano, *Fukugo Zairyo (Composite Materials)*, **1**, 70 (1972).
76. J D Walter in *Mechanics of Pneumatic Tires*, ed. S K Clark, US Department of Transportation, National Highway Traffic Safety Administration, DOT HS 805 952, August 1981, Ch. 3.
77. O Po'sfalvi, *Tire Sci. Tech.*, **4**, 219 (1976).
78. O Po'sfalvi, P Szor, *Periodica Polytechnica*, **1**, 189 (1973).
79. O Po'sfalvi, P Szor, *Muszaki Tudomany*, **48**, 401 (1974).
80. H P Patel, J L Turner, J D Walter, *Rubber Chem. Tech.*, **49**, 1095 (1976).
81. T Akasaka, Plenary Lecture, Eighth Annual Tire Society Meeting, Akron, Ohio, March 1989.
82. J D Walter, G N Avgeropoulus, J L Janssen, G R Potts, *Tire Sci. Tech.*, **1**(2), 210 (1973).
83. J D Walter, *Rubber Chem. Tech.*, **51**, 524 (1978).
84. J L Turner, J L Ford, *Rubber Chem. Tech.*, **55**(4), 1078 (1982).
85. A Y C Lou, J D Walter, *Rubber Chem. Tech.*, **52**, 792 (1979).
86. D O Stalmaker, R M Kennedy, J L Ford, *Experimental Mechanics*, **20**, 87 (1980).
87. M Hirano, T Akasaka, *Fukugo Zairyo (Composite Materials)*, **2**(3), 6 (1973).
88. J DeEskinazi, R J Cembrole, *Rubber Chem. Tech.*, **57**, 168 (1983).
89. J M Whitney, *J. Composite Mater.*, **6**, 426 (1972).
90. K Boustany, A Y Coran, US Patent 3697364 (1972).
91. A Y Coran, K Boustany, P Hamed, *J. Appl. Polymer Sci.*, **15**, 2471 (1971).
92. A Y Coran, K Boustany, P Hamed, *Rubber Chem. Tech.*, **2**, 396 (1974).

93. A P Foldi, *Rubber Chem. Tech.*, **2**, 379 (June 1976).
94. A P Foldi, *Rubber World*, **196**(2), 19 (1987.
95. K R Watson, A Frances, *Rubber World*, **198**(5), 20 (1988).
96. J E O'Connor, *Rubber Chem. Tech.*, **50**, 945 (1977).
97. M Ashida, T Naguchi, S Mashimo, *J. Appl. Polymer Sci.*, **29**, 661 (1984).
98. S Mashimo, S Nakajimo, T Noguchi, Y Yamaguchi, M Ashida, *Rubber World*, **200**(1), 28 (1989).
99. T Akasaka, S Yamazaki, K Asano, *Tire Sci. Tech.*, **11**, 3 (1984).
100. T Akasaka, M Katoh, S Nihei, M Hiraiwa, *Tire Sci. Tech.*, **18**(2), 80 (1990).

CHAPTER 9

High performance fibres 1: aramid fibres

V GABARA

9.1 Introduction

By the 1960s polymers had achieved significant penetration of various markets through substitutions for natural rubber, fibres and wood in nonstructural applications. At that point the interest of scientists and engineers turned toward the substitution for metals to capitalize on resistance to corrosion, ease of fabrication as well as lower energy consumption in their fabrication. For these applications high strength, high stiffness and thermal stability become essential.

The realization that a nearly perfectly oriented polymer and a fully extended polymer chain are necessary to achieve high strength and stiffness dates back to the early 1930s.[1,2] Calculations of modulus based on bond force constants,[3-7] or lattice extension,[8-12] and of strength based on the strength of the weakest bond in a chain, led to tensile property estimates substantially higher than those observed experimentally. The breakthrough occurred when Kwolek, Blades and colleagues discovered nematic solutions of poly(p-benzamide) and poly(p-phenylene terephthalamide)[13-16] that were processable into fibres with a highly oriented, extended chain configuration.

The discovery came only 15 years after Flory[17] had defined conditions leading to the formation of anisotropic solutions in which rigid chains pack with a great degree of local orientation. The importance of this discovery went beyond the introduction of the first commercial p-aramid fibre (Kevlar) in 1971. This discovery demonstrated that the theoretical estimates are correct and that products with properties approaching the theoretical values can be synthesized. The most elegant confirmation came later when Galiotis and co-workers described polydiacetylene single crystal[18,19] and its properties.

While unusual tensile properties were the driving force behind the development of these materials, several other characteristics were equally responsible for their success. In this chapter we will attempt to summarize how some of these properties are utilized in applications. Aromatic polyamides will be used as a model because they are the largest volume material on the market and their performance is understood best. But first we will take a brief look at the range of materials that have been developed.

9.2 Polymers and processes for high performance fibres

Kwolek and Blades' discovery led to the development of a whole series of other rigid polymers. It also provided an impetus for the development of other processing methods yielding fibres with fully extended chains. These include drawing of semirigid and flexible polymers. Figure 9.1 shows the chemical structures of materials discussed below.

9.2.1 Rigid polymers

The initial work focused on poly(p-benzamide) and poly(p-phenylene terephthalamide). The rigidity and linearity of these molecules leads to organized packing in solutions, with formation of anisotropic solutions. Good local orientation of chains within anisotropic domains is combined

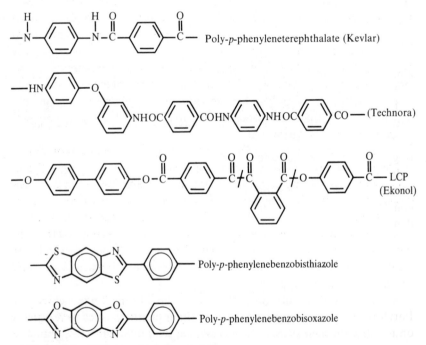

Figure 9.1   Polymers for advanced fibres.

with orientation of these domains by an elongational flow during air-gap spinning to produce very well oriented, extended chain organization.

The concepts developed in this work were extended to other materials. In addition to series of aromatic polyamides, aromatic heterocyclic polymers are the most interesting examples among lyotropic polymers. Most work in this area focused on poly(*p*-phenylene benzobisthiazole; PBZT)[20–22] and poly(*p*-phenylene benzobisoxazole; PBO).[23] Similar concepts underlie developments in the area of liquid crystalline polyesters (LCP). These thermotropic materials are difficult to process as homopolymers at the desired high molecular mass. A very broad range of materials was developed to deal with this issue of tractability.[24–28] In general, these materials are spun at low molecular mass and as a result exhibit only moderate strength at that point. To obtain high strength, they have to be treated for prolonged times at high temperatures (close to their melting points) to continue polymerization in the solid phase.[29,30]

9.2.2 Semirigid and flexible polymers

Semirigid polymers represent an attempt to balance the properties and the tractability of rigid polymers. Work of Ozawa and his co-workers from Teijin[31] led to an aramid copolymer fibre called 'Technora'. The fibre is spun from an isotropic solution and achieves its orientation during drawing at very high temperature. The chemical composition (Fig. 9.1), aimed at tractability, allows for high tenacity (>230 cN tex^{-1}) but limits stiffness of the material to ~ 4850 cN tex^{-1}.

The work of Pennings[32,33] and later of Smith and Lemstra[34] led to the development of fibres based on flexible polymers. Most of the work focuses on ultra-high molecular mass polyethylene (see Chapter 10) and the use of gel spinning to produce a series of commercial products.

In summary, over the past 20 years there has been an avalanche of materials developed with high strength and stiffness in mind. Figure 9.2 compares some of them, normalizing their tensile properties for their density.

9.3 Structure and properties

Early introduction of Kevlar obviously focused a major portion of work on the structure–property relations on that material. We will use it as an example, with only occasional references to other materials.

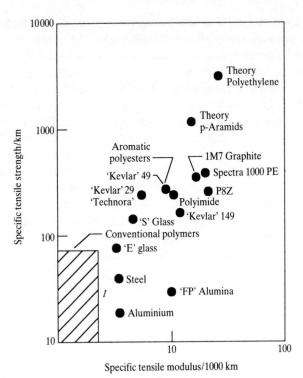

Figure 9.2   Specific properties of advanced materials. (From Tanner *et al.*[35])

9.3.1 Crystallinity

Poly(*p*-phenylene terephthalamide) is a highly crystalline material. The estimates vary from 68 to 85 per cent for Kevlar-29 and 76 to 95 per cent for Kevlar-49.[36,37] Two structures have been identified: the first is due to Northolt[38] and the second to Haraguchi *et al.*[39] Several mechanisms describing the formation of these structures have been proposed.[39] The most recent results of Roche *et al.*[40] point to a complex interaction of solution and coagulation processes. To obtain the Haraguchi polymorph, Roche suggests that the polymer solution must be allowed to crystallize into a crystallosolvate (described previously[41–43]), after which sulphuric acid is exchanged with water to form a hydrate, as proposed by Xu *et al.*[44] Drying of the hydrate leads to the Haraguchi polymorph, which, under heat, rearranges into the Northolt one. Any coagulant other than water, or direct coagulation without the formation of crystallosolvate, leads to the Northolt polymorph. Roche also suggests that domination of the Northolt polymorph in fibres is due to the fact that, in spite of relatively low temperatures of coagulation, no crystallosolvate is formed owing to super-cooling. On the other hand, very low temperature and slow coagulation will lead to the Haraguchi polymorph as the crystallosolvate is allowed to crystallize before coagulation is accomplished. Heat treatment of the fibres further improves orientation and crystalline perfection.[14,37,45–47] The orientation angle (2θ) obtained from azimuthal X-ray diffraction is about 12–20° for Kevlar-29 and <12° for Kevlar-49. Experimental limits of the procedure do not permit differentiation of Kevlar-149 even though the very high modulus points to further improvement in orientation.

9.3.2 Morphology

Studies (by electron diffraction and dark field transmission electron microscopy) of supramolecular organization of poly(p-phenylene terephthalamide) fibres point to an unusual radial orientation of hydrogen-bonded sheets[48] and pleated structure. Figure 9.3 shows schematically radially arranged sheets of hydrogen-bonded molecules that exhibit a regular, cooperative misorientation with the fibre axis of $\pm 5°$. The pleated structure has been observed both for Kevlar-29 and Kevlar-49, and is absent only in the product of highest modulus, Kevlar-149.[49,50] Hagege et al.[51] has shown a radial distribution of voids in an experimental high modulus aramid and some indication of the same for Kevlar-29, while Dobb et al.[52] estimated their aspect ratio at ~5. Kevlar also exhibits skin–core differentiation. Panar et al.'s work[36] through fibre etching, and Morgan et al.'s[53] from observation of separation of skin, represent early studies of the phenomenon. More recent work[54,55] shows a measurable differentiation of orientation between the skin and the core. A model proposed by Panar et al.[36] points to better orientation of the skin than of the core.

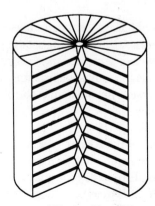

Figure 9.3 Pleated structure of Kevlar fibre. (From Dobb et al.[48])

9.3.3 Properties

The Kevlar structure finds its reflection in fibre properties. We will attempt to elucidate this connection.

Thermal properties

Bond dissociation energies of chemical bonds in aromatic compounds are significantly higher than those in aliphatic ones. While this is true of all chemical bonds, it is especially important for C–C and C–N bonds, which frequently represent the main chain of polymers. In this case the dissociation energies are ~ 20 per cent higher for aromatic molecules. This fact is responsible for a substantially higher thermostability of aromatic materials in general and *p*-phenylene terephthalamide specifically, which is stable up to about 550°C. Pyrolysis of the poly(*p*-phenylene terephthalamide) fibre leads to products (*p*-phenylene diamine, aniline, benzonitrile, etc.) that indicate that the mechanism involves homolytic cleavage of the amide bond.[56] The outstanding example of superior thermostability is given by poly(*p*-phenylenebenzobisthiazole). Its stability is due to a fully aromatic character and highly rigid molecule. Introduction of flexible groups such as –O– into the main chain, as in Technora, leads to a more flexible chain and lower thermal stability.

The C–N bond of the amide group in poly(*p*-phenylene terephthalamide) has a double bond character. The conjugation between the amide group and the aromatic ring in *p*-aramids is responsible not only for the yellow colour of the polymer but also for the further increase in chain rigidity, which translates into excellent retention of physical properties at elevated temperatures. At 300°C Kevlar still retains about 50 per cent of its strength at room temperature, while the modulus remains at 70 per cent of this level.[55]

Excellent orientation of chain-extended molecules in high performance fibres results in an anisotropy of the thermal expansion coefficient. The linear expansion coefficient for Kevlar is negative at $(-2 \text{ to } -4) \times 10^{-6}$ cm/cm∗deg C while the radial coefficient is 60×10^{-6} cm/cm∗deg C. Its high crystallinity results in negligible shrinkage both at high temperature in air (<0.1 per cent at 177°C) and in hot water (<0.1 per cent at 100°C).

Aromatic polyamides of the Kevlar type, when exposed to flame, form a large amount of char, with relatively small amounts of combustible fragments being released into the gas phase. This results in a limiting oxygen index (LOI) of 29 per cent and a relatively small release of toxic gases during decomposition.

Chemical properties

When one compares materials with similar chemical compositions but different structures, the crystalline ones usually offer better resistance to attack of organic solvents and aqueous solutions. Table 9.1 shows that Kevlar-49 does have a very good chemical resistance. However, strong acids and bases do attack the fibre at elevated temperatures, with hydrolysis of the amide linkage as the most likely mechanism.

The hydrophilicity of amide groups leads to absorption of moisture by all aramids. In addition to chemical composition, fibre structure plays a critical role in determining moisture absorption. Thus, different Kevlar products absorb moisture to a different extent: from ~ 1 per cent for Kevlar-149 to ~ 7 per cent for Kevlar-29 (55 per cent RH at 20°C). Kawai and co-workers have recently published extensive work on the relationship between structure and moisture absorption and transport in aramids.[57]

The aromatic nature of p-aramids is responsible for a substantial absorption of ultraviolet light which, in turn, leads to a change of colour

Table 9.1. Chemical resistance of Kevlar-49 aramid fibre

	Retained tensile strength[a]/%	Retained tensile modulus/%
Acetic acid (99.7% aqueous)	100	99
Formic acid	88	98
Hydrochloric acid (37% aqueous)	100	97
Nitric acid (70% aqueous)	40	95
Sulphuric acid (96% aqueous)	0	Too Weak to Test
Ammonium hydroxide (23.5% aqueous)	100	98
Potassium hydroxide (50% aqueous)	74	97
Sodium hydroxide (50% aqueous)	90	95
Acetone	100	100
Benzene	100	98
Carbon tetrachloride	100	100
Dimethylformamide	100	98
Methylene chloride	100	100
Methyl ethyl ketone (MEK)	100	98
Trichloroethylene ('Triclene')	98	99
Chlorothene (1,1,1-trichloroethane)	100	100
Toluene	100	100
Benzyl alcohol	100	99
Ethyl alcohol	100	98
Methyl alcohol	99	98
Formalin	99	97
Gasoline (Regular)	100	100
Jet fuel (Texaco 'Abjet' K-40)	96	99
Lubricating oil ('Skydrol')	100	99
Salt water (50% solution)	100	92
Tap water	100	100

From reference 59.

[a] Yarns were tested at 10-inch gauge length with 3 turns per inch twist.

Table 9.2. Typical tensile properties of aramid fibres (as yarn bundles)

Fibre	Density/g cm^{-3}	Tenacity/N tex^{-1}	Elongation/%	Modulus/N tex^{-1}
Kevlar-29	1.44	2.0	3.6	49.0
Kevlar-49	1.44	2.0	2.4	78.1
Kevlar-129	1.44	2.3	3.3	68.9
Kevlar-149	1.47	1.6	1.5	97.1
Kevlar-119	1.44	2.1	4.4	38.0
Technora	1.39	2.3	4.3	50.3

due to oxidative reactions, as well as a decrease of fibre properties when uncovered yarn is subjected to substantial exposures.[58,59] On the other hand, this high absorption leads to a self-screening phenomenon, with thin structures more sensitive to degradation than thicker ones.

Mechanical properties

Table 9.2 shows typical tensile properties of high strength aramid fibres. The values quoted are those of yarn bundles which are lower than values determined on individual filaments measured at a short gauge length. A substantial body of knowledge exists that deals with modelling of the impact of strength variability of fibres on the strength of their assemblies. We will omit it here, while focusing on a consideration of the structure–property relationship.

The idealized model of fibre strength described in the introduction relies on fully extended chains and perfectly oriented molecules. This has been analysed by Termonia *et al.*[60–62] Their kinetic model is in reasonable agreement with experimental data for high strength polyethylene as well as poly(*p*-phenylene terephthalamide). The model points to the importance of molecular mass, molecular mass distribution, as well as intermolecular interaction as factors in controlling the strength of these fibres. The analysis also pointed out that chain end segregation will lead to a substantial decrease in the strength of the material, with only marginal changes in its stiffness.

Transition from this idealized, perfectly oriented model to one closely related to fibre morphology can be found in the work of Allen *et al.*[63–65] On the basis of cyclic loading studies of poly(*p*-phenylene terephthalamide) they have been able to separate the impact of local orientation of crystallites from that of supramolecular organization on the tensile properties of the fibre. The initial modulus of poly(*p*-phenylene terephthalamide) fibre was found to have no correlation with fibre strength (Fig. 9.4a). This is due to a complex interaction between the local and global orientation of the material. As discussed above, pleated morphology of

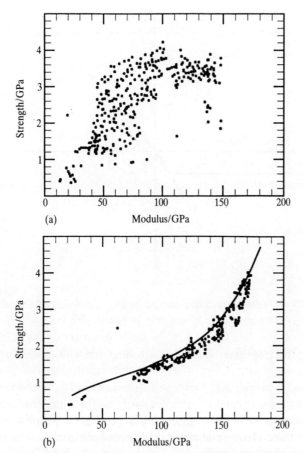

Figure 9.4 Strength versus modulus of poly(*p*-phenylene terephthalamide):
(a) initial modulus; (b) asymptotic modulus. (Redrawn from
Allen *et al.*[45] by courtesy of Butterworth-Heinemann Ltd ©)

the fibre is responsible for a global misorientation. The opening of pleats
during fibre deformation explains well the nonlinearity of the elastic
deformation. Only after the pleats are pulled out does the underlying local
orientation demonstrate itself in fibre properties. Allen introduced the
concept of an asymptotic modulus as a measure of local orientation. This
is the modulus of the fibre stretched almost to breaking point, when pleats
are pulled out (with an additional correction for time-dependent effects),
and thus reflects the local orientation. A modulus determined in this way
correlates extremely well with fibre strength (Fig. 9.4b). The steep
dependence of strength on the asymptotic modulus (or local orientation)
follows shear failure criteria (actually the solid line in Fig. 9.4b represents calculated values with shear modulus of 2 GPa). Thus, one can
conclude that inherent orientation and secondary interaction between

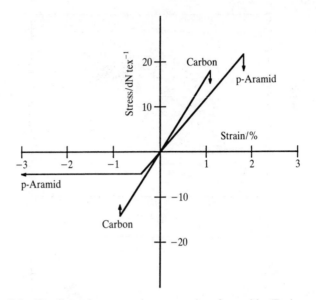

Figure 9.5   Tensile and compressive properties of aramids. (Redrawn from Tanner *et al.*[66] by courtesy of Marcel Dekker Inc.)

chains are the dominant factors controlling the tensile strength of the fibre.

Compressive properties of high performance fibres have been of interest to researchers since the very first discovery of these materials. Excellent tensile properties were not matched by the compressive ones. Figure 9.5 shows that the poly(*p*-phenylene terephthalamide) fibre exhibits a yield point at a compressive strain of ~0.5 per cent. The yield is accompanied by the formation of kink bands (a morphological deformation at 45–60° to the fibre axis). In spite of significant morphological changes, the fibre loses only ~10 per cent of its tensile strength, even after experiencing compressive strain of 3 per cent.[67] This behaviour is due to the buckling phenomenon, which in turn reflects weak lateral properties of these highly anisotropic materials. The presence of some lateral interaction due to hydrogen bonding in aramids results in significant improvement of compressive properties in comparison to polyethylene as shown in Table 9.3. Development of new tools for the direct measurement of lateral deformation of a single filament will help to elucidate these phenomena further.[69]

Fatigue behaviour of aramids is also dependent on the mode of fibre deformation (tension versus compression). Tension/tension fatigue is quite good, with no failure observed at loads as high as 60 per cent of the breaking strength, even after 10^7 cycles.[54] However, compressive fatigue of organic high performance fibres is significantly worse. The

Table 9.3. Recoil compressive strengths

Fibre	Strength/N tex^{-1}
Kevlar-49	25.8
Kevlar-29	25.8
Asahi-A5	24.9
AMP1	15.1
AMP2	13.3
PBT	17.8
PBO	13.3
Cellulose triacetate	13.3
Polyethylene	7.5
'Magnamite' AS4	80.8
'Thornel' P-55	23.0

From Allen[68] with permission of Chapman and Hall Ltd.

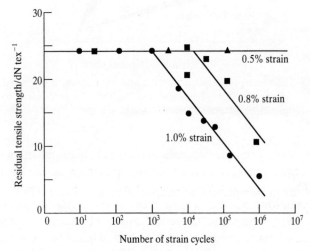

Figure 9.6   Compressive fatigue of Kevlar. (Redrawn from Tanner *et al.*[66] by courtesy of Marcel Dekker Inc.)

retention of tensile strength decreases significantly with increased compressive strain (Fig. 9.6). At 0.5 per cent strain no strength loss is observed, even after 10^6 cycles, while at 1 per cent strain loss is observed after 10^3 cycles.

Synthetic fibres break at loads lower than their tensile strength, provided the load is applied for a long time. Figure 9.7 shows results for the lifetime of poly(p-phenylene terephthalamide) and polyethylene, together with values calculated by Termonia[60] based on his kinetic model. Differences between these materials are quite substantial. Polyethylene fails in ~ 2.5 min when strained to half of its 1 s breaking strain. Under analogous conditions poly(p-phenylene terephthalamide) will support the

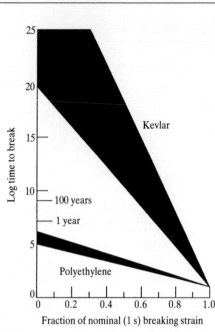

Figure 9.7   Lifetime properties of Kevlar and high strength polyethylene. (Redrawn from Termonia and Smith[60] by permission of the publishers, Butterworth-Heinemann Ltd ©)

load for 100 years. Termonia's calculations point to high activation energy of primary bond breaking as a major factor controlling the lifetime of materials. Thus, longer lifetime is a characteristic of high-melting polymers. Similar conclusions can be derived from study of the creep of these materials. Poly(p-phenylene terephthalamide) fibres exhibit very low creep, even at elevated temperatures. At a load of 0.5 of the breaking load after 10^7 s, creep strain for Kevlar-29 is only 0.3 per cent.[70] Creep strain is, in general, linear with log time and increases with load and temperature. Ericksen showed[71] that creep rate increases with temperature (activation energy of the order of 6.3–18.8 kJ mol^{-1}, depending on the load).

9.4 Applications

Since Kwolek and Blades' discoveries, strength and stiffness have been major driving forces behind research in the field of compositions, structures and processes yielding these unusual materials. On the other hand, efforts to translate these discoveries into practical application very quickly brought about a realization that a balance of all properties is essential to meet this task. We will attempt to illustrate that temperature resistance, flame resistance, creep and chemical resistance are essential for many applications whose development began with strength and stiffness as the driving forces. Moreover, a very fast growing range of applications, using short fibre or pulp type materials, underscores that there is more to advanced fibres than strength and stiffness. In many of the applications these new materials replace traditional ones. The anisotropy of properties of the advanced organic fibres is only one of the reasons why an attempt at simple substitution frequently leads to failure. A systems approach is essential to maximize the value of these new materials.

9.4.1 *Composites with soft matrices*

This was one of the earliest applications for *p*-aramids and it encompasses a very broad range of products. Here we will discuss exclusively uses of continuous fibre. Owing to their specificity, short fibre applications are reviewed together in section 9.6. Rubber reinforcement is the best example of this application. The first end-use for these materials was in tyres. High strength, high modulus, high fatigue resistance and good adhesion to rubber are especially important in this application.

In designing composites with soft matrices the goal is to provide such adhesion between the reinforcing fibre and matrix that the failure occurs in the matrix rather than at the interface. While poly(*p*-phenylene terephthalamide) fibres have polar groups that facilitate good adhesion, their very high crystallinity makes it impossible to achieve good adhesion with an RFL (resorcinol–formaldehyde–latex) dip only. This necessitates a two-layer approach. The treatment used by most producers at present is first to provide an epoxy resin layer that has good adhesion to aramids (it is likely that epoxy resin reacts with terminal amine groups). This is followed by the RFL dip, which ensures very good adhesion of rubber. The literature is full of examples of various forms of surface treatment, from introduction of reactive groups into the chain to plasma treatment, being developed to improve the situation further.

We have already pointed out that the compressive properties of advanced organic fibres represent one of their weaknesses, while compressive fatigue has been determined to be the major mechanism for strength loss in tyres. With tyre life and durability being one of the critical factors for consumers, it is clear that axial compressive strain in a tyre should be below ~ 0.8 per cent (Fig. 9.6). This is accomplished by twisting yarns and combining them into a cord with a twist in the opposite direction. This structure allows filaments to move so as to relieve the stress and thus increase fatigue resistance. While the increase of twist increases fatigue resistance, it decreases cord modulus and tenacity (the transverse properties of the fibre are significantly worse than the axial ones). The balancing of these two factors constitutes an important factor

Table 9.4. Fatigue-resistant *p*-aramid properties

	Kevlar-119[a]	Kevlar-29
Tenacity/N tex^{-1}	2.1	2.0
Elongation/%	4.4	3.6
Tensile modulus/N tex^{-1}	38.0	49.0
Loop tenacity/N tex^{-1}	1.1	0.99
Loop elongation/%	2.7	2.1
Relative strength retention/% (disc fatigue at 15% compression; twist multiplier 6.5)	154	100
Belt growth/% (after 2×10^8 cycles)	0.48	0.65
Relative cycles to failure/% (Mallory Tube Fatigue Test)	247	100

From reference 72.
[a] 1666 dtex yarn.

in tyre design. Moreover, other design modifications, such as folding tyre belts, place a large amount of fibre in especially critical regions. Renewed interest in reducing the weight of automobiles in order to decrease fuel consumption is likely to rekindle interest in advanced fibres as a reinforcement for tyres.

Automotive products (belts, hoses) and conveyor belts are another area in which rubber reinforcement is important. The development of Kevlar-119 illustrates the integration of the fibre preparation process, fibre properties and functionality in the final application (strength retention 50 per cent higher than Kevlar-29; Table 9.4). This improved fatigue resistance, when combined with better dimensional stability (belt growth), yielded an excellent product for automotive belts. The cyclic nature of deformation (pressure fluctuation) in hoses leads to fatigue failure as well. Table 9.4 shows that Kevlar-119 performs better than Kevlar-29 when this deformation is simulated in a Mallory Tube Test. In conveyor belt reinforcement, Kevlar is replacing steel, especially in belts rated above 500 pounds per inch width. In these applications the new material not only offers better fatigue resistance but also is not susceptible to corrosion, and has only one-fifth the weight at comparable strength. The significantly better energy absorption leads to better wear properties and longer belt life.[73]

9.4.2 Advanced composites

High specific strength and stiffness are the most important attributes of *p*-aramid fibres as a reinforcement for advanced composites. Applications that depend on weight reduction are those in which these fibres offer advantages over the less expensive glass-fibre-reinforced composites or metals.

Clearly, aerospace applications represent a primary area. Wound pressure vessels represent one of the applications where *p*-aramids have established a dominant position. Performance measurement in this application involves burst pressure times volume of the vessel normalized for its weight (PV/W). In this application the fibre is exposed to stresses both parallel and normal to fibre axes. As already pointed out, lateral properties are significantly lower for highly oriented organic fibres. To maximize performance in this application, a product with reduced adhesion to the resin was developed to minimize stresses normal to fibre axes. The result was a product that gives PV/W 25–40 per cent higher than glass or carbon fibres.[74]

While strength and stiffness are critical factors in advanced composites, impact damage tolerance is an important factor in conventional aircraft. An impact test based on a falling weight showed that high modulus aramid-reinforced composites can absorb 2–4 times as much energy as carbon-fibre-reinforced materials.[75] In applications where the very high stiffness of carbon fibre is required, good results can be achieved by hybrid composites reinforced by a mixture of carbon and *p*-aramid fibres. In general these materials offer properties which are intermediate between those of single fibres.

Even that which in most cases represents a severe negative for *p*-aramid fibres as a structural material, their low compressive properties, can be utilized in special situations. The fact that *p*-aramids yield at compressive strain slightly higher than 0.5 per cent leads to very high energy absorption during crushing. Moreover, the material maintains its integrity after failure; thus it does not generate fragments, as brittle materials do (e.g. carbon-fibre-reinforced parts).

The discussion above clearly shows that tensile anisotropy of organic high performance fibres necessitates special approaches in designing their applications. From our discussion of other properties of these materials it is clear that their values also depend on the direction in which they are measured. Anisotropy of the thermal expansion coefficient leads to a problem in minimum gauge thickness structures in aircraft. Kevlar-49 has a negative linear expansion coefficient of -2×10^{-6} cm/cm $*$ deg C but a positive one in the radial direction $(60 \times 10^{-6}$ cm/cm $*$ deg C). Special care needs to be taken in designing these structures, otherwise the thermal cyclic will lead to microcracks in the matrix which will result in water ingression and significant added weight to the structure.

9.4.3 Ropes and cables

High specific strength and specific modulus were the major characteristics that opened this application to aramid fibres. Other properties (corrosion resistance, low creep, nonconductivity) provided important added value for users.

In submerged applications the incumbent product was steel. In comparison to steel the specific strength ratio of 5–7 in air becomes greater than 20 in seawater. This advantage becomes critical where very long ropes are used. It allows either a significantly higher payload or reaching depths not possible with steel ropes that cannot support their own weight under similar circumstances. This combination of properties led to the development of nonrusting mooring lines from Kevlar-29, which enable oil rigs to probe depths down to 1200 metres. Low creep permits applications such as yacht lines, tow ropes and antenna guys. Electromechanical cables and fibre optics reinforced with Kevlar have been introduced successfully into telecommunication systems.

While capitalization on the above advangages of *p*-aramids requires the tailoring of their properties to specific applications, this is even more true when dealing with the deficiencies of these materials. Gibson[76,77] showed that internal loads in twisted ropes increase rapidly with increasing size. This leads to heating, internal abrasion and shear fatigue as elements move by each other. Relatively weak lateral properties of well-oriented organic fibres lead to modest abrasion and shear properties. To deal with this problem, rope structures that minimize cross-over points were developed. This was combined with jacketing of each strand and lubrication to reduce friction. External jacketing provided the solution, not only to abrasion but also to the UV stability of aramids.

9.4.4 Protective apparel

The thermal and flame resistance of Kevlar, combined with its dimensional stability and tensile properties, formed the basis for the application of this material in a broad range of end-uses involving the protection of people and equipment.

The very low shrinkage of Kevlar in flames prevents garments made from blends with Nomex from breaking open during exposure to flame and ensures insulating protection by the material. This allows Nomex III to be used in a broad range of applications from firemen's uniforms and turnout coats, NASA take-off/re-entry suits, race car drivers' uniforms to oil refinery uniforms.

Kevlar's mechanical properties translates into excellent cut and puncture resistance. For example, a knit/felt fabric made from Kevlar (1/1 ply with weight 7.5/3.5 oz yd^{-2}) withstands a slashing force of up to 9 kg, while a traditional glove from leather (60 oz yd^{-2}) is cut at a force of less than ~3.5 kg.[78] Similar results are obtained when one considers puncture resistance. The same materials show almost 5.5 kg of puncture force for

Kevlar and 2.3 kg for leather. Kevlar exceeds cotton in these tests by a factor of 2. The above properties, when combined with excellent thermal characteristics, are responsible for quite universal applicability of this material in protective gloves (from surgical to industrial uses in the sheet metal industry, the glass industry and meat processing plants).

9.4.5 Ballistics

A special application in protective equipment is the use of fibres in both soft and hard ballistics protection. The properties of advanced fibres such as *p*-aramids make them especially suited for this application. The need to consider a broad range of projectiles, and the materials from which the projectiles are made, as well as the structure of ballistic protection barriers transforms this problem into a separate field of study of its own. Full analysis of the performance of advanced fibres in a ballistic event is beyond the scope of this chapter. We will focus only on the fundamental principles and approaches that dominate the field.

Until advanced fibres entered the field, nylon fibres were the materials of choice and most of the basic studies focused on a comparison of these new materials with nylon. Very early it became obvious that, while absorption of energy is what is involved the ballistic event, simple fibre toughness is not adequate to differentiate materials. In this case nylon would absorb twice the amount of energy as *p*-aramids. Measurements of actual performance point to the opposite relationship. An analysis of the protective systems focuses on mechanisms that control the amount of material involved in the event.

To understand the phenomena involved in this deformation, we will explore issues at different levels of complexity of the system: from a single yarn to a two-dimensional fabric and, finally, to a three-dimensional multilayer structure. Figure 9.8 shows the model of deformation when a single yarn is impacted by a projectile. The impact leads to a longitudinal wave propagation and deflection (or transverse wave).[79] The longitudinal

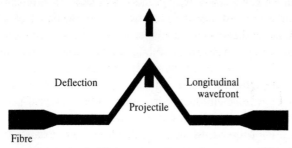

Figure 9.8 Model of a ballistic event. (From Tanner *et al.*[66] by courtesy of Marcel Dekker Inc.)

wave propagates at speed C (the speed of sound in the material) and the transverse wave at speed U, while the projectile speed is V.

$$C = (E/\rho)^{0.5} \qquad [9.1]$$

where E is the initial modulus and ρ is mass per unit length.

$$U = C\left(\frac{\varepsilon}{1 + \varepsilon}\right)^{0.5} \qquad [9.2]$$

where ε is fibre strain.

$$V = C\big(\varepsilon(1 + \varepsilon) - \{[\varepsilon(1 + \varepsilon)]^{0.5} - \varepsilon\}^2\big)^{0.5} \qquad [9.3]$$

Thus, if the strain represents strain at break, then V is the critical velocity of the projectile above which the material will never stop it because interaction between yarn and projectile is minimal. For Kevlar-29 at 4 per cent elongation and 48.5 N tex^{-1} modulus this critical velocity is ~900 m/s^{-1}. The longitudinal wave is the one that determines the amount of material involved in interaction with the projectile. The result of the difference between the specific moduli of Kevlar and nylon is that the amount of material involved in the interaction is 3–4 times higher for the higher modulus fibre.

Obviously the situation becomes more complex when we consider a two-dimensional woven fabric with cross-over points. What happens with wave propagation is of critical importance if one considers that a typical fabric has ~10^6 cross-over points in each square metre. At these points we can expect redistribution of the signal. A fraction will continue along the same yarn, part will be reflected back (if the node is fixed, essentially all of the signal will be reflected back) and part will be redistributed along a yarn which is perpendicular to the one in question. This complex situation was the subject of several analytical and modelling approaches.[80–82] Transverse deflection of the fabric is responsible for the fact that about 50 per cent of the total energy is absorbed by secondary yarns. The above considerations indicate that, in addition to the selection of material, one needs to optimize fabric construction. Thus, too tight a weave (almost fixed cross-over points) will lead to a reflection of the longitudinal wave and stress concentration at the point of impact. On the other hand, if the fabric weave is too loose, the projectile can penetrate by just pushing the yarns apart. While a significant amount of energy is absorbed owing to the out-of-plane deformation of layers, one obviously has to limit this deformation in order to avoid injury without penetration of the protective garment (blunt trauma). When material properties are modified by the presence of water, a significant decrease of ballistic performance is observed. In addition to the factors discussed above, the shape of the projectile plays a very important role.

All of the above factors are considered in designing the final protective garment. The garments are designed to protect from several levels of threat. A ballistic vest is truly a complex system. In addition to an appropriate number of layers and an appropriate hydrophobic treatment, a combination of ceramic panels with fabrics is used for the most severe threat. Today this protection is widely used by both civilian and military organizations around the world.

p-Aramids are also used in protection against fragments. In this case they are used in the form of composites. The design in this application is further complicated by influence of resin type, interaction between resin and fibre, and so on. Some of the applications include military helmets, spall liners and other fragment protection uses on vehicles and ships.

9.4.6 Short fibre applications

Short fibres and pulps represent one of the fastest growing applications of advanced fibres. The original driving force behind these applications was the replacement of asbestos, but by now a whole range of possibilities has been explored. Pulp products are short (0.05–8 mm), highly fibrillated fibres. One of the methods of their production involves a mechanical action on the cut fibre. The shear failure mechanism discussed earlier is responsible for the formation of shapes of high surface area. The end-use application determines which of the short fibre forms is to be used. This includes both the product itself and ease of processing of the material into the desired final product. For example, short fibre is used when increased strength of the product is desired, while pulp is easier to mix and mould. So far major areas of applications include friction products (brakes and clutches), gaskets and thixotropes, as well as additives that improve wear resistance of thermoplastic composites.

Introduction of Kevlar pulps into friction products in place of asbestos requires a reformulation of mixes. Because of its very low price, asbestos was generally used both as the 'active' ingredient and as filler. Kevlar pulp is used only in quantities dictated by processing and final product needs. As a result, while 50–60 per cent of asbestos was added to a mix, usually less than 10 per cent of Kevlar is used as shown in Table 9.5. In general, friction products containing *p*-aramid pulp show stable frictional characteristics, high heat stability, reduced drum and rotor wear and 3–5 times longer pad wear. Figure 9.9 shows a comparison of wear for asbestos-based pads with mixes containing 5 per cent Kevlar and dolomite stone as filler.

In gaskets Kevlar pulp is also a substitute for asbestos. As for friction products, less than 10 per cent Kevlar is used instead of 85 per cent asbestos in typical mixes. Here, too, a systems approach to the development of applications was required. Both new mixing conditions and new

Table 9.5. Examples of mixes with Kevlar pulps

Material	Composition/%
Friction products mix	
Dolomite filler (<200 mesh)	50
BaSO$_4$	15
Cashew friction particles (NC104-40[a])	15
NC126[a] resin	15
Kevlar pulp	5
Gasket material	
Elastomer binder	10–18
Curing agent, antioxidant, accelerator, etc.	1
Kevlar pulp	10
Structured silica	10
Talc	69–61

From references 83, 84.
[a] 3M trademark Cardolite.

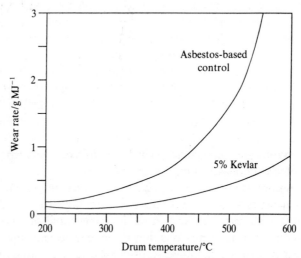

Figure 9.9 Wear of brake pads at high temperature. (From reference 84)

filler had to be identified. An example of such a composition is given in Table 9.5. Depending on the type of application, such compositions yield gaskets with tensile strength, creep resistance, compressibility and recovery equal to or better than those of asbestos. In many applications the lower chloride content compared with that of asbestos is an important advantage. The primary need in the sealants and adhesives industry is for pseudo-plastic material that offers a high viscosity at low shear rates and low viscosity at the high shear rates at which the material is applied. The

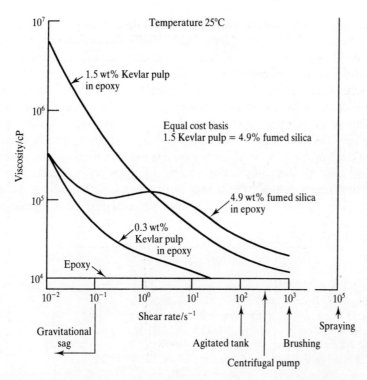

Figure 9.10 Viscosity versus shear rate of Kevlar pulp in epoxy resin. (From Frances and Dottore[85] by courtesy of Adhesive Age).

incumbent material is fumed silica, which when added to epoxy resin at 4.9 per cent increases its viscosity at low shear (10^{-2} s^{-1}) to 0.3×10^6 cP, while only 1.5 per cent of Kevlar pulp leads to a viscosity of 4×10^6 cP. At high shear rate (10^3 s^{-1}) the viscosity of both decreases, to 2.5×10^4 and 1.4×10^4 cP, respectively.[85] A complete match of performance between the two materials can be attained by decreasing pulp content to 0.3 per cent with substantial cost advantage (Fig. 9.10). The high toughness of these pulps ensures that performance does not change as length of exposure to high shear increases. Addition of ~2.5 per cent pulp into plastisol adhesives, silicone sealants and castable polyurethanes significantly improves tensile and tear strength as well as abrasion resistance.

The last example to be considered here is that of reinforcement of thermoplastic materials such as nylon and polyphenylene sulphide. Addition of Kevlar short fibres to these thermoplastic materials brings about several changes in properties. Addition of 20 per cent of 6-mm Kevlar-49 increases tensile strength of nylon-6,6 by ~70 per cent and more than doubles its modulus. A similar effect is observed in flexular properties. At this fibre content, heat-deflection temperature increases from 88°C to 255°C at 1.82 MN m^{-2}.[86] High toughness and increase in heat-deflection temperature are critical in achieving improved wear characteristics, especially at higher use temperatures. Wu[86] summarizes the results of wear tests (thrust washer test, ASTM D3702) at different conditions of pressure and velocity. Addition of short Kevlar fibre decreases the wear factor by a factor of 4–8 when compared with pure nylon-6,6, and by a factor of 2 when compared with nylon-6,6 containing 33 per cent of glass fibre. The results are even more dramatic with polyphenylene sulphide, which is a brittle material. The use of Kevlar fibre decreases the wear factor 17 times when compared with material containing glass fibre.

References

1. H Mark in *Physik und Chemie Der Cellulose*, Julius Springer, Berlin, 1932, p. 61.
2. H Mark, *Trans. Faraday Soc.*, **32**, 143 (1936).
3. K H Mayer, W Lotmar, *Helv. Chim. Acta*, **19**, 68 (1936).
4. L R G Treloar, *Polymer*, **1**, 95, 179, 290 (1960).
5. W J Lyons, *J. Appl. Phys.*, **29**, 1429 (1958).
6. L Holliday, J W White, *Pure Appl. Chem.*, **26**, 545 (1971).
7. G S Fielding-Russell, *Textile Res. J.*, **41**, 861 (1971).
8. W J Dulmage, L E Contois, *J. Polymer Sci.*, **28**, 275 (1958).
9. I Sakurada, Y Nakushina, Y T Ito, *J. Polymer Sci.*, **57**, 651 (1962).
10. I Sakurada, T Ito, K Nakamae, *Makromol. Chem.*, **75**, 1 (1964).
11. I Sakurada, T Ito, K Nakamae, *Bull. Inst. Chem. Res., Kyoto Univ.*, **42**, 77 (1964).
12. I Sakurada, K Kaji, *J. Polymer Sci.*, **C31**, 57 (1970).
13. S L Kwolek, US Patents 3600350 (1971), 3671542 (1972).
14. H Blades, US Patents 3869429 (1975), 3869430 (1975).
15. S L Kwolek, P W Morgan, J R Schaefgen, L W Gulrich, *Macromolecules*, **10**(6), 1390 (1977).
16. P W Morgan, *Macromolecules*, **10**, 138 (1977).
17. P J Flory, *Proc. Roy. Soc. (London), Ser. A*, **234**, 6, 73 (1956).
18. C Galiotis, R J Young, *Polymer*, **24**, 1023 (1983).
19. C Galiotis et al., *J. Polymer Sci.: Polymer Phys. Ed.*, **22**, 1589 (1984).
20. J F Wolfe, B H Loo, F E Arnold, *Polymer Prepr. ACS Polymer Chem. Div.*, **19**, 1 (1978).
21. J F Wolfe, B H Loo, F E Arnold, *Macromolecules*, **14**, 915 (1981).
22. S R Allen, *Macromolecules*, **14**, 1135 (1981).
23. J F Wolfe, P D Sybert, J R Sybert, US Patent 4533693 (1985).
24. G W Calundann, US Patent 4184996 (1980).
25. J R Schaefgen, US Patent 4118372 (1978).
26. C R Payet, US Patent 4159365 (1979).
27. R S Irwin, F M Lugullo, US Patent 4500699 (1985).
28. K Ueno, H Sugimoto, K Hayatsu, US Patent 4503005 (1985).

29. R R Luise, US Patent 4184996 (1980).
30. W J Jackson, *Br. Polymer J.*, **12**, 154 (1980).
31. S Ozawa, *Polymer J. (Tokyo)*, **19**, 119 (1987).
32. A J Pennings, K E Meihuizen, *J. Polymer Sci., Polymer Phys. Ed.*, **3**, 117 (1979).
33. A J Pennings, *Proc. Fiber Producer Conf.* 1-1, Greenville, SC, 1986.
34. P Smith, P J Lemstra, US Patents 4344908 (1982), 4422993 (1983), 4436689 (1984).
35. D Tanner, V Gabara, J R Schaefgen in *Polymers for Advanced Technologies*, ed. M Lewin, VCH, New York, p. 384.
36. M Panar *et al.*, *J. Polymer Sci., Polymer Phys. Ed.*, **21**, 1955 (1983).
37. A M Hindeleh, N A Halim, K Ziq, *J. Macromol. Sci. Phys.*, **B23**(3), 289, 383 (1984).
38. M G Northolt, *Eur. Polymer J.*, **10**, 799 (1974).
39. K Haraguchi, T Kajiyama, M J Takayanagi, *J. Appl. Polymer Sci.*, **23**, 915 (1979).
40. E J Roche, S R Allen, V Gabara, B Cox, *Polymer*, **30**, 1776 (1989).
41. M Arpin, C Strazielle, A Skoulios, *J. Phys.*, **38**, 307 (1977).
42. M M Iovleva, S P Papkov, *Vysokomol. Soeyed.*, **A24**, 233 (1982).
43. K H Gardner, R R Matheson, P Avakian, Y T Chia, T D Gierke, *Polymer Prepr., ACS Polymer Chem. Div.*, **24**(2), 469 (1983).
44. D Xu, K Okuyama, F Kumamura, M Takayanagi, *Polymer J.*, **16**, 31 (1984).
45. A M Hindeleh, Sh M Abdo, *Polymer*, **30**, 218 (1989).
46. A M Hindeleh, Sh M Abdo, *Polymer Commun.*, **30**, 184 (1989).
47. S J Krause, D L Vezie, W W Adams, *Polymer Commun.*, **30**, 10 (1989).
48. M G Dobb, D J Johnson, B P Saville, *J. Polymer Sci., Polymer Phys. Ed.*, **15**, 2201 (1977).
49. E J Roche *et al.*, *Mol. Cryst. Liq. Cryst.*, **153**, 547 (1987).
50. S J Krause, D L Vezie, W W Adams, *Polymer Commun.*, **30**, 10 (1989).
51. R Hagege, M Jarin, M Sotton, *J. Microscopy*, **115**, 65 (1979).
52. M G Dobb, D J Johnson, A Majeed, B P Saville, *Polymer*, **20**, 1284 (1979).
53. R J Morgan, C O Pruneda, W J Steele, *J. Polymer Sci., Polymer Phys. Ed.*, **21**, 1757 (1983).
54. R E Wilfong, J Zimmerman, *J. Appl. Polymer Sci. Polymer Symp.*, **31**, 1 (1977).
55. E I Du Pont de Nemours Inc., *Bulletin* **X-272** (1988).
56. J R Brown, A J Power, *Polymer Degrad. Stabil.*, **4**, 379 (1982).
57. M Fukuda, M Ochi, M Miyagawa, H Kawai, *Textile Res. J.*, **61**, 668 (1991).
58. E I Du Pont de Nemours Inc., *Bulletin* **K-4** (1979).
59. E I Du Pont de Nemours Inc., *Bulletin* **K-5** (1981).
60. Y Termonia, P Smith, *Polymer*, **27**, 1845 (1986).
61. Y Termonia, P Meakin, P Smith, *Macromolecules*, **18**, 2246 (1985).
62. P Smith, Y Termonia, *Polymer Commun.*, **30**, 66 (1988).
63. S R Allen, *Polymer*, **29**, 1091 (1988).
64. S R Allen, E J Roche, *Polymer*, **30**, 996 (1989).
65. S R Allen *et al.*, *Polymer*, **33**, 1849 (1992).
66. D Tanner *et al.* in *High Technology Fibers, Part B*, ed. M Lewin, J Preston, Marcel Dekker, New York, 1989, Ch. 2.
67. S J Deteresa, R J Farris, R S Porter, *Polymer Composites*, **3**, 57 (1982).
68. S R Allen, *J. Mater. Sci.*, **22**, 857 (1987).
69. S Kawabata, *J. Textile Inst.*, **81**, 432 (1990).

70. M H Lafitte, A R Bunsell, *Polymer Eng. Sci.*, **25**, 182 (1982).
71. R H Ericksen, *Polymer*, **26**, 733 (1985).
72. E I Du Pont de Nemours Inc., *Bulletin* **H24224** (1990).
73. E B Hoffman, *Bulk Solids Handling*, **10**, 283 (1990).
74. G E Zahr, An improved aramid fiber for aerospace applications, in *Progress in Advanced Processes: Durability, Reliability and Quality Control*, ed. G Bartelds and R J Schliekelmann, Elsevier Science Publishers, Amsterdam, 1985.
75. M W Wardle, E W Tokarsky, Drop weight testing of laminates reinforced with Kevlar aramide fibers, *E*-glass and graphite, *Composites Technology Review*, ASTM, Philadelphia, PA, Spring 1983.
76. P T Gibson, Analytical and experimental investigation of aircraft arresting gear purchase, NTIS Report AD 852074, Battelle Memorial Institute, Columbus, OH, July 3, 1967.
77. P T Gibson, Continuation of analytical and experimental investigation of aircraft arresting gear purchase, NTIS Report AD 904263, Battelle Memorial Institute, Columbus, OH, April 8, 1969.
78. E I Du Pont de Nemours Inc., *Bulletin* **H-05500** (1990).
79. J R Vinson, J A Zukas, *J. Appl. Mech.*, **42**, 263 (1975).
80. D Roylance, *Fibre Sci. Tech.*, **13**, 385 (1980).
81. C M Leech, J Mansell, *Int. J. Mech. Sci.*, **19**, 93 (1977).
82. C M Leech, B A Adeyefa, *Computers and Structure*, **15**, 423 (1982).
83. E I Du Pont de Nemours Inc., *Bulletin* **E-69541** (1991).
84. E I Du Pont de Nemours Inc., *Bulletin* **E-65333** (1984).
85. A Frances and J Dottore, *Adhesives Age*, 27 (April 1988).
86. Y T Wu, *Modern Plastics International*, 67 (April 1988).

CHAPTER 10

High performance fibres 2: high performance polyethylene fibres

D C PREVORSEK

10.1 Introduction

The high performance polyethylene (HPPE) fibres available to date are made by a solution spinning process that evolved from the Couette flow surface growth experiments of Pennings.[1] By studying the mechanisms of fibre formation in the Couette flow experiments, Pennings and his students Smith, Lemstra and Kalb[2,3] developed the principles for solution spinning of ultrahigh molecular mass polyethylene as well as a process disclosed in the patent that represents the foundation of DSM technology for producing Dyneema fibres. Some of the fibres used in the present study are made by an Allied-Signal solution spinning process developed by Kavesh and Prevorsek.[4] These fibres are commercially available under the trade name of Spectra 900 and Spectra 1000.

It should be noted that the solution spinning used to produce ultra-strong polyethylene fibres using high molecular mass polyethylene is frequently referred to as gel spinning. This term is misleading. The process, as it is practised to date, involves spinning a solution that, on cooling to ambient temperatures, is transformed into gels that have sufficient strength and integrity to be handled by conventional fibre processing equipment.

The commercially available solution-spun polyethylene fibres exhibit outstanding mechanical properties. Nevertheless, it is not only their specific modulus and strength, illustrated in Fig. 9.2, that contributed to their rapid acceptance in a variety of applications. Their commercial success must also be attributed to their unmatched damage tolerance and fatigue resistance. The ability of these fibres to fail in shear and/or compression without losing a great deal of tensile strength is uniquely characteristic of polyethylene and represents a distinct advantage over other reinforcing materials. This is particularly important when the products are used in the technology of survival, where the primary function of a part is to protect primary structures, people or equipment from the blasts of explosions, impact of fast moving projectiles, etc. The scope of this chapter is to identify the properties or combination of properties that make HPPE fibres particularly suitable for the applications that represent the foundation for their successful commercialization. Focusing on their unique damage tolerance, fatigue and abrasion resistance, and the capability for delayed recovery after an apparently plastic deformation, the molecular and morphological origins of these properties will be identified.

10.2 Morphology of HPPE fibres

Ultrastrong polyethylene fibres produced by solution spinning exhibit a morphology that is typical for all organic fibrous materials. They exhibit a well-defined aggregate structure on the macro (50 nm) as well as the microfibrillar (5 nm) level. Fundamental differences that may exist between HPPE and other high performance organic fibres, polyaramids, rigid rod and thermotropic liquid-crystal (LC) fibres, must therefore originate in the longitudinal characteristics of these fibres and especially in the microfibril and interfibrillar matter.[5]

It has been established that high strength, high modulus polyethylene fibres consist of microfibrils whose lateral dimension is well defined and amounts to about 4 nm. The studies of Schaper *et al.*[6] also indicate that the microfibrils have a finite length of about 1000 to 2000 nm. This yields a microfibril aspect ratio L/D of 250–500. In addition, small-angle X-ray diffraction studies by Grubb[7] indicate the presence of a long period of about 200 nm. This means that the structure of the microfibril is not uniform in density, but contains areas of different density that could be attributed to domains containing a high concentration of crystal defects. These findings are supported by the electron microscopy work of Kramer.[8] Based on the analysis of mechanical properties and degrees of crystallinity measured by wide-angle X-ray scattering (WAXS), it is proposed that these disordered domains, whose longitudinal dimensions appear to be 4–5 nm, are 'amorphous' domains that may contain a substantial amount of chain ends. It is important to note that this 'amorphous' domain is covalently bonded to the adjacent nearly perfect, needle-like crystalline domains whose L/D is 40.[5] The needle-like nearly perfect crystalline domains exhibit a crystal size in the (002) direction that is detected by WAXS but does not give rise to small-angle X-ray diffraction. Since the periodicity does not involve density fluctuations, the crystallite size in the (002) direction is attributed on the basis of the work of Reneker[9] to twist boundaries of the needle-like crystals.

10.3 Modulus

The crystalline modulus of polyethylene has been the subject of numerous experimental and theoretical studies. Nakamae *et al.*[10,11] recently published a comprehensive summary of reported values for the crystalline modulus in the direction parallel to the chain axis. Theoretical and experimental results vary a great deal. Most of the experimental data fall in the range between 230 and 340 GPa while the calculated values are generally higher and fall mostly in the 280–420 GPa range, with a few exceptions on the lower side (Treloar,[12] 182 GPa in 1960) and on the high side (Dewar et al.,[13] 494 GPa in 1979).

Considering that the distribution of stress is nonhomogeneous and that the disordered regions amount to about 25 per cent of the volume fraction of the fibre, the value of 235 GPa reported by Sakurada[11] and Nakamae[10] must be considered as the lower bound and the actual value must be signficantly higher. Considering that the low temperature ($-155°C$) measurement of Nakamae[9] yielded a crystalline modulus of 254 GPa, it can further be inferred that the actual modulus of the crystalline domains is higher than 254 GPa at $-155°C$. Therefore, 300 ± 20 GPa seems the most realistic range consistent both with experimental and theoretical analyses. This is also in agreement with the theoretical value of 290 GPa proposed by Boudreaux.[16]

Based on the structural model presented in section 10.2, in which roughly 25 per cent of the fibre volume is in the rubbery state, and the existence of nanometre defects that can transport the chain under the effect of stress and temperature, it is anticipated that the modulus of ultrastrong polyethylene fibres is strain rate dependent, and this is indeed the case.

The modulus of HPPE fibres was investigated by a variety of techniques and equipment, including transverse impact using high velocity projectiles.[5] By combining all the measurements, it was found that there was a fivefold increase in the modulus as the stress rate increased by about five orders of magnitude from 4×10^{-2} to 1×10^3. Based on these data, the room temperature modulus approaches 300 GPa at ballistic rates of deformation. It should be noted that the data of transverse impact analysed by the method of Smith[5] also yield a value of fibre modulus of about 300 GPa, which agrees well with the above data obtained using high speed testing equipment.[5] Transverse impact on fibres using direct measurements of strain wave propagation yields for the same fibre significantly lower values of modulus, between 160 and 200 GPa depending on the rate of deformation and pre-tension.[17] This shows the difficulties in selecting proper modulus data in the analysis of the behaviour of structures made of ultrastrong polyethylene fibres.

The effect of temperature on the modulus of ultrahigh modulus polyethylene fibres was investigated by Barham and Keller.[18] Their estimate of a modulus of 288 GPa at 77 K falls in the range of high strain rate modulus determined by Prevorsek et al.,[5] confirming that at low temperature and/or high strain rates the modulus of HPPE fibres falls fairly close

Table 10.1. Theoretical strength σ^*, modulus E, elongation-at-break ε_B, and activation energy ΔF^* of the ethylene repeat unit

$\sigma^*/$GPa	$E/$GPa	$\varepsilon_B/\%$	$10^{19}\,\Delta F^*/$J bond^{-1}	Reference
66.00	405	43	3.390	14
34.60	405	20	1.015	29
19.25	59	33	0.467	16
106.66	297	33	2.566	1
Not determined	Not determined	Not determined	2.260	30

to the theoretical modulus of the crystalline domains. Considering the high degree of crystallinity of about 80 per cent, high aspect ratio of crystallites and crystalline orientation function close to unity, it would be expected that the low temperature, high deformation rate modulus of HPPE fibres should be above 250 GPa.

10.4 Strength

A great deal has been written about the theoretical strength of polyethylene fibres.[19,20] The unresolved issues exist at several levels: the strength of an isolated molecule, the strength of an ensemble of perfectly oriented molecules, and the analysis of the achieved strength based on morphological analysis.

Considerable differences already exist with regard to the strength of a single molecule. Based on the work of Crist *et al.*,[14] Table 10.1 lists the theoretical strength, modulus, elongation at break, and activation energies for bond breakage of ethylene repeat units. Using these data, Prevorsek[21] calculated the theoretical strength of a perfect ensemble of extended polyethylene molecules as a function of the molecular mass. The calculations assume a random distribution of chain ends and a fracture mechanism consistent with Griffith's crack stability criteria. Several authors consider values of the theoretical strength of 20 GPa acceptable. Prevorsek, on the other hand, considers values of about 50 GPa realistic. It should also be noted that this theory neglects stress relaxation effects and hence is suitable only for conditions of low temperature and high deformation rate. With regard to actual achieved strengths, it should be noted that commercially available products have a strength of 3–4 GPa at ambient temperature and standard rates of deformation.

Several relationships are of scientific and technological interest. For example, a frequently reported linear relationship between modulus and strength has recently been confirmed by van der Werff.[22] This work also reports some of the highest room-temperature values of strength (7 GPa) and modulus (240 GPa) obtained at the low deformation rate of about

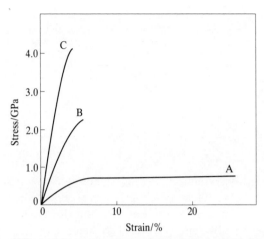

Figure 10.1   Effect of draw ratio λ on stress–strain relationship of HPPE
fibres. A, λ = 15; B, λ = 30; C, λ = 70. (From ref. 22 with
author's permission)

0.37 min^{-1}, but using relatively short samples (32.5 mm) and single
filaments. Differences in strength and modulus reported are usually
obtained by maintaining the spinning conditions constant but changing
the draw ratio. An example of the effect of draw ratio on strength is shown
in Fig. 10.1.[22] The important findings are (1) the low strength (0.8 GPa)
and ductility of the sample having a draw ratio of 15, and (2) that an
increase of draw ratio from 15 to 70 increases the strength of the fibre by
a factor of 5.[22]

The effect of deformation rate on strength is more difficult to investigate
than the effect on modulus. This is because, with increasing rate of
deformation, the improper breakage of fibres becomes more frequent and
a great many data have to be discarded. The effect of strain rate expressed
as cross-head speed is shown in Fig. 10.2 for a fibre having a draw ratio of
15.[22] The author proposes that the levelling of the strength indicates a
ductile–brittle transition at about 32 mm min^{-1} cross-head speed, beyond
which the rate effects are much smaller.[22]

Because of testing problems, there are at present no reliable data on
strength for the technologically very important deformation rates encountered in explosions or impact of fast projectiles. In the absence of such
data, the temperature dependence of strength is used, assuming that
the time–temperature relationship reflected in strain rate–temperature
relationships, and observed with modulus, holds also for strength. The
tensile strength–temperature relationship for an HPPE fibre of draw ratio
100 is shown in Fig. 10.3.[22]

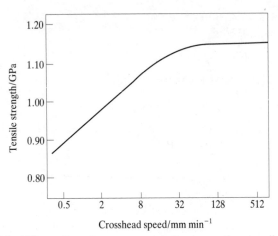

Figure 10.2   Effect of strain rate on tensile properties: tensile strength vs cross-head speed of HPPE fibre having draw ratio λ of 15. (From ref. 22 with author's permission)

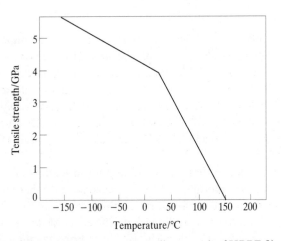

Figure 10.3   Effect of temperature on tensile strength of HPPE fibre having draw ratio λ of 100. (From ref. 22 with author's permission)

The fibre shown in Fig. 10.3 has a strength of about 4 GPa at room temperature. On cooling, the strength increases linearly with decreasing temperature and reaches 5.6 GPa at about $-150°C$. Based on this and similar results obtained by others, tensile strengths involving ballistic impact (i.e. deformation rates exceeding $10^3 \ min^{-1}$) are about 40 per cent higher than those measured at standard rates of deformation of about $1 \ min^{-1}$. In the analysis of projectile armour interactions (deformation and penetration) or confinement of explosions, Prevorsek et al.[5] therefore use for Spectra 1000 fibres a tensile strength of 5 GPa for deformation rates exceeding 10^3 min^{-1}.

10.5 Energy absorption potential in armour

10.5.1 Straining in tension

Considering the large strain rate effects discussed above and the fact that these effects are unique for HPPE, the relative merits of HPPE versus other reinforcing fibres increase with the rate of deformation. Thus it is not surprising that HPPE is rapidly gaining acceptance in the technology of survival where the principal role of a component is to protect people, apparatus and equipment from blasts of explosions, fast moving projectiles, and so on. In the analysis and design of structures involving fast moving fragments, the properties corresponding to the deformation rates caused by such impact must be used. Under such conditions the HPPE modulus and strength are about 280 GPa and 5 GPa. This increases the energy absorption of HPPE products substantially above the values expected on the basis of testing experiments conducted at the rate of deformation used in standard testing procedures. However, this factor alone is insufficient to establish the ranking of HPPE with respect to other materials. In ballistic mechanics, where impact velocities approach the velocities of the strain propagation, it is also necessary to take into consideration the amount of material that experiences the stress during the duration of the impact. The times are of the order of several microseconds and the amount of strained material is proportional to the velocity of strain propagation.

To determine the strain propagation rate in HPPE transverse impact was employed and two methods were used to treat the data. The results, which depended a great deal on the data treatment, are summarized in Table 10.2. In column A are shown the results obtained by treating the experimental data using the analytical procedure of Smith *et al.*[2] Column B, on the other hand, contains more recent data of Field *et al.*[17] based on the outputs of sensors placed at the points where the fibres are supported. The difference in HPPE/Kevlar strain propagation rate ratio, 1.48 vs 1.30, is somewhat larger than expected. The difference may be caused by a much larger strain amplitude and pre-tension dependence of strain propagation rate in HPPE than in Kevlar. Work is in progress to resolve this problem.

Consider that at very fast rates of deformation the energy absorption potential W_{max} of a fibre broken in tension is given by

$$W_{max} = K \cdot W_f \cdot C \qquad [10.1]$$

where K is a constant, W_f is the energy to break fibres at a given rate

Table 10.2. Strain propagation rate C and modulus E

	A		B	
	C/m s^{-1}	E/GPa	C/m s^{-1}	E/GPa
HPPE	17 800	310	13 000	230
Kevlar	12 000	170	9 300	130
Elastomer	95–200		NM	
HPPE/Kevlar ratio	1.48		1.30	

NM = not measured.

of deformation, and C is the strain propagation velocity. This expression enables us to estimate the relative energy absorption potentials of HPPE and aramid fibres at ballistic rates of deformation. In the absence of the energy-to-break data we assume that both fibres exhibit linear response up to their breaking point. Then the energy to break equals strength/ modulus. Using the data in Table 10.2 and the estimated high deformation rate strengths of HPPE and Kevlar as 5 GPa and 3 GPa, respectively, we obtain the value

$$\frac{W_{max}(\text{HPPE})}{W_{max}(\text{Kevlar})} \approx 3.0 + 0.75 \qquad [10.2]$$

Relative penetration resistance falling in this range is indeed observed for the impact of large projectiles of about 1.5 cm diameter, at impact velocities ≥ 610 m s^{-2} with armour structures that are thicker than about 1.5 cm. These conditions of armour design and impact ensure that the predominant failure mode in the plate is straining of fibres in tension, a mode of failure that is relatively unimportant with small size impactors and thin plates.

A great deal of work, both experimental and analytical, has been dedicated to this issue. Studies in which the penetration resistance of HPPE and aramid composites were compared frequently confirmed the expected energy absorption capacity ratio of about 3. There are, however, numerous cases where HPPE products are only marginally better than or about equal to aramid ones. Those cases were relatively easy to explain but, surprisingly, there are also cases where the advantages of HPPE over other products are much greater than expected on the basis of the energy absorption ratio of 3. To explain these data, it is necessary to review the principles of penetration dynamics as currently applied to fibre structures. In discussions of the relative energy absorption potential of HPPE versus other fibres, consideration is usually limited to straining in tension. In this mode of failure HPPE indeed has a very large advantage over

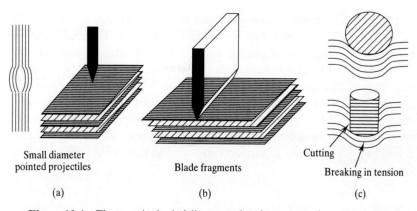

Small diameter
pointed projectiles

Blade fragments

Cutting

Breaking in tension

(a)   (b)   (c)

Figure 10.4 Three principal failure modes in penetration. (a) Lateral displacement. (b) Cutting. (c) Breaking in tension.

aramids. However, failure frequently also involves shear, and in this failure mode the advantage of HPPE is much less. It must therefore be stated that the threefold advantage of HPPE over aramids is expected only if the impact and composite design are such that the penetration involves primarily straining and breaking in tension.

10.5.2 Penetration mechanisms

The analysis of the penetration of fibre-reinforced composites considers three primary mechanisms: lateral displacement of fibres, breaking of fibres in shear, and straining and breaking of fibres in tension.[23] These failure modes are represented schematically in Fig. 10.4. A small, pointed projectile can penetrate armour without breaking the fibres and in this case the penetration resistance depends on the composite construction: weave, yarn and fibre count, matrix characteristics, fibre, adhesion, etc. Failures in shear and cutting are very important with sharp fragments. As the projectile diameter increases, and with increasing thickness of the armour, the tension mode becomes increasingly important. The penetration resistance of various composites in these modes of failure can be studied using high-speed testing equipment by firing projectiles of various geometries against targets.

Systematic experimental and analytical work showed, however, that in actual cases all penetration modes exist and that their relative importance depends on the design of the target (fibre layup) and the geometry, size and speed of the penetrator. In this respect it should be noted that the relative values of shear strength or cut resistance were measured both by instrumented impact as well as by firing of blade-like projectiles at high speed against suspended filaments. Both methods indicate a substantially higher cut resistance of HPPE compared to aramids. It should also be noted that the analyses of penetration of soft body armour and military helmets show that in these cases the penetration is primarily by cutting and lateral displacement of filaments because of specified projectile geometries and speeds. Hence, with soft body armour and helmets the advantage of HPPE over aramids is due primarily to its superior cut resistance. With projectiles larger than 1 cm diameter and armour thicknesses exceeding 1.5 cm, failure and straining in tension becomes a major factor.

10.5.3 Damage tolerance

On the basis of these considerations it is possible to explain and analyse quantitatively a great deal of experimental data. There are, however, observations and experiments that can easily be reproduced showing that, in some cases, the energy absorption potential of HPPE products exceeds the threefold advantage over other composites by a wide margin. The following experiments illustrate these points.[23]

To compare various fibres, composites were constructed for which the following parameters were held constant: fabric plain weave construction, fabric weight 50–60 oz yd^{-2}, number of layers (5), resin system (araldite 6010 epoxy resin with hardness Hy56), and fibre content volume (60 per cent). A Dynatap 8200 MST instrumented impacted tester was used, which yielded the following data: load curve, maximum load value, energy absorption curve, energy absorbed at maximum load and total energy absorbed upon completion of impact. The properties of fibres from which these composites were made are listed in Table 10.3.

The composite samples were tested under conditions indicated in Table 10.4. The key observation is that S-2 glass, Kevlar and carbon fibre composites were penetrated at much lower impact energies. The load and energy data show that the absorbed energy and peak loads were substantially higher than anticipated on the basis of the threefold advantage discussed above. To some extent, this discrepancy could be attributed to differences in sample preparation, namely the HPPE (Spectra 900) sample quality may have been superior to that of the other samples. We consider this to be very unlikely, but it must not be ruled out.[23]

To broaden the data base and develop design criteria for penetration in damage tolerant structures, it was necessary to study the behaviour of these composites on repeated impact. To accomplish this the HPPE 900 panel was subjected to the same impact conditions (namely 6.6 times the energy to penetrate the Kevlar 49 panel) five times, and the HPPE panel survived all five impacts without failure. Knowing that an impact energy

Table 10.3. Fibre properties

Property	HPPE (Spectra 900)	Aramid HM (Kevlar 49)	S-2 Glass	Graphite
Density/g cm^{-3}	0.97	1.44	2.49	1.86
Filament diameter/μm	27	12	9	7
Elongation/%	2.7	2.8	5.4	0.6
Tensile strength/GPa	3.0	2.8	4.6	2.34
Specific strength/GPa/g cm^{-3}	3.09	1.94	1.85	1.26
Tensile modulus/GPa	172	124	90	379
Specific modulus/GPa/g cm^{-3}	177	86.1	36.1	203

Table 10.4. Instrumented impact single fibre systems

	Maximum load (kg)		Energy total (ft-lb)		
	Actual	Index	Actual	Index	Observations
Spectra 900	753	100	73	100	No penetration
S-2 glass	168	22	11	15	Penetration
Kevlar 49	115	15	9	12	Penetration
Graphite	60	8	3.4	5	Penetration

of 6.6 times that to penetrate the Kevlar panel must be very close to the energy required for the HPPE panel, it is concluded that these panels can survive, without an apparent weakening, repeated impact at levels that are close to the energy required for their penetration.[23]

To determine what happens in these composites on severe impact, a series of experiments was conducted in which the impact energy was progressively increased from 6.6 to 9.6 times the energy to penetrate the Kevlar panel. The panel rigidity increased on impact. This indicates that contrary to expectation, namely a weakening of the test specimen and decrease of the panel modulus, the HPPE panels actually underwent structural changes that enhanced their properties.[23]

In the analysis of the behaviour of composites under transverse impact, it must be recognized that the fibres on the impact side experience compression and shear and that the tension is developed only on the back side of the panel. Also, according to current estimates, the compression strength of HPPE is significantly lower than that of aramids (by about 30–60 per cent), the shear strength of HPPE is about 30 per cent higher than that of aramids, and the tensile strength at the deformation rates of these experiments is possibly about 40 per cent higher than that of aramids. It is obvious that this set of properties cannot explain large energy absorption differences between HPPE and S-2 glass, aramid and carbon fibre composites. Based on the cited differences of compressive shear and tensile strength, only a 50 per cent increase in the penetration resistance would be expected. However, an almost tenfold advantage recorded in these experiments shows that other energy absorption mechanisms must play a key role in the penetration resistance of these composites.

What then is the source of this energy absorption capacity that greatly exceeds the levels expected on the basis of strength data? Some of the answers are provided by inspection of targets after impacts that did not lead to penetration and after which the targets still retained the energy absorption capability of the original sample. Microscopical examination of such samples shows on the impact side large plastic deformations of the fibres, flattening, and kink bands indicating shear as well as compressive failure. That these shear and compressive failures of the fibres in the impact zone produced no readily detectable detrimental effects on the impact resistance of the panels means that polyethylene can, under certain conditions, fail in shear and compression and still retain its tensile properties essentially unchanged. Since it is the energy associated with the tensile mode of deformation into which the bulk of kinetic energy of the impacting body is transformed, these findings and observations explain, at least qualitatively, the observed behaviour.

This also means that the key factor in the impact characteristics of HPPE composite discussed above is not the difference in the tensile strength, but the damage tolerance and capability of HPPE to undergo a variety of solid-state deformation mechanisms discussed above without losing its ability to sustain large tensile loads and absorb energy in tension. Although the massive formation of kink bands is experimentally recorded as compressive failure, it can be inferred on the basis of these experiments that these compressive failures do not involve a great deal of breakage of primary bonds and that, upon subjecting fibres that failed in compression to tension, the original tensile properties can to a large degree be restored.

While a quantitative analysis of impact data is not yet available, many fundamental investigations are currently dedicated to these phenomena and some progress has been made towards elucidating this technologically important behaviour. The issue in this case is the deformation energetics at large deformations. To simplify the case, consider the properties in tension, focusing on the absorbed energy or hysteresis during cyclic deformation. The following experiment was carried out by van der Werff and Pennings.[22] Figure 10.5 shows two deformation cycles for HPPE fibre made from a 2.5 wt% gel with a draw ratio λ of 60, which corresponds closely to the maximum draw ratio in one step at 148°C. After the first cycle, the cross-head was returned to its original position and after a recovery period of two minutes the second cycle was recorded.

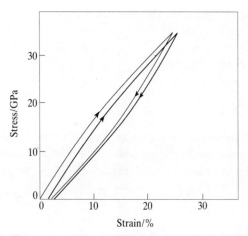

Figure 10.5   Two consecutive deformation cycles of an HPPE fibre having draw ratio λ of 60. (From ref. 22 with author's permission)

Note that the hysteresis is nonelastic in the two deformation cycles. The strain at which the stress becomes zero in the unloading curve is in both cycles larger than the strain at which the loading cycle started. In addition, the strain at which the loading curve in the second cycle started is smaller than the strain at which the unloading curve of the first cycle stopped. This indicates that, after unloading to zero stress and zero initial strain, a recovery process must have taken place during which the fibre contracted. In the case of long delay times between the two cycles (about 1–2 days) the stress–strain characteristics of the first cycle are completely recovered and the second cycle becomes completely identical to the first.

The process that causes nonelastic hysteresis during a cyclic deformation is therefore reversible and under these experimental conditions no plastic deformation, i.e. permanent elongation after deformation, occurs. Deformation of a fibre causes a recoverable, time-dependent change from its initial morphology, thus providing a recoverable source of energy absorption. To quantify various energy absorption processes taking place during cyclic straining, van der Werff and Pennings consider the energy needed to elongate a fibre from a strain ε_1 to a strain ε_2:

$$U = \int_{L_1}^{L_2} F \, \mathrm{d}L = V \int_{\varepsilon_1}^{\varepsilon_2} \sigma \, \mathrm{d}\varepsilon \qquad [10.3]$$

where L is the fibre length, V is the fibre volume, F is the tensile force and σ is the tensile stress. From Fig. 10.5 it can be seen that the area under the first loading curve is larger than the area under the first unloading curve. The difference between them is the dissipated energy. The same holds for the second cycle, but the dissipated energy is smaller than that in the first cycle. The energy recovered under the unloading curve, however, is nearly the same for both cycles. The fact that total deformation energy and the dissipated energy differ between the two cycles indicates that the properties of the fibre have changed during deformation. Van der Werff and Pennings[22] propose that the following energies can be determined on the basis of cyclic deformation:

U_{tot} Total deformation energy (the area under the loading curve)
$U_{elastic}$ Apparent elastic energy (the area under the unloading curve)
U_{dis} Dissipated energy ($U_{tot} - U_{elastic}$)

By performing a second deformation cycle, the dissipated energy can be split up into two parts:

$U_{dis,2}$ Dissipated energy in the second cycle, i.e. the energy that can be dissipated again immediately

$U_{dis,\Delta}$ Dissipated energy in the first cycle, i.e. the energy that cannot be dissipated again (difference in U_{dis} between the first and second cycle)

According to van der Werff,[22] $U_{dis,\Delta}$ reflects changes in the fibre structure. When this value is zero (which is the case when the recovery period between the two cycles is sufficiently long), the stress–strain behaviour in both cycles is identical.

This energy analysis was implemented by van der Werff[22] for the fibre shown in Fig. 10.5. The plot of the energies as a function of strain in the first cycle is given in Fig. 10.6. At the largest strain at which the energies could be determined, still 72 per cent of the deformation energy is elastic. This percentage goes up for smaller strains, although there is no region where the fibre is completely elastic. The highest value measured was 79 per cent elastic energy at a strain of 1.09 per cent. Considering the dissipated energies, $U_{dis,2}$ is substantially higher than $U_{dis,\Delta}$. This means that, on this timescale, most of the dissipated energy is dissipated as heat and that only a relatively small amount is used to change temporarily the structure of the fibre. When the change of structure is recovered with time, however, this small amount of energy will also be dissipated gradually as heat. The amount of dissipated energy per volume, U_{dis}, can be related to an elongational viscosity through the relation

$$U_{dis} = \eta \dot{\varepsilon}^2 t \qquad\qquad [10.4]$$

where $\dot{\varepsilon}$ is the deformation rate and t is the elapsed time. The fibre in Fig. 10.6 has dissipated 26.3×10^6 J m^{-3} at a cycle strain of 3.37 per cent ($\varepsilon = 1.67 \times 10^3$ s^{-1} and $t = 40.4$ s). Using the above relation for U_{dis} it follows that $\eta = 2.33 \times 10^{-11}$ N m^{-2} s. This value compares very well

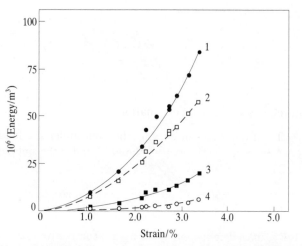

Figure 10.6 Deformation energies vs strain of an HPPE fibre having draw ratio λ of 60. $1 = U_{\text{tot}}$, $2 = U_{\text{elastic}}$, $3 = U_{\text{dis},2}$, $4 = U_{\text{dis},\Delta}$. (From ref. 22 with author's permission)

with the elongational viscosity of HPPE fibres at higher temperatures,[24] and this viscosity must then be assigned to disordered domains in the fibre morphology. Such domains do not seriously contribute to the tensile stress, but elongation and contraction of these viscous regions results in dissipation of deformation energy as heat. Van der Werff and Pennings[22] also noted that the results of their energy analysis on HPPE fibres, presented in Fig. 10.6, differ markedly from similar studies by Frank and Wendorff on polyamide fibres.[25] Their electron spin resonance (ESR) experiments showed that deformation above a critical strain level leads to chain rupture. Such tensile deformation also irreversibly changed the stress– strain behaviour of the polyamide fibres, and the irreversible dissipation of energy during deformation was accounted for in terms of energy release after chain rupture. The breaking of chains was confirmed by a decrease in molecular mass.

From the results of the preceding energy analysis of HPPE fibres, Pennings concludes that during the tensile deformation of these fibres, even close to breaking strain, no detectable amount of chain breaking occurs. Irreversibly dissipated energy is not present, except for yielding fibres where this energy is associated with work consumed in plastic deformation. Stoeckel et al.[26] found no relevant decrease in the molecular mass of a polyamide sample after it had been fractured in tensile deformation. ESR studies on HPPE fibres in Pennings' laboratory have have shown that under tensile stress no significant amounts of free radicals due to chain rupture could be detected.[22] These data, together with the

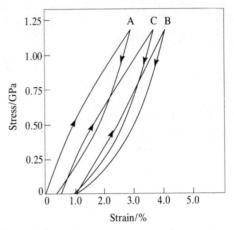

Figure 10.7   Three deformation cycles from a consecutive cycling experiment on an HPPE fibre having a draw ratio λ of 15. A, first
cycle; B, 35th cycle; C, 36th cycle. Recovery time between
cycles B and C is 75 minutes. (From ref. 22 with author's
permission)

results from the work of van der Werff *et al.*, indicate that tensile
deformation of high modulus and high strength HPPE fibres does not
involve significant breaking of chains. Viscoelastic processes in the
disordered phase and solid transformation of the crystalline phase play a
major role, as shown by experiments involving repeated cyclic loading.[22]
  In Fig. 10.7 three loading cycles are presented for a fibre of draw ratio
15 made from a 6.0 wt% gel (a nonyielding fibre at the applied deformation
rate). In cycle A, the fibre is strained up to 1.20 GPa stress, after which
the cross-head speed is reversed. When the tensile stress reaches zero, the
cross-head is again immediately reversed in a similar way and a number
of cycles are performed. Cycle B is the 35th cycle. The comparison of
cycles A and B shows that cycle A corresponds to the cycles shown in
Fig. 10.5 and that cycle B is distinctly different. Note that the tangent
modulus to the loading curve of B does not decrease monotonically as
in A but, after an initial decrease, goes through a minimum and then
increases again. The shapes of the unloading curves in A and B are similar.
The end of the unloading curve in B, however, falls very close to the
beginning of its loading curve. The hysteresis has become almost elastic
and the dissipated energy in cycle B is much less than in cycle A. After
performing cycle B, the cross-head was returned to the position corresponding to zero initial strain and the sample was allowed to recover
during the 75 minutes before cycle C (36th) was recorded.[22] The
stress–strain behaviour changed dramatically after recovery. The beginning

of the loading curve C shifted to about halfway between the beginnings of A and B, showing that the sample length had decreased. Typically, the hysteresis has become nonelastic again after recovery and the energy dissipated in cycle C is much larger than in cycle B. Furthermore, the shape of the loading curve in C again resembles that of cycle A.[22]

This work of Pennings *et al.*[22] explains quantitatively the unique capability of polyethylene fibres to withstand repeated impact under loads that are close to the breaking loads without an accumulation of damage and hence without an appreciable chain breakage. The finding that the energy absorbing structural rearrangements taking place in the disordered amorphous domains and solid crystalline phase are frequently almost completely recoverable demonstrates a characteristic not encountered with other high performance fibres used in composites (S-2 glass, aramids and carbon fibres). This in turn explains what is the source of the damage tolerance and energy absorption capability of ultrastrong polyethylene fibres and composites. It is believed that the same recoverable energy absorption mechanisms are responsible for the outstanding fatigue and abrasion resistance of ultrastrong polyethylene fibres.

10.6 Dimensional stability under load-creep

Despite very high levels of crystallinity (75 per cent) in comparison with other fibres from flexible polymers, polyethylene fibres exhibit dimensional instability under load. The creep of polyethylene fibres has been a topic of numerous studies because it was clear that, unless the problem is solved, the prospects for a successful commercialization of HPPE fibres would be rather small. The most important studies of creep behaviour of polyethylene fibres were conducted by Ward and his associates.[27,28] His work led to several important conclusions relating to the magnitude of creep in polyethylene fibres:

1. Creep and recovery depend strongly on applied stress and temperature.
2. Temperature effects can be reproduced using Eyring's activation rate theory.
3. Creep decreases with increasing draw ratio and increasing molecular mass.
4. Creep is reduced by modifying linear polyethylene to include small amounts of co-monomers containing bulky groups.
5. Creep is reduced by cross-linking of fibres.

The effects of load, draw ratio and molecular mass are illustrated in Figs 10.8 and 10.9. The data in Fig. 10.9 are particularly interesting because they show the magnitude of the improvements in the dimensional

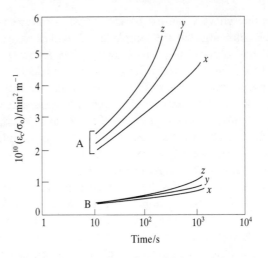

Figure 10.8　Creep compliance (ε_c/σ_0) of ultrahigh modulus polyethylene fibres, $\bar{M}_n = 6180$, $\bar{M}_r = 101\,450$. Effect of draw ratio, $A = 10$, $B = 30$. Effect of load, $x = 0.1\,\mathrm{GPa}$, $y = 0.15\,\mathrm{GPa}$, and $z = 03.2\,\mathrm{GPa}$. (Redrawn from I M Ward and M A Wilding, *Polymer*, **19**, 969 (1978); and M A Wilding and I M Ward, *Polymer*, **22**, 870 (1984); by courtesy of Butterworth-Heinemann Ltd ©)

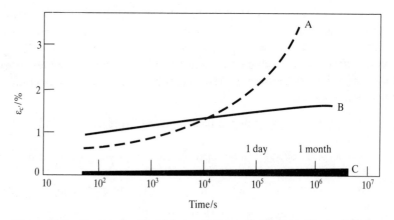

Figure 10.9　Creep of ultrahigh modulus polyethylene fibres. Load = 0.1 GPa. Effect of molecular mass at draw ratio of 20: A, $\bar{M}_n = 6180$, $\bar{M}_r = 101\,450$; B, $\bar{M}_n = 33\,000$, $\bar{M}_r = 290\,000$; C, draw ratio of 100, $\bar{M}_n = 290\,000$, $\bar{M}_r = 2\,200\,000$. (Redrawn from I M Ward and M A Wilding, *Polymer*, **19**, 969 (1978); and M A Wilding and I. M. Ward, *Polymer*, **22**, 870 (1984); by courtesy of Butterworth-Heinemann Ltd ©)

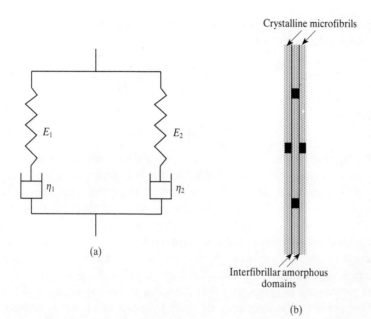

Figure 10.10 (a) The two-process model for permanent flow creep. (b) Structural model of HPPE fibre. The interfibrillar domains are the sites of large activation volume process 1, and the crystalline microfibrils are the sites of the large activation volume process 2. (Part (a) redrawn from M A Wilding and I M Ward, *Polymer*, **22**, 870 (1984) by courtesy of Butterworth-Heinemann Ltd ©. Part (b) from ref. 22 with author's permission)

stability that have been achieved by increasing the molecular mass and draw ratio through process modifications, namely, the change from melt to solution spinning.

To discuss the effects that factors 1 to 5 have on the process mechanisms of creep, the analytical work of Ward is complemented here by structural interpretations. Ward showed that creep and recovery data can be reproduced by a mechanical model consisting of two parallel Maxwell models shown in Fig. 10.10, where E_1 and E_2 are the moduli of the springs and η_1 and η_2 are the viscosities of the dashpots. On the basis of the structural model that represents well the morphological characteristics of HPPE fibres (Fig. 10.10b), it follows that the responses of the two Maxwell models can be attributed to (1) the amorphous domains that exist between the microfibrils and (2) the long needle-like crystals of length-to-diameter (L/D) ratio about 40. These crystals are covalently joined by a short 'amorphous' domain with L/D ratio about 1–2.

Assuming that the plastic deformation of each dashpot can be represented by the Eyring relationship, Ward finds that the total stress of this model is given by

$$\sigma = \frac{2.3kT}{v_1}\left[\log \varepsilon_p - \log\left(\frac{\varepsilon'_{01}}{2}\right)\right] + \frac{kT}{v_2}\sinh^{-1}\left(\frac{\varepsilon_p}{\varepsilon_{02}}\right) \qquad [10.5]$$

where

$$\varepsilon'_0 = \varepsilon_0 \exp(-\Delta U/kT)$$

ε_p is the rate of plastic flow, ΔU is the activation energy of the process, v_1 and v_2 are the activation volumes of the two processes, ε_{01} and ε_{02} are the pre-exponential factors associated with these two processes, σ is the applied stress, k the Boltzmann constant, and T the absolute temperature.

The use of the constant 'plateau' creep rates enables one to determine the activation volumes v, and the pre-exponential factors. By applying the context of the Eyring reaction rate theory to the processes involved in creep, it is postulated that the activation volume reflects the 'size' of units that participate in creep, and that the value of the pre-exponential factor corresponds to the frequency of creep events and probably reflects the population size of potential creep sites. This background allows one to interpret the effects of factors affecting the macroscopic creep in terms of nanometre-scale micromechanisms involved in creep on the basis of their activation volumes and their pre-exponential factors. Because of the nature of the experimental and analytical work involved in this analysis, a twofold change in the process parameters is required to indicate a significant process change.

Let us consider the effect of cross-linking at a constant draw ratio of 20 and constant relative molecular mass ($M_r = 101\,450$). The corresponding parameters characterizing the two microcreep processes are listed in Table 10.5. Based on this analysis, the large reduction in the macroscopic creep on cross-linking can be attributed solely to the changes in the amorphous phase. It can then be inferred that cross-linking took place only in the amorphous regions and resulted in a nine-order-of-magnitude decrease in the number of the potential creep sites in this phase. All other changes appear to be insignificant.

The effects of copolymerization with hexane co-monomer, leading to about 1.5 butyl groups per thousand carbon atoms at about constant relative molecular mass ($M_r = 155\,000$) and a constant draw ratio of 20 are shown in Table 10.6. The analysis shows that almost all the butyl groups are excluded from the crystalline domains. The net effect is a very large (eight orders of magnitude) decrease in the potential creep sites in the amorphous phase and a relatively smaller (fourfold) decrease in the potential creep sites in the crystalline phase. It should be noted that the effects of cross-linking and copolymerization on the micromechanisms of

Table 10.5. Effect of cross-linking on creep processes

	Process 1 (amorphous)		Process 2 (crystalline)	
	Activation volume/nm^3	Pre-exponential factor/s^{-1}	Activation volume/nm^3	Pre-exponential factor/s^{-1}
Control	0.46	2.2×10^{-8}	0.077	1.0×10^{-6}
Cross-linked	0.49	1.8×10^{-17}	0.110	0.9×10^{-6}

Table 10.6. Effect of copolymerization on creep processes

	Process 1 (amorphous)		Process 2 (crystalline)	
	Activation volume/nm^3	Pre-exponential factor/s^{-1}	Activation volume/nm^3	Pre-exponential factor/s^{-1}
Control	0.37	2.4×10^{-13}	0.123	1.5×10^{-6}
Copolymer	0.58	5.2×10^{-21}	0.086	3.8×10^{-7}

Table 10.7. Effect of draw ratio on creep processes

	Process 1 (amorphous)		Process 2 (crystalline)	
	Activation volume/nm^3	Pre-exponential factor/s^{-1}	Activation volume/nm^3	Pre-exponential factor/s^{-1}
Draw ratio 10	0.58	1.2×10^{-13}	0.37	2.8×10^{-6}
Draw ratio 30	0.51	2.4×10^{-17}	0.08	6.0×10^{-7}

creep that are indicated by this analysis are also expected on the basis of common knowledge. Thus these two examples not only illustrate the utility of this analysis but also confirm the validity of its results.

The effect of increase of draw ratio from 10 to 30, at constant $M_r = 135\,000$, yields the results shown in Table 10.7. Draw ratio affects both the structure of the predominantly crystalline microfibril and the interfibrillar amorphous domains. With increasing draw ratio, the structure of the microfibril becomes tighter, as indicated by about a fivefold decrease in the activation volume v and the number of potential creep sites. The most important changes, however, occur in the structure of the interfibrillar amorphous domains. These changes are attributed to the increases in the amorphous orientation and degree of chain extension that produce a four-order-of-magnitude decrease in the number of potential creep sites.

The effect of the molecular mass change from $M_r = 101\,450$ to $M_r = 132\,000$ at a constant draw ratio of 20 is presented in Table 10.8. The fact

Table 10.8. Effect of molecular mass on creep processes

	Process 1 (amorphous)		Process 2 (crystalline)	
	Activation volume/nm^3	Pre-exponential factor/s^{-1}	Activation volume/nm^3	Pre-exponential factor/s^{-1}
$M_r = 101\,450$	0.46	2.2×10^{-8}	0.077	1.0×10^{-6}
$M_r = 312\,000$	0.26	1.1×10^{-17}	0.106	3.1×10^{-7}

Table 10.9. Effect of spinning on creep processes

	Process 1 (amorphous)		Process 2 (crystalline)	
	Activation volume/nm^3	Pre-exponential factor/s^{-1}	Activation volume/nm^3	Pre-exponential factor/s^{-1}
Melt-spun	0.46	2.2×10^{-8}	0.077	1.0×10^{-6}
Solution-spun	0.25	7.0×10^{-15}	0.010	1.8×10^{-9}

that the only significant change resulting from the increase in the molecular mass involves a nine-order-of-magnitude reduction in the potential creep sites in the amorphous phase shows that the chain ends are heavily concentrated in the amorphous phase. This could be a result of fractionation of molecular mass through crystallization.

Finally, let us consider the differences between creep micromechanisms of a melt-spun low draw ratio (20) low molecular mass ($M_r = 101\,450$) fibre and a solution-spun HPPE fibre having a molecular mass of $M_r = 2\,220\,000$ and a draw ratio of 60. (See Table 10.9.) The comparison of the microprocesses of a dimensionally fairly unstable melt-spun fibre with those operating in HPPE fibres shows that the improvements in the dimensional stability result from major changes in structure of both the amorphous interfibrillar domains and the crystalline microfibrils. The activation volume of amorphous phase creep is decreased by a factor of 2, and the reduction in the potential creep sites of the amorphous interfibrillar domains amounts to seven orders of magnitude. In addition, the macroscopic creep of the melt-spun product is further reduced by the changes in the structure of the microfibrils that result in a sevenfold decrease in the activation volume and a three-order-of-magnitude decrease in the number of creep sites.

In summary, polyethylene, although known for its dimensional instability under load, can be converted into a product that exhibits remarkable dimensional stability that is sufficient for use in cordage, fish nets, sails, parachutes, etc. Analysis of creep data supported by morphological and mechanistic analyses allows the macroscopic creep to be interpreted in terms of nanoscale processes that occur in the crystalline microfibrils and amorphous interfibrillar domains.

References

1. A J Pennings, K E Menniger in *Ultra High Modulus Polymers*, ed. A Cifferi, I M Ward, Applied Science, London, 1979, p. 117.
2. P Smith, P J Lemstra, B Kalb, A J Pennings, *Polymer Bulletin*, **1**, 733 (1979).
3. P Smith, P J Lemstra, US Patents 4344908 (1982), 4422993, 4430383 (1984).
4. S Kavesh, D C Prevorsek, US Patents 4413100 (1983), 4536536, 4663101 (1987).
5. D C Prevorsek, H B Chin, Y D Kwon, J E Field, *J. Appl. Polymer Sci.: Appl. Polymer Symposium*, **47**, 45 (1991).
6. A Schaper, D Zenke, E Schultz, A Hirte, M Taege, *Phys. Stat. Sol. (a)*, **116**, 179 (1989).
7. D T Grubb, private communication.
8. V Kramer, Allied-Signal Inc., unpublished results.
9. D H Reneker, J Mazur, *Polymer*, **24**, 1387 (1983); *ibid*, **29**, 3 (1988).
10. K Nakamae, T Nishino, H. Ohkubo, *J. Macromol. Sci.-Phys. B*, **30**, 1 (1991).
11. I Sakurada, I Ito, K Nakamae, *Macromol. Chem.*, **75**, 1 (1964); *J. Polymer Sci.*, **C15**, 75 (1966).
12. L R G Treloar, *Polymer*, **1**, 95 (1960).
13. M J S Dewar, Y Yamaguchi, S H Suck, *Chem. Phys.*, **43**, 145 (1979).
14. B Crist, M A Ratner, A L Browder, J R Sabin, *J. Appl. Phys.*, **50**, 6047 (1979).
15. B J Kip, M C P van Eijk, R J Meir, *J. Polymer Sci.: Polymer Phys. Ed.*, **29**, 99 (1991).
16. D S Boudreaux, *J. Polymer Sci.: Polymer Phys. Ed.*, **11**, 1285 (1973).
17. J E Field, unpublished data.
18. P J Barham, A Keller, *J. Polymer Sci.: Polymer Lett. Ed.*, **17**, 591 (1979).
19. D C Prevorsek, in *Encyclopedia of Polymer Science and Engineering*, supplementary volume, 2nd edn, Wiley, New York, 1989, p. 803.
20. Y Termonia, P Meakin, P Smith, *Macromolecules*, **18**, 2246 (1985).
21. D C Prevorsek in *Polymers for Advanced Technologies*, ed. M Levine, VCH, New York, 1988, p. 557.
22. H van der Werff, PhD thesis, University of Gröningen (A. Pennings, supervisor), April 26, 1991, pp. 31–58.
23. D C Prevorsek, H B Chin, A Bhatnagar, *Composite Structure*, **23**, 137 (1993).
24. J Smook, A J Pennings, *J. Appl. Polymer Sci.*, **27**, 2209 (1982).
25. O Frank, J H Wendorff, *Colloid Polymer Sci.*, **259**, 1047; *ibid*, **266**, 216 (1988).
26. T M Stoeckel, J Blasius, B Crist, *J. Polymer Sci.: Polymer Phys. Ed.*, **16**, 485 (1978).
27. I M Ward, M A Wilding, *Polymer*, **19**, 969 (1978).
28. M A Wilding, I M Ward, *Polymer*, **22**, 870 (1981).
29. P Morse, *Phys. Rev.*, **34**, 56 (1929).
30. S N Zhurkov, V E Korsukov, *J. Polymer Sci.: Polymer Phys. Ed.*, **12**, 385 (1974).

CHAPTER 11

Thermostable and fire-resistant fibres

G DESITTER AND R CASSAT

11.1 Introduction

In 1862 Harbordt first introduced thermostable polymers with poly(*m*-benzamide). However, it was only in the late 1950s, with the development of the aeronautics and space industries, that the use of thermostable polymers became essential. For example, a plane flying at Mach 2 has a surface reaching 250°C in the low strata of the atmosphere and 100°C at an altitude of 10 000 metres. When space shuttles enter the atmosphere, the temperatures reached are over 500°C, despite the use of braking systems. At that time there was only a small range of materials available, such as organic silicones and fluoropolymers, products with weak mechanical properties.

The emergence of such needs resulted in the definition of ambitious programmes by major companies. The necessary materials had to satisfy specific mechanical characteristics at lower densities than metals and alloys, while assuring good retention of these characteristics under heat. The very first successes were by Du Pont, with the discovery of polybenzimidazoles and the development of polyimides. Following these precursors, many laboratories ended up with polymers based on sulphur or nitrogen and on aromatic structures, e.g. polyphenylenes, polyaryloxides, polyarylsulphides, polysulphones, etc.

However, many of these polymers had a short lifespan because of their very intricate preparation or difficult utilization. Industrially speaking, only a limited number have proceeded to more significant development, such as phenols, polysulphones, polysulphides, aromatic polyethers, polybenzimidazoles, polyimides, aromatic polyamides and polyamideimides.

11.2 Nonflammable polymers

Nonflammable polymers, by definition, possess significant fire-resistant properties because of their chemical structure or due to additives. All thermostable polymers belong to the broad nonflammable polymer group, but a number of polymers within the nonflammable group possess performance characteristics that are different from those of thermostable polymers. Polymers that are halogenated by nature, such as poly(vinyl chloride), fluoropolymers, or polymers containing fire-proofing agents such as polyesters or polyamides, are classed as FR (fire-resistant) polymers. In fact, every polymer can be improved by the addition of fireproofing agents, but each combination has a specific performance. This field is still very empirical.

11.3 Thermostability

The concept of thermostability is not easy to define. It basically comprises three independent factors: (a) physical properties (mechanical, electrical, etc.); (b) the level of thermal exposure; and (c) the time of this exposure. The idea of a 'thermal index' establishes a relation between these factors. It is the temperature that a fibre may bear for 20 000 hours without losing more than half of its mechanical properties, whilst also ensuring that certain other proprties are retained, depending on the field of application. For instance, for aeronautical uses it is important that properties are comparable to those of light metal alloys.

The thermal index relates to thermostability through a temperature value only. More interesting information is provided by the Arrhenius curve, which involves time as well as temperature (again for a loss of less than half of the initial properties). The degradation of a polymer depends upon both time of exposure and temperature:

$$f(P) = -kt \qquad [11.1]$$

where $f(P)$ is a function of a specific property, k is a velocity constant, and t is time. But the variation of the velocity constant with temperature is given by

$$k = A\,e^{-E/RT} \qquad [11.2]$$

where E is the energy of activation, T is the temperature in kelvins, and A, e, R are constants.

$$\frac{f(P)}{t} = A\,e^{-E/RT} \qquad [11.3]$$

$$\log f(P) - \log t = -\frac{E}{RT} + \log A \qquad [11.4]$$

It follows that the logarithm of the 'time limit' for a property to fall to a certain level at various temperatures is a linear function of the inverse of temperature (for this 'time limit' $f(P)$ is constant):

$$\log t = \frac{a}{T} - b \qquad [11.5]$$

where a and b are constants. The greater the value of a, the lower is the speed of thermal degradation (long-term thermostability). When $t = 1$, $T = a/b$. For a given value of a, the initial thermostability is inversely proportional to b, which is a measure for short-term thermostability or instantaneous protection.

In practice it is easy to measure the short time t that corresponds to a high temperature T. However, the Arrhenius curve can be extrapolated to obtain long times of exposure that are hard to measure. These are the exploitable values of T. The use of the Arrhenius curve to define the thermostability of a polymer does suffer from a certain lack of precision, since experimental errors are increased by extrapolation. However, the method remains the quickest way of determining the thermal index of a polymer.

11.4 Nonflammability

The limiting oxygen index (LOI) roughly defines 'thermostability' from the flammability aspect. This index gives the minimum volume of oxygen in an oxygen–nitrogen mixture required to maintain the combustion of a particular polymer. A list of LOIs is given in Fig. 11.1, which shows the similarities between thermostability and nonflammability. For instance poly(acrylonitrile oxide), where the organic fraction has been partially replaced by nitrogen, has an excellent fire classification. So do the heterocyclic polymers, which have an improved thermostability in relation to aromatic polymers, particularly when carbon is replaced by nitrogen or sulphur. By contrast, halogenated polymers, whose fire performance is rather good, have a poor classification as far as their thermostability is concerned. For example, poly(vinyl chloride) with an LOI of around 40, has a thermal index of around 80°C.

Following the definition of thermostability by the use of the thermal index and its practical exploitation by the Arrhenius curve, plus the definition of flammability through the LOI, different aspects of the thermal performance of polymers will now be examined.

11.5 Irreversible degradation of polymers

Among the phenomena that result from rise in the temperature of a polymer, some are reversible and others are irreversible. For the reversible phenomena, such as glass transitions, the resulting variations of properties

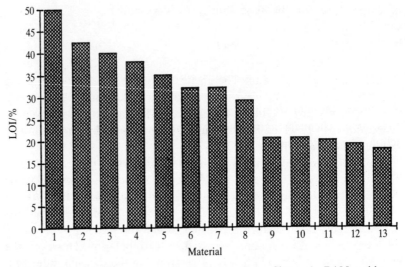

Figure 11.1 Limiting oxygen index for various fibres. 1, PAN oxide; 2, polyacrylate; 3, PBI; 4, chlorofibre; 5, PPS; 6, novoloid: 7, polyamideimide; 8, aramid; 9, viscose; 10, polyester; 11, polyamide; 12, polypropylene; 13, acrylic.

are not permanent. Once the original temperature is regained, the variations of properties are reversed. But irreversible modifications definitely change the chemical composition and the structure of the polymer. These modifications can be classified into three types:

1. Those that lead to reorganization of the original macromolecule. This type of modification is exploited in the making of thermostable products by cyclization or controlled pyrolysis.
2. Those that lead to cross-linking and the creation of bi- or tridimensional configurations. This kind of modification, although mainly produced thermally, is usually helped by chemical or ionic agents. The cross-links created reduce the original plastic or elastic characteristics and reinforce the resistance of the material to thermal decomposition.
3. Those that bring about a split of the original molecule into segments of lower molecular mass. This is the kind of irreversible modification that relates directly to the thermal stability of the polymer.

There are two types of reaction of decomposition by pyrolysis, random splitting of the chain and depolymerization. During pyrolysis, these two kinds of decomposition occur simultaneously but, according to the nature of the polymer concerned, one type distinctly prevails over the other.

Although it is not the most common mode, 'thermostable' polymers can decompose by depolymerization. This is particularly true of polytetrafluoroethylene.

Practically speaking, thermal decomposition is rarely the result of heat only, but arises from various factors combined together, implying a synergy. To estimate the thermostability of a polymer, the notion of complex thermal stability must be introduced. It takes into account physical or chemical agents acting together with the purely thermal action. The photochemical effect of electromagnetic radiation such as UV has to be considered. Among the chemical agents, the element most usually associated with thermal effects is oxygen (in the air). Oxygen preferentially attacks the carbon-carrying hydrogen, leading to thermooxidation.

Thermogravimetric analysis (TGA) is used to determine the thermostability of polymers. This method consists of following the loss of weight of a sample with temperature in various environments (air, nitrogen, vacuum, etc.) with a thermobalance. Of course, this method does not give complete information about the modification of properties. Thus TGA of polystyrene does not show any loss before 350°C, whereas the reduction of the modulus shows that polystyrene cannot be used over 100°C. Thus TGA has to be used simultaneously with other methods of evaluation in order to determine the level of thermal application as well as thermostability. A typical TGA curve is shown in Fig. 11.2. T_0 is the beginning of decomposition; T_c indicates the first important loss of mass, and T' and T'' show other major changes. TGA under oxygen is really accelerated ageing.

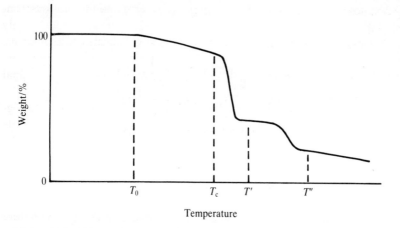

Figure 11.2 Theoretical thermogravimetric curve for thermostable fibres.

11.6 Chemical mechanisms of polymer degradation

The mechanisms of thermal degradation of polymers have essentially a radical origin initiated by symmetric breaking of covalent links by homolytic reactions. These homolytic reactions are triggered photochemically as well as thermally: thus, as mentioned before, the importance of radiothermolysis. Among the simplest cases, the initial result of degradation in air is a peroxide:

$$R\cdot + O_2 \rightarrow ROOH\cdot$$

This then oxidizes a hydrogenous moiety:

$$RO_2\cdot + RH \rightarrow RO_2H + R\cdot$$

As the products of degradation are radicals able to produce further oxidation, degradation by oxidation becomes autocatalytic at high temperatures. Also, decomposition into radicals accelerates the process of degradation:

$$RO_2H \rightarrow RO\cdot + HO\cdot$$

In order to stop the development of these chain reactions, antioxidizing agents have been developed (the best are phenols and aromatic amines). They react with the peroxidic radicals $RO_2\cdot$, and have some influence on the other free radicals.

Because of the high thermal energy associated with the thermal degradation of a thermostable polymer, there is competition between the reaction of the peroxidic radicals with the antioxidizing agent and dissociation into radicals, and it is virtually impossible to reinforce the thermal performance of a thermostable polymer by adding antioxidizing agents.

Typical mechanisms of degradation involving radicals are as follows:

$$RCH_3 + \cdot OH \rightarrow RCH_2\cdot + H_2O$$

$$RCH_2\cdot + O_2 \rightarrow RCHO + HO\cdot$$

$$RCHO + O\cdot \rightarrow RC\cdot O + OH$$

$$RC\cdot O \rightarrow CO + R\cdot$$

$$CO + HO\cdot \rightarrow CO_2 + H\cdot$$

with simultaneous reactions of coupling between radicals which limit further degradation development.

11.7 Factors influencing thermal performance

Factors that influence the thermal performance of polymers are first the strength of the covalent bonds, and second those that determine the degree of intermolecular attraction (secondary links). Molecular geometry is also important since it can affect intermolecular attractions. The factors that determine the melting resistance in general are the secondary links and the molecular geometry.

11.7.1 Dissociation energy

The bond dissociation energy depends on the nature of the other parts of the molecule. Thus the bond energy is not necessarily constant for different polymers. Average energies are considered satisfactory, but if the covalent bond has a polar characteristic or if it is involved in a resonant system, its energy can be greatly above the average value. One of the best methods of determining bond dissociation energy is by electronic bombardment in a mass spectrometer. The energy of the electrons produced is increased progressively until the appearance of molecular fragments. Values are given in Table 11.1.

In a complex molecule, the weakest bond energy determines the thermal resistance. It is therefore best to use uniform chemical structures that have no weak point. Molecules with bond energies greater than $400 \, \text{kJ mol}^{-1}$ can be used for the synthesis of thermostable polymers ($400 \, \text{kJ mol}^{-1}$ corresponds to the energy that can be produced from UV radiation of 300–400 nm). Table 11.1 shows that the strongest C–H and C–C bonds are found in aromatic components. However, C–H remains inferior to C–C and it is useful, for example, to replace C–H by C–F, or better to eliminate these bonds by using heterocycles with nitrogen or sulphur. On the other hand, the high dissociation energy of the aromatic C–C bond relates only to the first broken bond. The aromaticity of the whole system is then disturbed, which makes the other bonds weaker. This is also true for heterocycles with aromatic properties. Thus the concept of thermal stability based on bond dissociation energy is strictly valid only for systems where the bonds are independent. Completely aromatic polymers may be undesirable, in being insoluble, nonsoftening, brittle materials. It may be convenient to break up the complex bonds with small linkages in the structure, e.g. polyphenylene sulphide, poly-*p*-xylylene, etc., but this can harm the thermostability.

11.7.2 Rigid and tridimensional structure

A tridimensional architecture reduces the mobility of the chains and gives a polymer with a high melting point. This architecture is produced by

Table 11.1. Bond dissociation energies

Bond	Type of linkage	Energy (kJ mol^{-1})
C–C	Alkane	334
C=C		606
C≡C		836
C–C	Aromatic	518
C–C	C_6H_5–CH_3	372
C–C	$C_6H_5CH_2$–CH_3	263
C–H	Primary	410
C–H	Secondary	393
C–H	Tertiary	372
C–H	Aromatic	426
C–H	$C_6H_5CH_2$–H	326
C–H	Ethylene	443
C–H	Acetylene	505
C–H	CF_3–H	430
C–H	CCl_3–H	372
C–N		276
C≡N		874
C–O		364
C–O	Ether	330
C=O		727
C–S		276
C=S		539
C–F	CF_4	506
C–F	CH_3–F	447
C–F	C_6F_6	606
C–Cl	Alkane	334
C–Cl	Aromatic	359
O–H		460
S–H		364
N–H		351
Si–H		313
Si–O		372
Si–C		293
N–N		155

chemical cross-linking between chains. Partial cross-linking increases the melting temperature (and also modifies other properties) while complete cross-linking leads to a nonsoftening polymer. The rigidity of polymeric chains depends on the level of energy that prevents chain bending, which is comparable to the cohesion energies. Large or polar bonds, or aromatic portions, can prevent such bending. The existence of strong intermolecular attractions is also favourable to thermostability.

11.7.3 Cracking and depolymerization

The creation of free radicals when a high temperature is reached begins with the breaking of the weakest bonds, but continues via a chain mechanism. In addition-polymers these free radicals can catalyse a rapid chain depolymerization at around 280–300°C. The simplest way to avoid this chain depolymerization is the use of polycondensation, because these polymers can only depolymerize if they are in contact with the products of polycondensation.

11.8 Combustion

The factors discussed above govern the thermal performance of polymers that have not reached the point of combustion; this aspect of thermal degradation will now be discussed. It is convenient to consider polymer flammability not as a phenomenon resulting from exposure to intense heat flux in the sequence of events described before, but rather as a particular phenomenon governed by different reaction mechanisms. While thermostability is always an inherent characteristic of the polymer structure, nonflammability may be linked to the use of nonconstitutive elements in the specific polymeric chain. These additives are fire-resistant agents. Nevertheless, some polymers exist in which nonflammability is due to the constitution of the polymer itself. This is particularly the case of halogenous polymers. It is important to note that although thermostable polymers generally have acceptable fire performance, the converse is far less applicable in that only a few fire-proofed polymers can be considered as thermostable.

When heat or a flame raises the polymer to a sufficient temperature, the weakest bonds are broken and thermal degradation starts, increasing more or less quickly according to the composition. When a temperature specific to the particular material is reached, the strongest bonds also break. This is the decomposition stage, accompanied by volatilization. In contact with the combustion agent and at a temperature that depends on their specific nature, the volatile gases produced during decomposition catch fire: the combustion stage starts and then flame propagation. The maintenance of combustion requires that the available energy (heat of formation of combustion products, minus radiated energy) is sufficient to secure the decomposition and volatilization of more decomposable material. It also requires adequate supply of the combustion agent. Factors that influence the different stages of the combustion process are given in Table 11.2.

The process of combustion is easier if melting occurs. Depending on whether the polymer is thermoplastic or cross-linked, when treated it will either melt to generate droplets or simply deform. A strong tendency for the formation of droplets can increase the susceptibility to the action of heat in the first stage, particularly for highly crystalline polymers with a relatively low melting point.

In the second step (thermal degradation) there may occur reorganization

Table 11.2. Factors that influence different stages of the combustion process

	Combustion stage	Influencing factors
I	Heating	Specific heat
		Thermal conductivity
II	Degradation/decomposition	Critical temperature of decomposition
		Latent heat of decomposition
III	Volatilization	Critical temperature of volatilization
		Latent heat of volatilization
IV	Flammability	Characteristics of polymer
		(flammability temperature, nature of volatile gas, etc.)
		Combustion agent
		Source of flame
		(temperature, sparks, flame, etc.)
V	Combustion	Heat transfer rate to the material
		Kinetics of degradation
		Nature of volatile gas
		Availability of combustion agent

of the structure with the formation of stronger bonds (polyacrylonitrile), depolymerization (PTFE), or total destruction (polybutadiene). In certain cases, the processing of polymers can endow specific behaviour (for example, shrinkage of fibres that have been drawn).

During combustion a bulk piece of polymer exhibits four regions: (1) an area of virgin polymer heated by conduction and radiation; (2) an area undergoing degradation which can be divided into two parts—one of initial degradation in which the polymer aspect may still be preserved and the other damaged, with a charred aspect; (3) the flame area in which volatile products stemming from the solid area of degradation undergo oxidation and complete their degradation, resulting in combustion products of a low molecular mass; and (4) the area above the flame in which by-products of the combustion are diluted and cooled and where recombination reactions occur that destroy free radicals.

It is impossible to obtain a completely nonflammable polymer, because of the organic nature of the material. What can be done is to reduce its ability to ignite and to propagate flame. Methods of improving fire performance can be grouped as follows: (1) changing the structure of polymer in order to make it more thermostable; (2) covering the polymer with a fireproof product after processing; and (3) introducing a fireproofing agent in the polymer before processing. These fireproofing agents help by prevention of decomposition and oxidation reactions, prevention of the evolution of combustible gas, the presence of protective charred residues and a fall in temperature by endothermic reactions.

Of these three methods of improving fire resistance, the first, concerning the reinforcement of thermostability, has already been discussed. The second one, concerning fireproofing surface treatments, relates essentially to natural fibres. The third method, consisting of adding fireproofing agents to the polymer, will be further elaborated below.

The function of fireproofing agents in relation to the stages of combustion in Table 11.2 is as follows.

— *Stage I* (*Heating*) Development of a vitreous film of low thermal conductivity; endothermic reaction at low temperature producing noncombustible by-products; production of a swollen layer providing protection by its low thermal conductivity.

— *Stage II* (*Degradation/decomposition*) Transformation in the degradation process, restricting the development of chain reactions; formation of heavy gas, slowing down the supply of combustion agent.

— *Stage III* (*Volatilization*) Entrapment of free radicals, restricting the development of chain reactions.

— *Stage IV* (*Inflammation*) Formation of heavy gas, slowing down the supply of combustion agent.

— *Stage V* (*Combustion*) Formation of particles and development of fumes that reduce thermal radiation and fire propagation.

Fireproofing agents behave in two ways:

1. Reactive fireproofing agents (products with a reactive group). These are introduced into the structure of the macromolecule via the synthesis of monomers or in the polymerization process. Their action is thus additional to fire resistance itself since they also affect other characteristics (mechanical, thermal, etc.). On the other hand, they offer absolute certainty of a permanent effect.
2. Fireproofing agents added as blending agents. These can be added to the raw materials, before the production of chips for example, i.e., during formulation. But they can also be added later in production, i.e., during the use of the polymer. The second case is more flexible than the first; however, as they are not chemically fixed, their stability is not so great. Organic and inorganic materials are used, and in some cases, particularly to develop synergies, several can be added at the same time.

11.8.1 Fireproofing agents with physical action

This range of agents mainly includes hydrates and hydroxides (alumina, magnesia, etc.), borates, silicates and carbonates.

Alumina trihydrate decomposes on reaching 200°C according to the reaction

$$2Al(OH)_3 \rightarrow Al_2O_3 + 3H_2O \qquad (-297 \text{ kJ})$$

Thus its activity is linked to the endothermic nature of the reaction, which uses part of the heat of combustion; to the release of water, which dilutes the combustible gas; and to the conjunction of both actions in creating charred residues that are not completely burned, to form a protective layer.

Magnesium hydroxide works in the same way as hydrated alumina, but decomposition occurs at a higher thermal level (around 340°C). It is consequently used with polymers having higher temperatures of degradation.

Zinc orthoborate also releases water during transformation into metaborate and boric anhydride. This anhydride, which is thermally highly stable, is formed at a high temperature. Moreover, it leads to the creation of a protective glass, which has the property of dissolving metallic oxides by forming borates. Thus it functions during degradation by using radicals initiated by the oxidation–reduction reactions of metals (for example, metal introduced at the polymerization stage which may also help depolymerization when heated).

Silicates as well as borates can form a protective glass on the surface of the decomposing polymer, slowing down the spread of the flame and limiting the supply of combustion agent in the spreading area.

Carbonates, which are widely used, are often added in large quantities to the polymer in order to reduce its costs. In terms of fire resistance they produce a dilution of combustible elements in the solid as well as the gaseous state.

11.8.2 Fireproofing agents with chemical activity

Phosphorus. Organophosphorus products act in the pyrolysis area, where they encourage the formation of a charcoal-like layer (limitation of combustion agent supply, formation of CO to the detriment of CO_2). At a high temperature, phosphorus products decompose into phosphorus oxides and phosphoric acid. Phosphorus encourages the direct decomposition of the polymer into heavy and nonvolatile fragments that play no part in keeping the flame burning. The last stage of the degradation of these fragments is a residue with a high carbon content, mentioned above, that protects the lowest layers. Phosphoric esters (such as isodecyldiphenylphosphate) are substituted for the usual plasticizing agents when nonflame characteristics are needed.

Halogens. The fireproofing activity of the halogens ranks as follows:

$$I > Br > Cl > F$$

They display two types of fireproofing action:

1. Formation of heavy and nonflammable halogenated products that dilute the combustible gas resulting from the degradation of the polymeric chain. These nonflammable gases slow down the supply of combustibles that might keep the flame burning.
2. Formation of halogen hydrides, which inhibits the radical reactions.

$$HO\cdot + HBR \rightarrow H_2O + Br\cdot \quad \text{(Initiation)}$$

$$Br\cdot + RH \rightarrow HBr + R\cdot \quad \text{(Regeneration)}$$

The disappearance of OH. radicals in the degradation products (even if they transform into other radicals such as $R\cdot$) is very important because their concentration quickly increases according to the following process:

$$HO\cdot + H_2 \rightarrow H_2O + H\cdot$$

$$H\cdot + O_2 \rightarrow HO\cdot + O\cdot$$

$$O\cdot + H_2 \rightarrow HO\cdot + H\cdot$$

In contrast, the $R\cdot$ radicals play a smaller role in the propagation of degradation because they take part more easily in molecular reorganization. For example,

$$R\cdot + R'\cdot \rightarrow R\text{--}R'$$

$$R\cdot + HR_1R_2\cdot \rightarrow RH + R_1 + R_2$$

In the pyrolysis area, the presence of halogenated fireproofing agents helps the formation of CO to the detriment of CO_2, which reduces thermal emission by $290\,kJ\,mol^{-1}$. The conversion is shown in Fig. 11.3. The pyrolysis area is covered with a tar-like layer, reducing the flow of

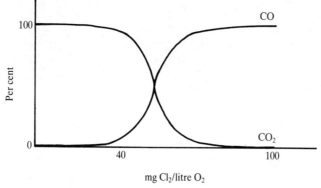

mg Cl_2/litre O_2

Figure 11.3 Conversion of CO/CO_2.

combustive material. Among halogenated products, aromatic structures (decabromodiphenyl, tetrabromobisphenol A, etc.) possess good heat and hydrolysis stability.

Synergy of antimony/halogen. Antimony used alone has no fireproofing effect, but when it is associated with halogen a synergy is developed with the formation of antimony halides. These halides inhibit the action of free radicals with a greater efficiency than halogen hydrides or phosphorus oxyhalide. The formation of halides and oxyhalides, depending on the temperature, is given by the following reactions:

$$10RCl + 5SB_2O_3 \longrightarrow 10SbOCl + 5R_2O$$

$$5SbOCl \xrightarrow{245-280°C} Sb_4O_5Cl_2 + SbCl_3$$

$$4Sb_4O_5Cl_2 \xrightarrow{410-475°C} 5Sb_3O_4Cl + SbCl_3$$

$$3Sb_3O_4Cl \xrightarrow{475-565°C} 4Sb_2O_3 + SbCl_3$$

The action of $SbCl_3$ produces breaks in the degradation mechanism where radicals assist propagation:

$$SbCl_3 \rightarrow SbCl_2{\cdot} + Cl{\cdot}$$

$$R-CH_3 + OH{\cdot} \rightarrow R-CH_2{\cdot} + H_2O$$

$$R-CH_2{\cdot} + Cl{\cdot} \rightarrow R-CH_2Cl$$

The effect of the coupling of antimony and halogen may be considered to be due to the disappearance of OH· radicals according to $RCl + OH{\cdot} \rightarrow ROH + Cl{\cdot}$. The synergy developed by Sb_2O_3 in association with halogenated agents is evident only at low contents of 3–5 per cent in the polymer.

Synergy of phosphorus/halogen. The simultaneous use of phosphorus and halogenated fireproofing agents produces new substances containing phosphorus–halogen linkages. As already noted, phosphorus products help the appearance of a charcoal-like surface which limits the formation of volatile materials in the flame. The existence in this protective layer of halogenated products allows the formation, when heated, of phosphorus trihalides, pentahalides and oxyhalides. These are heavier and less volatile than halohydrides, and are therefore more effective. Moreover, these products are efficient in trapping free radicals. However, phosphorus agents can be inconvenient in that they are not easily retained by the polymer, which is a considerable limitation when fibres must be subjected to washing.

11.9 General properties of thermostable fibres

Polymers that are thermostable usually have a polyaromatic or hetero-cyclic structure and the process of formation into fibres requires the use of a highly polar solvent. When it is possible to colour them (although this is not always possible, e.g. Kevlar, oxidized PAN, etc.) the resulting colours cannot be pale. This is true even for those that are dyeable (e.g. Kermel). The fibres usually have a higher modulus than conventional synthetic fibres. In some cases this modulus is much higher (e.g. *p*-aramids). It is not possible to melt them. At high temperature or when a flame is used, degradation usually occurs (except for Ryton, Teflon, etc.). In a flame the fibres do not usually release fumes or toxic gases (apart from nonflammable fibres like chlorofibres and fireproofed polyesters). They have good fire resistance and their LOI level is high. They maintain their mechanical properties over a large range of temperatures and keep their length in wet conditions and when heated, because shrinkage is usually low. Moisture regain is comparatively low, except for Kynol, and, above all, for PBIs (13–15 per cent at 20°C and 65 per cent RH). Their resistance to chemical agents is quite good, especially to solvents; behaviour in acidic or basic environments is more variable.

11.10 Thermostable and fire-resistant fibre groups

11.10.1 Meta-aramid fibres

Nomex, a member of the aramid family of fibres, was developed by Du Pont in the 1960s for applications requiring good textile properties, dimensional stability and heat resistance. It is available as staple. The fabrics are used in protective clothing for firemen and policemen, for gas and electric utility operatives, for workers with rocket and petroleum based fuels, molten metals, etc. Nonwovens are used for hot gas filtration and thermal insulation. Nomex paper is made entirely of aramid polymer in two forms: short fibres (flock) and microscopic fibrous binder particles (fibrids). They are combined into a sheet structure by normal papermaking methods. During this process the fibrids join together to form filmy webs in the spaces between the flock fibres. This paper may be subsequently calendered at a high temperature and pressure to lock the constituents permanently together and produce a relatively impermeable sheet with high dielectric strength and good mechanical strength. Because of its combination of electrical properties, temperature resistance and flexibility, Nomex paper is widely used in electrical insulation. Another interesting use for paper is honeycomb core for composite materials when a high strength-to-weight ratio is needed. Nomex filament yarns are used in some applications where good aesthetics, or the higher strength of filament yarns compared with staple, are required. They include industrial coated fabrics, electrical insulation, hot gas filter bags, rubberized belting and hoses.

Nomex in an aromatic polyamide, more precisely a poly(*m*-phenylene isophthalamide), which results from the condensation of *m*-phenylene diamine with isophthaloyl chloride [I]. The fibre is made by dry spinning.

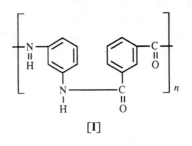

[I]

It has useful mechanical properties at temperatures above the melting point of most synthetic fibres. Even after longer periods of exposure to high temperatures in dry air, Nomex retains a relatively high strength. It has a low level of flammability, and does not melt and flow at a high temperature. However, at temperatures above 370°C, the fibre degrades rapidly to a fragile charred state at a rate proportional to the intensity of the heat source. Fibres, fabrics, and paper made of Nomex may be flammable in the presence of oils, resins or sizing materials.

Mixing Nomex with flammable fibres makes the blended Nomex itself flammable (this is true for all the thermostable fibres). Laundering or dry-cleaning does not significantly change the flammability characteristics of 100 per cent Nomex fabrics.

Another meta-aramid fibre is Conex. It is manufactured by Teijing from poly(*m*-phenylene isophthalamide), similarly to Nomex, but uses a wet spinning process. It has the same properties and the same end-uses.

The addition of Kevlar to *p*-aramids (e.g. Nomex III is a blend of 95 per cent Nomex and 5 per cent Kevlar) increases the strength during exposure to flame, and provides lightweight fabrics. These blends are used, for instance, as a fireblocker liner in aircraft seating, for protective apparel, and in barrier fabrics for home furnishing.

11.10.2 Polyamideimide

Rhône-Poulenc has produced a polyamide fibre, Kermel, since 1972. It is made by the condensation of trimellitic anhydride and diisocyanatodi-phenyl methane. It has alternative amine and imide groups [II]. The fibre is produced by wet spinning from a highly polar solvent. In order to

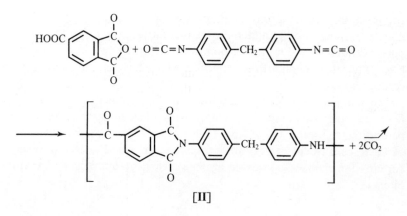

[II]

give the filaments their final properties, they are stretched under favourable plasticity conditions (the correct level of solvent and temperature). The colour of the fibre depends on the choice of solvent and on the stretching temperature. A light-coloured fibre is now available instead of the former straw-yellow fibre.

This fibre has a good tensile strength and a low modulus of elasticity, and does not pill. Its high LOI value of 32 per cent is well suited for textile uses such as protective clothing for heat and fire resistance. It retains its mechanical properties at temperatures far above the melting point of most synthetic fibres. Extremely high temperatures would be required to produce complete physical degradation of the fibre. At 400°C in air, with a temperature increase of 5°C per minute, there is a weight loss of only 5 per cent.

Kermel fibre does not melt and form droplets, but carbonizes. If a Kermel fabric is exposed to a flame, it first chars and then stiffens slightly without significant shrinkage or the formation of droplets. The fabric loses its mechanical properties but totally retains its physical structure—a very important consideration for protective garments. The carbonization of Kermel (type V) fibre is a function of temperature and exposure time to the heat source: 420°C, 1 min; 400°C, 3 min; 350°C, 8 min; 300°C, 20 min; 250°C, no carbonization.

Kermel woven and nonwoven fabrics offer a high level of thermal insulation against flame and radiant heat. With a protective garment made of 250 g m^{-2} Kermel fabric and under a heat flux of $Q = 80 \text{ kW m}^{-2}$ (temperature 1040°C) the pain protection is 5.5 seconds and the second-degree burn 8 seconds (giving an escape time of 2.5 seconds). With the complementary protection of an undergarment, these values become 14.5 and 24 seconds (i.e. escape time of 9.5 seconds). Kermel has a high LOI of 32 per cent.

Fire risks from textile fabrics can arise not only through the combustibility of the fabrics themselves but also through their ability to initiate burning of other materials. Kermel fibres have a very low level of post-incandescence. During carbonization, Kermel releases a very low level of fumes, in comparison with other fibres such as polyester, viscose, cotton and acrylics.

The main use for Kermel fabrics is in fire protective apparel. This fibre has the advantage of being easily blended with other fibres, permitting the production of fabrics that embody the respective advantages of each constituent fibre. Other uses are in carpets, wall-coverings, curtains, fire blankets, nonwoven insulation elements and high temperature filters.

11.10.3 Polyimide

Fully imidized polyimide fibre has been developed by Lenzing AG. It is produced from benzophenonetetracarboxylic acid dianhydride and a mixture of aromatic diisocyanates (toluene diisocyanate and methylene diphenyldiisocyanate) [III].

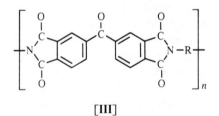

[III]

For many years this polymer was used in powdered form for making pressed and sintered components. However, the polymer is soluble in some strong polar organic solvents like dimethylacetamide, dimethylsulphoxide, N-methylpyrrolidone and dimethylformamide, and so can be dry spun. The fibre does not melt and starts to carbonize at temperatures over 450°C with low generation of smoke and toxic gases. Its property of non-flammability is due to the aromatic, halogen-free polyimide structure, which gives it a high LOI of 36–38 per cent. In constant use it does not exhibit any major changes in air temperatures up to 260°C over a considerable period of time.

Its applications include protective clothing, high temperature filtration (needle-punched felts), sealing and packing material, braiding with impregnated yarns (PTFE dispersion, e.g. to reduce friction coefficient), replacement of asbestos, aerospace and aircraft uses.

11.10.4 Polybenzimidazole

Celanese developed PBI in 1963. The polymer is prepared from tetra-aminobiphenyl and diphenyl isophthalate [IV]. The PBI polymer in a

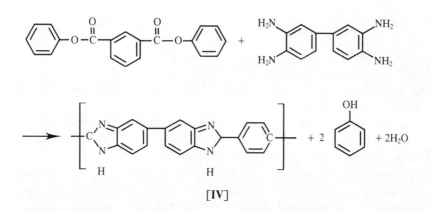

[IV]

solution of dimethylacetamide containing lithium chloride is spun into fibre via a dry spinning process in hot nitrogen. Washing and drying allow extraction of the solvent and lithium salt.

PBI fibre does not burn in air. Its LOI is greater than 41 per cent. It does not continue burning, and when charred it retains its integrity and flexibility with little shrinkage. It emits little smoke on exposure to flame. It has very good thermal stability and strength retention over a wide range of temperatures and environments. For example, PBI withstands temperatures as high as 600°C for short-term exposures (3–5 seconds). It does not melt or form droplets. It can survive for one week at 300°C, and shrinks only 3 per cent in a 24-hour exposure at 315°C. In low oxygen environments, PBI has remarkable stability and strength retention. For instance, under vacuum there is no change in mechanical properties after 300 hours at 350°C.

Since its origin in the NASA space programme, PBI has extended its usefulness to encompass a broad spectrum of applications in aerospace, and it is now used in the American space shuttle. A current area of interest is in the replacement of asbestos-reinforced rubbers used in rocket motors and boosters to control ignition. Classical applications include fire-blocking fabrics in aircraft seats, firefighters' protective clothing, industrial work apparel, racing-car driver suits and military flying suits. Aluminized PBI knits are used for crash rescue garments. It is also used in glass-handling fabrics, conveyor belts, molten metal applications and as an asbestos substitute in braided valve packing.

11.10.5 Novoloid fibres

In 1968, the American firm Carborundum fibre-formed a phenolic resin on a laboratory scale and named it Kynol. This phenolic resin fibre, generally known as novoloid fibre, was developed in response to a request from NASA for a cheap carbon fibre replacement. However, the resultant product did not fulfil the required characteristics so could not be applied to its intended use. Because of its strong flame resistance, research continued into its development, but industrial production technology was never established successfully in the United States, resulting in its transfer to Japan in 1972 to Gun Ei Chemical Industry. Industrial production of Kynol began in 1986.

In the first step of Kynol production, phenol and formaldehyde are combined with an acidic catalyst to produce a novolac type of phenolic resin. This has a very high stability to heat, making it well suited to melt spinning. The uncured novolac fibre is then cured with formaldehyde through an acidic reaction similar to the first stage, creating the final product. It is completely noncrystalline and is a three-dimensional network structure.

The recommended thermal limits are 150°C in air and 250°C in an inert atmosphere. The LOI is 30–34 per cent. Kynol does not melt under any circumstances when exposed to flame or heat over 600°C, and gradually carbonizes to form carbon fibre. The low smoke emission when it is burned reduces the dangers that result from obscured vision during a fire. The products of combustion are principally water vapour and carbon dioxide with virtually no toxic gas. Another novoloid fibre is Philene (Isover St Gobain).

Applications are for carbon fibres for packings, conductive textiles, composites, activated carbon for odour elimination, solvent recovery, contaminant removal and electrodes for batteries. Kynol is used in fire safety equipment for racing-car drivers, for firefighters and as a protection against chemicals or molten metals. Products include needled felts for insulation and fire protection in flame barriers and garment linings, yarns for insulation and cushioning of electric cables. The fibres are used to replace asbestos fibres in heat-resistant and chemical-resistant sealing materials (gaskets and braided packings), and as heat friction materials in solid-fuel rockets.

11.10.6 Polyacrylate fibres

Courtaulds produces Inidex fibres based on cross-linked polyacrylonitrile chemistry together with metal complexing. This combination of organic and inorganic chemistry leads to a fibre that combines flame resistance, very low emission of smoke and toxic fumes, nonmelting behaviour and resistance to a wide range of chemicals including acids and alkalis. Its LOI is 40 per cent. It is used for the production of fire protective fabrics or blankets, soundproofing, wadding, filtration of hot gases (owing to the good heat resistance of cross-linked polyacrylonitrile to numerous reagents), protective clothing, and in the reinforcement of plastics and cements. Inidex is well suited to use in confined spaces (low smoke emission) or areas rich in oxygen, because of its high LOI.

11.10.7 Fluoropolymers

The first fluoropolymer fibre, Teflon, was developed by Du Pont in 1953. It is characterized by chemical inertness, resistance to flexing, and excellent electrical insulation properties. Now many other types are produced that may have some, but not all, of these properties: for example, Halar, made from poly(ethylene–chlorotrifluoroethylene); Knar, made from poly(vinylidene fluoride); Tefzelk made from poly(ethylene–tetrafluoro-ethylene); and Hostaflon, among others.

Manufacturers other than Lenzing AG, who use a split film process, produce PTFE monofilament via a spinning process. It is difficult to make fibres from poly(tetrafluorethylene) since it is insoluble in all solvents and cannot be melt spun, as heating decomposes it before it becomes fluid enough to spin. These difficulties are overcome by spinning from an aqueous dispersion of the polymer. Spinning into an acid solution precipitates the PTFE fibre, which is washed, dried and sintered briefly by running over a surface held at 385°C. It is then cold-stretched to about four times its original length. Fluorocarbon fibres other than PTFE are produced by conventional monofilament extrusion equipment, since they can be handled as a melt.

Fluorocarbon fibres are used for filtrations that require maximum resistance to high temperatures and chemical resistance to acid and alkali; as substitutes for asbestos in packing and gasket tapes; for protective clothing; for packing for distillation columns; and for nonlubricated bearings for high loads.

11.10.8 Polyphenylene sulphide (PPS) fibre

PPS fibres are produced by Phillips from Ryton. The molecular structure is shown in [V]. Its LOI is 34 per cent.

PPS fibres are used for filtration, sewing threads, laundry materials, in the rubber industry, and in blends with other fibres such as carbon, in which the thermoplastic PPS fibres are used as binders.

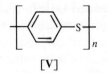

[V]

11.10.9 Poly(ether ether ketone) (PEEK) fibres

The molecular structure of PEEK is shown in [VI]. The fibre has an extremely high resistance to acids, alkalis and organic solvents, even at elevated temperatures. Its low flammability and LOI of 33–34 per cent make it flame-resistant.

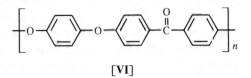

[VI]

PEEK fibres are used for filtration, dryer canvas, conveyor belts, sewing threads, sealing materials, and composite materials using fabrics made of PEEK and carbon or glass.

11.10.10 Polyetherimide (PEI) fibres

The molecular structure of PEI is shown in [VII]. PEI has a good resistance to a wide range of chemicals, except for highly concentrated

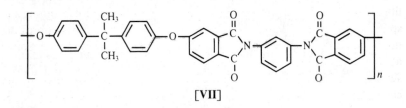

[VII]

alkalis and halogenated solvents. It is flame-resistant and produces little smoke. Its LOI is 47 per cent. The fibre is used for advanced thermoplastic composites via fabrics made of PEI fibres and carbon or glass fibres.

11.10.11 Oxidized polyacrylonitrile fibres

This fibre is produced by heating polyacrylonitrile (PAN) fibre in air. The carbon content is reduced from 67 per cent in PAN to 63 per cent in partially oxidized polyacrylonitrile (POA). POA fibres possess acrylic-like strength and textile processability, yet—like carbon fibres—they do not burn or melt. Examples of products are Dolamite, Sigrafil, Graphil and Avox.

POA fibres can be used as an asbestos replacement in friction materials such as brakes and clutch linings. In brake lining, 20–30 per cent of POA fibre is mixed with inorganic and metal particles and then with resin binders. These fibres are also used in protective garments for firefighters, steelworkers, etc., although the abrasion resistance is relatively low.

11.10.12 Chlorofibres

These fibres are based on poly(vinyl chloride) or poly(vinylidene chloride) polymer. The US Federal Trade Commission defines vinyon fibres as 'a manufactured fibre in which the fibre-forming substance is any long-chain synthetic polymeric composed of at least 85% by weight of vinyl chloride units $-CH_2-CHCl-$'.

Variants are homopolymer fibres (atactic and syndiotactic), postchlorinated homopolymer fibres and copolymer fibres.

Poly(vinyl chloride) fibres

Homopolymer fibres are made from a polymer containing 100 per cent vinyl chloride units and have a chlorine content of about 57 per cent. The atactic vinyon fibre is produced by a dry spinning method from a polymer solution, with a mixture of acetone and carbon disulphide (in Europe) or a mixture of benzene and acetone (in Japan). The syndiotactic vinyon fibre is produced by a wet spinning method, with hot cyclohexanone, and spun into a bath of water, cyclohexanone and ethanol. These fibres are nonflammable, which is their main characteristic. However, they soften at low temperatures (glass transition temperature of about 85°C), and swell in the presence of chlorinated solvents (such as those used in dry cleaning). Examples are Rhovyl, Viclon, Teviron, etc. The LOI is 35–37 per cent.

Homopolymer fibres are used for warmth retention (underwear, nightwear, tights) permeability (underwear, foundation garments, disposable nappies), negative electric charge (health wear), nonshrinkability (curtains, bath mats, hats) and nonflammability (curtains, toys, car mats, sheets, nighties, blankets).

Postchlorinated fibres are made from homopolymer that reacts with chlorine gas until it reaches a chlorine content of 62–65 per cent. This polymer may be dissolved in acetone, and fibres are produced by wet spinning. The solution is extruded into a bath of water.

Poly(vinylidene chloride) fibres

This polymer is composed of at least 80 per cent by weight vinylidene chloride units ($-CH_2-CCl_2-$). Fibres are melt spun, stretched, heat-treated and cut into staple, e.g. Saran (Dow).

Copolymer vinyon fibres

Copolymer vinyon fibre is made from a vinyl chloride–vinyl acetate copolymer. The copolymer is dissolved in acetone, and fibre is produced by dry spinning. The vinyl chloride content of the copolymer vinyon fibre must be over 85 per cent (which corresponds to a chlorine content over 47 per cent) to meet the Federal Trade Commission definition of vinyon fibres. A copolymer fibre is characterized mainly by its lower strength and softening point compared with the other types of vinyon. However, the low softening point makes it ideal for use in heat-sealable papers, accounting for 80 per cent of total consumption. Examples are Vinyon HH (Hoechst Celanese) and MP Fibre (Wacker Chemie).

Polychlal fibre

Polychlal fibre is the name given to a fibre based on poly(vinyl chloride) and poly(vinyl alcohol). Cordelan is 50 per cent poly(vinyl chloride) and 50 per cent poly(vinyl alcohol), but also contains a small amount of poly(vinyl chloride)–poly(vinyl alcohol) graft polymer. It is produced by emulsifying and polymerizing vinyl chloride in a poly(vinyl alcohol) solution. By extruding the spinning solution into an aqueous coagulating bath of sodium sulphate, a fibre having a poly(vinyl alcohol) substrate (in which poly(vinyl chloride) is dispersed) is produced. The graft polymer is at the interface of the two components of the matrix. The fibre is then stretched and heat treated to improve its tensile strength. Because the poly(vinyl alcohol) substrate is water-soluble, the fibre is treated with formaldehyde to form cyclic acetal groups along the polymer chain and cross-links between the polymer chains to insolubilize the poly(vinyl alcohol) fibre substrate.

The polychlal fibre derives its characteristics from each of its chemical components: the poly(vinyl alcohol), which is the continuous part, contributes to most of its basic physical properties, while poly(vinyl chloride) makes the fibre self-extinguishing when ignited. Polychlal fibre is fire-resistant. It is resistant to wet heat only up to 90°C and changes colour at 130°C when subjected to dry heat. These temperature limits are those for the syndiotactic homopolymer vinyon fibre. It is used when nonflammable qualities are required, such as in children's sleepwear, blankets, curtains, upholstery and some industrial applications.

Modacrylic fibres

Modacrylic fibres are defined by both the US Federal Trade Commission and the International Organization for Standardization as being composed of at least 35 per cent but less than 85 per cent by weight of acrylonitrile units. The remaining 15–65 per cent of a modacrylic fibre is not specially defined. However, commercially speaking, this consists primarily of halogenated co-monomers, such as vinylidene chloride, vinyl chloride, or vinyl bromide. The composition of modacrylic fibres makes them different from chlorofibres. Modacrylic fibres resemble acrylic fibres in most respects, but because of their halogenated co-monomer content, they have inherent flame-retardant characteristics. Examples are Velicren and Teklan.

Flame-resistant modacrylic fibres are becoming increasingly important, especially for use in home furnishings and simulated furs. Other uses are in carpet production, where a mixture of modacrylic and acrylic fibres may often be sufficient to meet flame resistance standards; in blankets for hotels and airlines, in toys (such as dolls' hair) and in air filters, military fabrics, overalls, and so on. The LOI is 28–32 per cent. A recent improvement (Lufnen) has an LOI of 35 per cent.

11.10.13 Flame-retardant polyester fibres

Producers of polyester fibres have been trying to improve the flame resistance of their fibres. The result of this research is the use of a phosphorus-containing co-monomer, such as a derivative of phosphonic acid, e.g. for Trevira [VIII].

$$\text{HP—}\overset{\overset{\displaystyle O}{\displaystyle |}}{\underset{\underset{\displaystyle Y}{\displaystyle |}}{\text{P}}}\text{—X—COOH}$$

[VIII]

In order to modify a polyester, the co-monomers must have two groups that are able to form esters. During manufacture of the polymer by polycondensation, temperatures up to 300°C and vacuum are required. A suitable co-monomer must be stable under these conditions. The vapour pressure must be low. The polycondensation reaction must not be affected by the co-monomer. The formation of by-products such as diethylene glycol should not be promoted. The phosphorus content is chosen to be sufficient to meet the different flammability tests but low enough to keep the textile properties almost unchanged.

Whilst Teijin (Extar), Toyobo (Heim or Toyobo GH), Toray (Uniflam), Du Pont (Dacron) and Hoechst (Trevira) have based the development of their flame-resistant fibres on the use of phosphorus, Snia Viscosa has chosen the use of co-monomers based on bromine and sulphur, present in percentages of 4.5 and 0.3, respectively, in their Wistel FR fibres.

The processing stages of polymerization, melt spinning and drawing of fire-resistant polyesters are almost the same as for conventional polyesters, and the physical properties are the same.

The most important end-uses are in furnishing fabrics for curtains, upholstery and wall coverings. The risk of fire in hotels and hospitals has stimulated a great interest in the use of flame-retardant polyester in mattresses, bed linens, pillows and blankets. For the same reasons, railway companies are using seat covers, curtains and carpets made of flame-retardant polyester fibres.

Further reading

H Breulet, *Ignifugation des textiles*, Institut Textile de France, 29–30 June, 1988.
Hi-Tech Textile 1987, JTN (1987).
J Preston (ed.), *High Temperature Resistant Fibres From Organic Polymers*, Applied Polymer Symposia, Chapter 9 3.20 (1969).
F Rocaboy in *Comportement thermique des polymères synthétiques*, Masson et Cie, Paris, 1972.
N Valentin, *Connaissance des fils composites et des fils à usage technique*, Institut Textile de France, 24/25 March, 1988.
K Wheeler, W Cox, M Tashiro, *Specialty Organic Fibers*, CEH Marketing Research Report, SRI International, 1988.

CHAPTER 12

Specialized uses 1: engineering uses of textiles

D W LLOYD

12.1 Introduction

It is appropriate to begin a consideration of engineering uses of textiles by defining what constitutes an 'engineering textile'. A useful working definition is that an engineering textile is a material composed to a significant extent of fibres, where the fibrous component is processed at some stage by a recognizable textile processing route, where the material is selected for an application primarily on the basis of performance rather than aesthetics, and where, at some stage of its life, use is made of its (mechanical) flexibility.

All definitions are exclusive; this definition excludes fashion apparel, domestic household textiles and decorative textiles. It also excludes some engineering products, such as injection moulded plastics reinforced by fibres too short to have been processed by textile machinery. The definition includes such a wide spectrum, however, that it still presents difficulties of classification. A simple division by form and structure is helpful in understanding the broad range of products that must be included. The primary form of the engineering textile may be one-, two- or nearly three-dimensional (see scheme 1). Thus, fabrics are primarily two-dimensional, ropes and cords one-dimensional. Such textiles may have a secondary form when considered in more detail; thus some fabrics have substantial thickness and are more nearly three-dimensional, as are some heavy ropes. Some products have components other than the textile, so that the engineering textile product has a different form from its textile component, as, for example, in the case of an industrial V-belt reinforced with textile cords.

Engineering textiles have a long history. Cave paintings in eastern Spain dating from late palaeolithic or mesolithic times show a person climbing down the face of a cliff to collect wild honey, using what appears to be ropes. However, it is only with the wide availability of synthetic fibres in the last half century that many applications have begun to reach their full potential, or even become possible at all. The dominance of synthetic

Secondary form

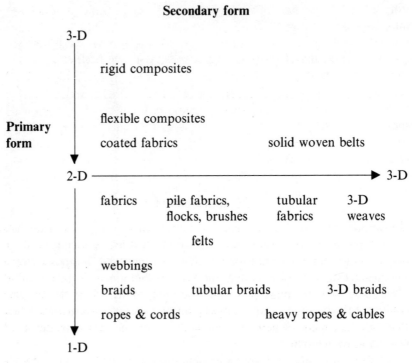

Scheme 1

Outline classification of engineering textiles by structural form

fibres may be attributed both to the superior properties that they offer compared to natural fibres, and to new structural forms, such as spun-bonded or melt-blown nonwovens, that they have made possible.

Constraints on space preclude any attempt at a comprehensive survey of engineering uses of textiles. Even attempts to classify applications by material, producer industry, or end-user sector, would be unsatisfactory, giving rise to ambiguities. For example, carpets, as pile fabrics, fit within the classification in the context of aircraft fittings, where fire resistance, light weight and durability are paramount, but carpets in the context of low-traffic areas in the home are excluded, as aesthetics are the dominant factor in their selection. As a consequence, a limited number of examples of engineering applications will be covered in greater detail to illustrate broader principles and to give some idea of the wide range of end-uses of engineering fabrics made from synthetic fibres.

However, before considering particular examples in detail, it is worth listing some of the applications of the different engineering textiles identified above. Rigid composites consist of a resin that becomes rigid on curing, reinforced by fibres, yarns or fabrics. Fibres may be chopped from a heavy tow and sprayed with the liquid resin, or be consolidated into a tissue sheet or heavier fibrous web to facilitate handling. Yarns are normally continuous filament bundles with little or no twist, whilst fabrics are either plain woven or sateen weave (to give straighter yarns and a stiffer composite). Glass fibres find the widest application on the basis of cost, in end-uses as diverse as aerospace, motor vehicles, ship and boat building, and street furniture. Carbon and aramids find application in specialist areas such as aerospace, high performance boats, and high performance cars and racing cars. Composite materials offer the ability to make complex, three-dimensional shapes without expensive press-tools, a combination of light weight, high strength and high stiffness, and greater fatigue resistance than metals. The transparency of composites (except carbon composites) to radar is of particular importance in the production of radomes.

Flexible composites involve the reinforcement of rubber, synthetic rubber, or other flexible polymer with one or more textile components. The role of the textile typically is to prevent excess extension or distortion. The most common flexible composites are automotive tyres, which are reinforced by cords made from polyamide, polyester or aramid. Other examples include power transmission belts, and hoses—from garden hoses, through automotive radiator hoses, to hoses used to load and unload supertankers.

Flexible composites merge, as the textile becomes the dominant component, into the category of coated fabrics. Base fabrics for coating are normally woven, though some warp knitted and nonwoven fabrics find application in this area. Polyamide, polyester, aramid, glass and poly(vinyl alcohol) are used as base fabrics, with poly(vinyl chloride), polyurethane, polytetrafluoroethylene, rubber and synthetic rubber coatings. Applications for coated fabrics include lightweight architectural structures, aircraft escape slides, inflatable boats and life-rafts, hovercraft skirts, awnings, salvage bags, protective clothing, vehicle airbags, airships, water-proof or weather-proof seals, as for example on aircraft loading ramps, flexible sides for cargo containers and lorries, intermediate flexible bulk containers, and conveyor belts in applications ranging from food processing to mineral extraction.

Fabrics probably form the largest and most diverse group of engineering textiles, with all synthetic fibre types represented. Woven fabrics find application as geotextiles (in polypropylene, polyester, and to a lesser extent, polyamide), and as filters (in polypropylene, polyamide, acrylic, polyester, aramid, PTFE, polyphenylene sulphide, or polyimide, depending on the chemical and temperature operating environment) for products as diverse as milk and flue gases. They are used to protect crops from sunlight, wind and birds. They are used to make sails (mostly in polyester, but in polyamide for spinnakers, and in aramid for racing), to make parachutes (polyamide), and to provide ballistic protection in flak jackets and bullet-proof vests (in aramid or UHM polyethylene).

Knitted fabrics are to be found in car upholstery, as geotextiles, in electromagnetic screening (using metal yarns), in filters and in some medical applications such as support hosiery. Nonwoven fabrics are to be found in applications where cost is a critical consideration, including geotextiles, horticulture, agriculture, packaging and in single-use medical and hygiene products. In the form of felts, nonwoven fabrics are used as thermal and mechanical insulators, sound absorbers, and in papermaking in the dewatering and drying stages.

Pile fabrics are used as synthetic turf for high-use sports pitches (polypropylene, polyamide or polyester), as artificial sheepskins (polyester or acrylic) for medical applications, high-performance sleeping-bags and outdoor pursuits clothing, and as brush-like seals in doors and windows. Brushes are used for end-uses from street cleaning to metal deburring, from painting to the application of makeup. Although fibres such as polyamide are widely used, special fibres find a niche here, such as poly(vinyl chloride) with an abrasive filler.

Heavy multicloths are used as conveyor belts in food processing (aramid) and as aircraft arrester tapes (polyester). Three-dimensional woven fabrics are beginning to find application as composite preforms for high technology applications in the aerospace industry (aramid, carbon and glass).

Webbings and narrow fabrics are used as seat-belts (polyamide and polyester), as slings for cargo handling by crane (polyamide and polyester), as tapes in cigarette manufacture (polyamide), and as webbings on flexible intermediate bulk containers (polypropylene). Braids are used as sutures, shoelaces, and covers for climbing ropes. In solid, three-dimensional form they are beginning to be used in the same way as three-dimensional woven fabrics for near-net-shape manufacturing in aerospace. Ropes of all sizes are an essential part of the marine transport and fishing industries, where polyamide, polyester and polypropylene predominate, with polypropylene having the particular advantage of being less dense than water.

12.2 Industrial V-belts and wedge-belts

Belts for power transmission provide an example of a flexible composite, in which the two components combine to provide characteristics not achievable by either alone. The belt is used to connect two pulleys, one driven by a motor, the other connected to the machinery to be driven. The driver pulley is normally smaller than the driven pulley. The torque supplied by the driver pulley is transmitted to the belt by friction

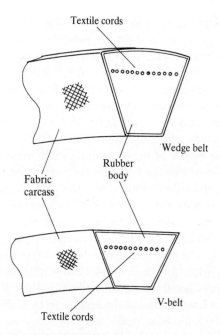

Textile cords

Wedge belt

Rubber body

Fabric carcass

V-belt

Textile cords

Figure 12.1 Construction of wedge-belt and V-belt, showing the rubber body reinforced by textile cords and wrapped in a woven fabric carcass.

between the angled sides of the grooved pulley and the sides of the belt. The torque is transmitted between the pulleys by the difference in tensions between the two sections of the belt, and by friction between the belt and the driven pulley. The system always includes some means of adjusting the separation between the pulleys, to allow belts to be fitted without stretching and to allow the belt tension to be adjusted.

The basic construction of a V-belt and a wedge-belt is shown in Fig. 12.1. The body of the belt is made from low-hysteresis rubber (to prevent excess power consumption and heat generation), with textile cords running around the length of the belt. The outside layer of the belt is reinforced with a layer of woven fabric scrim with its bias direction aligned with the axis of the belt. The reinforcing cords are placed from two-thirds to three-quarters of the distance from the inside of the belt, depending on the width of the belt.

The cords have a much higher modulus than the rubber of the body, so they impart inextensibility to the belt as well as strength. The belt is tensioned during installation so that all parts of the cords remain in tension when the belt is operating. This ensures that the cords experience a tension–tension fatigue regime, as this is very much less damaging to textile materials than the alternative tension–slack load cycling. Synthetic fibres offer higher modulus and fatigue resistance than natural fibres, as well as better resistance to creep. Polyester, with its better creep resistance than polyamide and its ability to bond well to rubber, is widely used, with aramids finding application for special purposes. Resistance to creep eliminates the need for frequent adjustment of the belt to maintain tension. The cords are separated from each other by a layer of rubber to prevent contact between cords and consequent damage from abrasion.

The inextensible nature of the cords forces the neutral axis of bending to lie in the plane of the cords; when the belt is forced by tension to conform to the curvature of the pulley, the rubber above the cords is stretched and the rubber below the belt is compressed. The compression results in a Poisson expansion of the rubber sideways; this increases the pressure between the sides of the belt and the sides of the pulley, increasing the frictional grip. As the belt straightens to exit the groove, the compression (and extra pressure) is released, helping to free the belt and reduce abrasion between the sides of the belt and the pulley.

The fabric outer layer increases the resistance to scuffing and abrasion damage. Woven fabrics have low shear modulus, so they offer little resistance to bias extension. Aligning the bias direction with the axis of the belt ensures that the fabric layer does not resist the longitudinal extension and compression of the rubber as the belt is operating.

12.3 Coated fabric buildings

Modern buildings constructed from coated fabrics are the direct descendants of the tents of antiquity. Synthetic fibres have allowed them to grow, where required, to previously unattainable sizes, sheltering many thousands of people—for example, the air-supported fabric roof on the Silverdome Stadium in Pontiac, Michigan, or the tensioned membrane roofs of the Haj Terminal at the King Abdulaziz Airport in Saudi Arabia. Modern fabric buildings fall into these two broad categories: air-supported structures and tensioned-membrane structures. In the former a 'bubble' of fabric is secured to the ground with an air-tight seal and inflated with low-pressure air; in the latter the fabric is supported by frames or poles and stretched to form the required shape. In both cases the fabric (normally referred to as the membrane) is stabilized by the biaxial stress state developed within it. Typical examples are shown in Figs 12.2 and 12.3.

Fabric buildings possess significant advantages over traditional buildings for some applications, offering large uninterrupted spans, simple, low-cost foundations, short construction times, comparatively low total costs, bold architectural styling, and swift and easy demolition at the

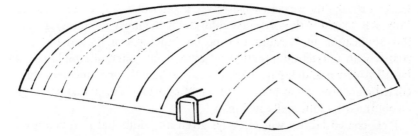

Figure 12.2 Typical air-supported structure of simple, cylindrical shape.

Figure 12.3 Typical tensioned-membrane structure, used to protect visitors to the 1988 Expo in Brisbane from inclement weather. (Photograph: D W Lloyd, University of Bradford, UK)

end of the building's life. Such buildings may be temporary, being erected for specific events (such as major open-air concerts or Expos), or even repeatedly re-erected. Sports stadia, leisure centres, swimming pools, shopping malls, warehousing and industrial units provide the main uses for fabric buildings, along with military applications such as temporary aircraft hangers and radomes.

Almost all modern architectural fabrics have a woven base in plain weave or panama weave (which gives a smoother surface for coating). Polyester coated with poly(vinyl chloride) is the most common fabric for this application in Western Europe for buildings of limited life; glass coated with polytetrafluoroethylene is used for longer-life buildings. In North America and Japan, PVC-coated polyamide and silicone-rubber-coated glass are used as architectural fabrics, with PVC-coated poly(vinyl alcohol) used additionally in Japan.

PVC-coated fabrics will often be given an additional thin coating of acrylic or poly(vinyl fluoride) compounds. These are intended to increase the durability of the fabric by protecting the coating from contaminants, reducing loss of plasticizers and slowing degradation. The PVC can have various additives incorporated in it to improve durability and aesthetics—coloured pigments, fillers, heat and light stabilizers, fungicides, flame-retardants, etc. PVC-coated fabrics can be sewn satisfactorily and can be heat-sealed, simplifying difficulties in obtaining strong joints that are impermeable to air and water and that do not allow water into the cut end of the fabric where wicking might occur.

PTFE coatings have the advantages of being translucent, minimizing the need for artificial lighting in daylight hours. PTFE is self-cleaning and, like PVC coatings, can be used in a wide range of climatic conditions. Small glass beads are sometimes incorporated into PTFE coatings as a filler. PTFE-coated glass fabrics cannot be sewn, as unacceptable damage would be caused to the glass yarns; instead, hexafluoropropylene film is used between the two fabric layers as a welding agent under heat and pressure, or a surface film of fluorinated ethylene propylene is applied to the PTFE to facilitate welding.

In use, the fabric membrane is subjected to constant biaxial stress from the inflation pressure or applied tension. The fabric must therefore be resistant to creep or stress relaxation. The instantaneous biaxial stress varies constantly as wind and snow loads vary and temperatures change, so the fabric must be resistant to fatigue loading also, but the fatigue regime is, as with V-belts, tension–tension. Clearly, seams constitute the weakest part of the membrane; these are required to have a tensile strength at least 90 per cent of that of the membrane material itself. The membrane is most vulnerable to tearing, so stress concentrations are avoided, and the opportunity for accidental damage is minimized, by careful detailed design. However, since some accidental damage (either mechanical or chemical) is almost inevitable during the lifetime of the membrane, the fabric should have high tear resistance so that tears do not propagate catastrophically from points of damage.

The safety of fabric buildings is largely dependent on their behaviour in case of fire. The inability of the fabric membrane to support combustion is critical; experience with full-scale structures suggests that such buildings are inherently safe, as a fire tends to burn a localized, self-extinguishing hole in the membrane that vents smoke and fumes, keeping exit routes clear and usable, and, in the case of air-supported structures, leading to a slow deflation of the membrane.

12.4 Geotextiles

Geotextiles are textiles that are incorporated into geotechnical or civil engineering works. The term geotextiles is normally taken to refer only to permeable textiles; textiles that are impermeable or an integral part of an impermeable layer are called geomembranes. Geotextiles, in the form of mud or soil reinforced by fibrous materials, have been known since earliest times, but it is only in the last 30 years or so, with the availability of a wide range of synthetic fibres and nonwoven fabrics in particular, that geotextiles have become accepted and a normal part of construction.

In use, geotextiles are required to perform one or more functions. The four basic functions are drainage, separation, filtration and reinforcement. Other functions that have been identified include tensioned membrane, cushioning, containment, erosion control, screening, tie, surfacing and slip-surface, but it can be argued that these are special categories of the basic functions. The combination of several functions allows innovative applications to be developed and has been an important factor in the rapid growth in the use of geotextiles.

Geotextiles are normally woven or nonwoven (needle-punched or heat-bonded), with a few knitted fabrics. In composition, polyester and polypropylene predominate, with high density polyethylene, polyamide and aramid being used in more specialized products. Fibres appear as staple fibres, continuous filaments, multifilament yarns, tape yarns and slit films, depending on the intended use of the fabric. Fabrics may be combined into composite structures; for example, preformed fin drains are made from a central layer of high in-plane permeability (such as a thick, heat-bonded web of coarse continuous filaments or coarse, open woven fabric) and outer layers of filter fabrics with small pore sizes (such as needle-punched nonwovens). It is worth remembering that geotextiles normally operate as part of a composite with the soil.

When a geotextile functions as a drain, it collects liquid or gas and provides a path for it to flow to another location, as in Fig. 12.4. The drains may be vertical or horizontal and normally collect water, but may collect leachate or gas from waste tips, and are often used in association with geomembranes. As a filter (Fig. 12.5), a geotextile is used to retain small soil particles while allowing liquid to pass; as, for example, when used to line the sides of a French drain and prevent it from becoming clogged. The fabric is chosen so that its distribution of pore sizes is

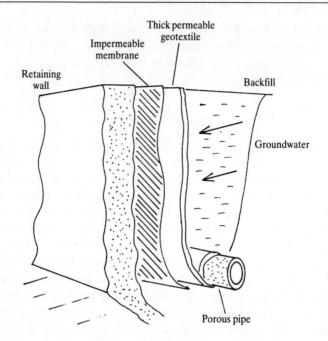

Figure 12.4   Thick geotextile used in combination with an impervious membrane to produce a vertical drain that protects a retaining wall by intercepting groundwater and carrying it to a porous pipe for removal.

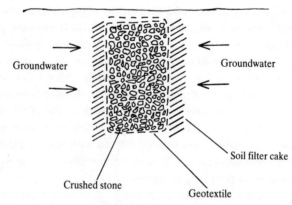

Figure 12.5   Geotextile used as a filter, to prevent a French drain becoming clogged with fine soil particles.

appropriate to the range of particle sizes in the soil, so that after a period of time, a 'filter cake' builds up behind the fabric.

Geotextiles as separators are placed between different types of materials to prevent them from becoming mixed. Geotextiles are placed on soft

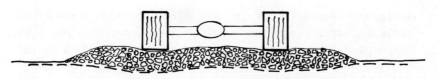

Figure 12.6 Geotextile used as a separator and reinforcement under a temporary road, to prevent loss of subbase into the soil and to inhibit the formation of ruts.

subsoil, for example, to prevent the granular subbase of a road or the ballast of a rail bed from mixing with the subsoil, as shown in Fig. 12.6. In this situation the geotextile enables the subbase to resist local failure and behave as a unit. If the geotextile in this situation also carries significant load, so that the soil is less stressed, the geotextile is providing reinforcement as a tensioned membrane. In general, whenever a geotextile improves the mechanical behaviour of an earth structure it is acting as a reinforcement. Thus, embankments are constructed on a geotextile base to improve the stability of the soil foundations, and may be used in the growing embankment to reinforce the embankment itself, as shown in Fig. 12.7. As the base to an embankment, the geotextile may be acting as a tensioned membrane, as a separator, and as a drain and filter facilitating the dewatering and consolidation of the soil under the embankment. Vertical geotextile 'tubes' may be driven into the soil to act as vertical drains to further aid dewatering and consolidation.

Geotextiles are sometimes used as flexible formwork in concrete construction, as, for example, in the construction of revetment mattresses for river or estuary bank stabilization. These fabrics may include frequent filter patches, to allow compensation for changing water levels. Ties are geotextiles that are used to connect different parts of a structure that might otherwise move apart, such as a bridge abutment and its facing wall.

Geotextiles are used as surfaces in, for example, temporary roads, to prevent damage to the ground and provide a sound working surface. This group of fabrics includes some of the heaviest geotextiles, incorporating galvanized steel wires and sprung steel bars. Slip-surfaces allow differential

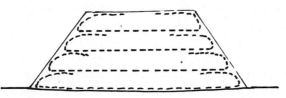

Figure 12.7 Geotextile used to reinforce an embankment, by enclosing layers of soil.

movement between layers, as can occur in multilayer river or canal bank linings, whereas cushioning fabrics absorb stress concentrations, preventing sharp stones from puncturing geomembranes or reflective cracking from damaging the top layers of repaired road surfaces.

The short- and long-term properties of a geotextile are both important to its successful use. The geotextile must be capable of withstanding the stresses of installation and of the building and consolidation of the structure of which it is a part. Thus tensile strength and stiffness, puncture resistance, tear resistance and permeabilities under compression must all be adequate for the initial phases. However, if the geotextile is designed to continue to operate as an integral part of the structure during its life, it must retain its properties, resist creep deformation, and not degrade with prolonged exposure to the chemicals and microorganisms in the soil. An important aspect is the resistance of the geotextile to UV light, as it is likely to be left exposed to sunlight for extended periods during installation, triggering degradation processes that will continue after the exposure ceases. For this reason, many geotextile fibres either incorporate UV stabilizers or are covered in a protective layer that screens out harmful UV light. Prediction of long-term degradation and consequent loss of properties is extremely difficult; accelerated testing has been used to provide guidance but only accumulated experience of the behaviour of real structures can provide firm data.

The success of synthetic fibres as geotextiles can be judged from the quantities of fabric involved, currently measured in hundreds of millions of square metres per annum worldwide. Geotextiles have become a mature industry with a wide range of developed products matched to particular applications and with a large, and growing, body of accumulated published research and experience. A selection of geotextiles is shown in Fig. 12.8.

12.5 Climbing ropes

Rock climbing and mountain climbing are outdoor pursuits enjoyed by thousands of people worldwide. Their safety and enjoyment is due, in large measure, to the availability of modern climbing ropes with an appropriate mix of properties. The primary purpose of a climbing rope is to arrest a fall, but in the gentlest way practicable so as to avoid serious injury to the climber.

Inevitably, there has to be a compromise between desirable, but conflicting, characteristics. The 'ideal' climbing rope should have low extensibility (high modulus) at low loads, so that only minimal extension is experienced during normal climbing. At high loads—the shock loads encountered during a fall on the rope—the rope should have higher extensibility (low modulus), as the kinetic energy of the falling climber is

Figure 12.8 Selection of geotextiles. The base fabric is a woven polyester double-cloth used as flexible formwork for producing concrete revetment mattresses for river and estuary bank protection. The grey and white samples are nonwoven polyester, the remaining samples are woven from a variety of polymers. The heavy fabrics on the top are woven polypropylene with additional steel reinforcement. (Photograph: D W Lloyd, University of Bradford, UK)

absorbed as strain energy by the rope (load × extension). Some hysteresis loss is desirable to give rapid attenuation to the inevitable bounces of the climber on the end of a rope following a fall.

Hysteresis is also important for another reason. The shock load of the climber 'hitting' the end of the rope travels back along the rope at the speed of sound in the rope, in the form of a tension wave. This is partly reflected at points of contact with the rock, knots, carabiners, etc. A complex pattern of waves can result with the risk of failure where tension waves combine. Hysteresis losses in the rope help to attenuate these waves.

After being fully loaded by a fall, the rope will suffer a permanent extension. The rope should be scrapped at this point, as it will have insufficient residual extension at high load to protect the climber in a subsequent fall.

The safe use of ropes in climbing demands that they should not snarl easily, as this would prevent the rope from paying out freely. Snarling is prevented by using a twist-balanced structure, and by ensuring a relatively high bending stiffness to prevent the rope forming small loops easily. However, the bending stiffness must not be so high as to make the rope difficult to handle or knot.

Climbing ropes are used in conditions that range from cold and saturated with water to hot and dry. The rope must be able to retain its properties under all conditions of use. Equally, the rope must retain its properties with time, so it should be resistant to rotting and microbial and fungal attack. Since climbing exposes a rope to strong UV light, the rope should be UV resistant, and, again as a result of its normal conditions of use, strongly resistant to abrasion. It should also possess high surface friction to prevent knots from slipping and to allow the climber to achieve a firm grip on the rope.

These characteristics are obtained through the rope structure and the synthetic fibre materials used. Modern climbing ropes have a core–sheath structure. The core is the main load-bearing element, with the sheath contributing no more than 20 per cent to the ultimate tensile strength of the rope. The core structure is twist balanced. Some ropes use a core of straight yarns (bundles of continuous filaments with little or no twist), though the preferred structure consists of twist-balanced cords. These cords, from 3 to 12 in number, are made up of fine continuous filaments. An odd number of cords is usual, the cords themselves being either three-strand cable-laid or braided structures. Fine fibres are used as they impart an appropriate bending stiffness to the rope and because the breaking of individual fibres is not catastrophic. Straight yarns give high initial modulus; cable-laid or braided cords give greater extensibility at high loads.

The sheath is formed by one or two layers of 2/2 braid formed over the core. Because the yarns in the braid can rotate easily relative to one another, the braid has little effect on the bending stiffness of the rope, except at small curvatures, when the structure 'jams' on the inside of the bend. The sheath provides the outer friction surface and screens the core from UV light. It also forms a sacrificial, abrasion-resistant layer. The sheath is normally brightly coloured; this makes the rope clearly visible against rock or snow surfaces and permits the easy identification of different components in a multirope system. The bright colouring also enables abrasion damage to be detected readily by visual inspection—

Figure 12.9 Climbing ropes, showing the twisted core strands and braided covers. Note the point of damage to the outer case of the light coloured rope, made visible by the dark inner braid showing through the hole. (Photograph: D W Lloyd, University of Bradford, UK)

this is assisted by using a different colour for the inner braid in a two-layer sheath.

The inner core of a climbing rope is always made of fine, continuous filament polyamides, as this provides the best shock absorption. The sheath, too, is always made from polyamide, though coarser fibres are used as these give greater abrasion resistance. The result is a highly engineered product, well matched to its intended use. Some typical climbing ropes are shown in Fig. 12.9.

Further reading

It has not been possible in the space available to present more than a brief summary of the engineering applications of synthetic fibres. The following publications are suggested for a fuller discussion.

General

The Design of Textiles for Industrial Applications, Papers of the Annual Conference of the Textile Institute, Rotterdam, 1977, edited by P W Harrison, published by the Textile Institute, Manchester.

High Performance Textiles, edited by P Lennox-Kerr, published monthly by Elsevier Advanced Technology.

Journal of Coated Fabrics, edited by W C Smith, published quarterly by Technomic Publishing Co. Inc.

Filtration and Separation, published by the Filtration Society, Uplands Press.

Coated fabric buildings
The Design of Air-Supported Structures, Proceedings of the Conference of the
Institution of Structural Engineers, Bristol, 1984.

Geotextiles
Geotextiles and Geomembranes, edited by T S Ingold, published bimonthly by
Elsevier Applied Science.
Geotechnical Fabrics Report, edited by Danette R Fettig, published monthly by
the Industrial Fabrics Association International.
Proceedings of the Second International Geotextiles Conference, Las Vegas, 1982,
published by the Industrial Fabrics Association International.
International Directory of Geotextiles and Related Products, edited by P R
Rankilor, published by Manstock Geotechnical Consultancy Services Ltd.

Ropes
K P Gilbert, *Ropemaking, A History of Technology*, Vol 1, Clarendon Press,
Oxford, 1954.
H F Microys, Climbing ropes, *American Alpine Journal*, 1977, published by the
American Alpine Club.

CHAPTER 13

Specialized uses 2: medical textiles

D W LLOYD

13.1 Introduction

13.1.1 Medical textiles and their functions

Materials containing fibres have been used for medical purposes since earliest times, most obviously as wound dressings. The huge growth of medical applications of textiles over the last 120 years or so has mirrored the development of modern medicine and surgery. This growth has not been limited to just the volume of materials used, but can be seen in the type of function and range of materials and structures used. Currently a great diversity of products is available commercially, with new products and applications being developed continuously. It is necessary to bring order to this diversity through some form of classification scheme. It would be possible to classify medical textiles by fibre type or processing route, for instance; however, for the present purposes, a classification based on function is more appropriate. Classification by function emphasizes aspects common to otherwise apparently different applications, different fibre types and different textile structures. It also emphasizes that some applications require several functions to be performed simultaneously. It is perhaps worth mentioning at this point that few, if any, textile structures are capable of meeting all the requirements of a medical application on their own, so the broad term 'medical textiles' includes many products with a composite structure, some containing nonfibrous components.

It is possible to identify six primary functions required of medical textiles. These are listed in Table 13.1 with examples of the types of product that perform each function. Some products appear against more than one function, indicating the complexity of some applications.

13.1.2 Regulation and legislation

Medical textiles are produced and sold within a strict framework of regulations and legislation. Such control is essential to protect patients,

Table 13.1. Primary functions of medical textiles and related products

Function	Product
Barrier	Operating-room apparel and drapes, dressings, face masks, caps, overshoes
Absorbent	Swabs, incontinence products, hygiene products, dressings
Prosthesis and reinforcement	Vascular grafts, artificial ligaments, artificial tendons, sutures, soft-tissue patches
Cushioning	Pile fabrics (for prevention of decubitis), sewing rings and pads on prosthetic devices, dressings
Thermal	Hypothermia prevention and treatment devices, bedding
Psychological	Modesty gowns, dressings, curtains and carpets

medical practitioners and manufacturers; for example, the approval of polyester for use in prosthetic devices has resulted in its almost universal use in this area. In the United Kingdom three sources are of primary importance, the British Pharmacopoeia, the British Pharmaceutical Codex and the DHS Guide to Good Manufacturing Practice for Sterile Medical Devices and Surgical Products 1981. Also, the European Pharmacopoeia will become increasingly important as standards for materials are agreed for inclusion in it.

The DHS Guide to Good Manufacturing Practice for Sterile Medical Devices and Surgical Products 1981 forms the basis of the DHS Registration Scheme. The Guide is based on the principles of BS 5750 Part 1 Quality Systems, and addresses those aspects of the manufacturing process that affect the quality, safety and performance of products. One of the requirements of the Guide—sterilization—interacts especially strongly with the materials and structures of medical textiles, and so requires separate description.

13.1.3 Sterilization

Many of the medical applications of fibrous materials demand a sterile product. However, many fibres suffer severe damage if sterilized by some routes, so different sterilization methods are used for different fibres and products. Four techniques are in use at present: steam, dry heat, ethylene oxide gas and irradiation.

Steam sterilization involves bringing steam into intimate contact with the organisms to be destroyed. The textile products are placed in an autoclave, which is then evacuated before steam pressure is raised to $220 \, \text{kN m}^{-2}$ (32 psi), at a temperature of 134°C. These conditions are maintained for at least 3.5 minutes, to allow the steam to penetrate the load fully, followed by a second, vacuum, cycle to remove the steam and to dry the load. Finally, filtered air is admitted to the auto-clave.

Dry heat sterilization is carried out in hot-air ovens or continuous infrared ovens, using 160°C for 1 hour or 190°C for 14 minutes. Clearly, these methods are inappropriate for fibres such as polypropylene that soften or melt at low temperatures.

Ethylene oxide gas is often used where the materials to be sterilized are sensitive to high temperatures. The gas alters proteins, killing bacteria, fungal spores and viruses. A thorough cleaning cycle is required before sterilization and a gas removal cycle is needed before use.

Irradiation is the other technique used on heat sensitive materials. Irradiation is an industrial-scale process, normally using cobalt-60 or caesium-137 as radiation sources. Many countries recommend a dosage of 2.5 Mrad, with 3.5 Mrad being recommended in Scandinavia.

The efficacy of a sterilization cycle is monitored using indicators placed throughout the load, including the most inaccessible places. After steriliz-ation, the product must be packaged so that it will preserve the sterile conditions until the product is used.

13.2 Fibres used in medical textiles

A wide range of fibres, both natural and synthetic, is used in medical textiles. A particular fibre will be used when its mix of properties represents the best compromise for the spectrum of properties required by the application. History also plays a part; fibres that have proved effective in the past may be slow to be replaced, even though newer fibres have a better mix of properties.

The natural fibres used in medicine include cotton, linen, silk and catgut. All find application as sutures, and cotton finds wide application in dressings, operating room apparel and drapes, absorbent devices such as swabs and gauzes, and in patient bedding systems. Cotton has the advantages of being easy to launder and to sterilize, and of being highly absorbent.

A greater range of properties is available from synthetic fibres and the range of fibres used is correspondingly greater. Viscose finds similar applications to cotton, in the form of single-use nonwoven fabrics for sheets and operating room apparel. In modified forms, the various superabsorbent forms of viscose find application in all the types of product that perform an absorbent function.

Polyester finds application in dressings, prostheses and sutures, where its good compatibility with body tissues, relatively low cost and good mechanical properties offer advantages over other fibres. Polyester and polyethylene are both used in spun-bonded nonwovens for disposable

operating room apparel. Polypropylene is used in sutures, as a hydrophobic layer in dressings and absorbent products, and, because of its low
cost and ease of processing, in single-use nonwoven fabrics in products
such as operating-room apparel and disposable sheets.

More specialized applications require special fibres. The need for so
many different types of suture leads to a wide range of different fibres
being employed. In addition to the fibres already mentioned, polyamide,
polyglycolide, polylactide, polydioxanone and steel are used. Carbon fibre
made from polyacrylonitrile precursor finds application as replacement
tendons and ligaments, and, in activated charcoal, woven fabric form, in
adsorptive dressings. Reconstituted collagen, in the form of fibrous
'fleeces' and partially fibrous 'sponges', and alginate fibres are used in
specialized dressings. X-ray-opaque yarns, normally made from poly(vinyl
chloride) or polypropylene containing at least 50 per cent by weight of
barium sulphate, are used as markers in surgical swabs and sponges.

13.3 Medical textiles as barriers

The use of a textile as a barrier is one of the most familiar medical
applications, as everyone, at some time or other, will have applied a simple
dressing to a cut finger 'to keep the dirt out'. Similarly, most people will
be familiar with the picture of operating room staff in gowns, caps and
masks, with the unconscious patient hidden under drapes. What is less
familiar is the range of routes of infection involved.

13.3.1 Operating-room apparel and drapes

Mechanisms of infection

Postoperative wound infection occurs in a significant percentage of
patients following a surgical procedure. Such infections may double the
length of the patient's stay in hospital, and inevitably increase the patient's
suffering and the cost of treatment. Two routes of wound contamination
can be identified, from environment to wound and from patient to wound
(self-infection). An additional infection route is of importance, that from
the operating-room staff to the patient.

Contamination of a wound from the operating-room environment
arises essentially from the people in it, as long as the operating room is
properly ventilated. People performing active tasks cause airborne contamination, as skin cells are shed continuously from the surface of the
body. Such cells may host colonies of the normal bacterial flora of the
skin and have been shown to be a major cause of wound infection. The
surgical staff and patient are both sources of dispersed skin cells and

contribute to the level of airborne contamination. One of the roles of operating-room apparel and drapes is to control this airborne contamination.

The wound itself is a source of potentially infectious body fluids, some of which may be transferred to the front of the surgeon's gown. If strike-through occurs, perhaps as the surgeon presses against the edge of the operating table, the potential exists for infection of the surgeon by the patient. It has also been established that a wet surgical gown provides a route for the transfer of bacteria from the surgeon's skin to the outside of the gown and thence to the patient.

Operating-room apparel

Normal operating room apparel comprises a scrub suit, consisting of trousers and short sleeved tunic, or a dress; this is clean but not normally sterile and will normally be worn all day. A cap, to contain the hair, a face mask and overshoes are also worn; the sterile gown is worn over the top, with at least one pair of sterile surgical gloves. The gloves are worn over the cuffs of the gown to close the sleeves. 'Unscrubbed' personnel, such as theatre runners, anaesthetists and technicians, wear a subset of the full apparel system, typically the scrub suit, cap and overshoes.

The traditional material for operating-room apparel is woven cotton dyed green, with, in some cases, waterproof panels produced by coating particular areas with polymer film. Woven cotton has the advantages of being easy to launder and sterilize, and of being relatively comfortable to wear. It has the significant disadvantages of linting—shedding fibre fragments—and hence of potentially contributing to wound contamination, and of offering a very poor barrier to skin cells or strike-through by liquids. These disadvantages are causing alternative materials to be used, especially for critical areas such as orthopaedic surgery.

Alternatives divide into single-use and multiple-use materials. Multiple-use fabrics include closely woven Ventile fabrics, developed to overcome the poor bacterial barrier properties of traditional cotton fabrics, and fabrics incorporating microporous materials, such as Gore-Tex. This is a two- or three-layer composite in which the outer layer is a traditional fabric (typically a polyester cotton blend) laminated to a microporous film of PTFE. An additional fabric layer may be laminated to the inside of the composite. The pores in the film are less than 0.2 μm in diameter and provide an effective barrier to liquids and bacteria while allowing the passage of water vapour. This helps to ensure the comfort of the surgeon. Such surgical gowns retain their effective barrier properties after repeated laundering and sterilizing cycles.

Single-use materials are based on nonwoven fabrics. Typical fabrics are spun-bonded polyolefins such as Tyvek, spun-bonded nonwovens made from polyester and cellulose (Fabric 450), and composite poly-propylene structures with an outer spun-bonded layer and an inner melt-blown layer. Gowns made from these materials are supplied in sterile packs and have the additional advantage that they can be stockpiled for use in the event of a major emergency. All fabrics used for operating-room apparel must have antistatic properties, as a risk of sparks cannot be tolerated because some anaesthetic gases are potentially explosive in air.

Disposable nonwoven fabrics are also used for the other single-use items. Spun-bonded polypropylene is used commonly for theatre caps and for overshoes. These may be made in one piece or may be made from two or three pieces sewn together to give a better fit. Elastic threads are sewn into the edges of the openings to provide a simple but efficient closure. Masks often have a multiple layer structure, to ensure more efficient filtration of the breath. Typically the inner layer will be melt-blown polypropylene, with outer layers of print-bonded viscose to provide strength and to prevent the loss of polypropylene fibres. Tapes are sewn to the mask to enable it to be tied firmly into place over the nose and mouth; in some instances a wire frame may be incorporated that can be bent to follow the contours of the individual face.

Drapes

Traditionally, drapes are made from woven cotton or linen, but increasingly synthetic fibre materials similar to those used in single-use gowns are being used. Drapes are used to cover the patient and reduce the risk of the wound becoming contaminated by skin cells shed by the patient. Drapes are supplied cut to a variety of different shapes appropriate to different surgical procedures and contain an opening corresponding to the position of the surgical site. Similar, smaller drapes are also used to cover instrument trays. The typical use of drapes is shown in Figs 13.1 and 13.2.

13.3.2 Dressings

The role of dressings

A shallow wound, one that involves the loss of the epidermis, produces a scab when allowed to heal 'naturally' without a dressing. The scab is rigid and dry, containing the dried blood and serous exudate that flowed from the wound initially, the dehydrated leukocytes that migrated across the wound after the injury and collagen fibres passing into the dermis. Regeneration of the epidermis occurs through a layer of liquid below

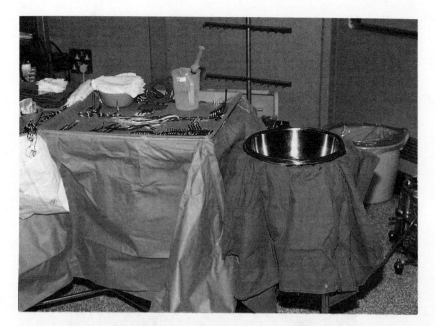

Figure 13.1   Disposable nonwoven synthetic fibre drape on an instrument
tray (compare with the traditional, nondisposable drape to
the right). (Photograph: T Browne, Killingbeck Hospital,
Leeds, UK)

the scab. The scab is porous to the oxygen required by the regenerating
tissue, but acts as a barrier to dirt and infection. The scab is shed when
regeneration of the epidermis is complete.

The natural scab has the disadvantages that its oxygen permeability is
limited, that it is not flexible, and that it is prone to cracking and damage
and hence is vulnerable to infectious microorganisms.

Dressings are used to try to overcome the disadvantages of the natural
scab as a wound covering. Dressings are normally classified into absorbent
and nonabsorbent types. There have been many attempts to specify a mix
of properties that would be possessed by an 'ideal' dressing; however, the
variation in types of wound and in wound management programmes
means that no single type of dressing is universally applicable. All
dressings should provide a barrier against infection and against the ingress
of foreign material that might cause adverse tissue reactions. This implies
that the dressings should conform well to the patient's body and its
movements and that it should not itself be a source of infection,
inflammation or foreign material. The dressing should protect the wound
from further injury by acting as a cushion, and it should promote rather
than interfere with healing.

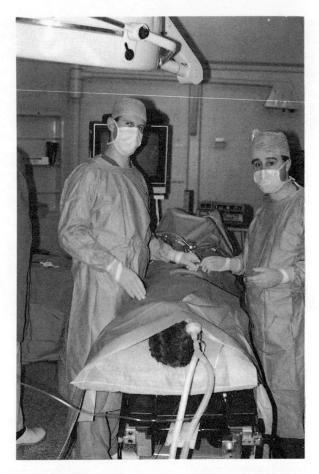

Figure 13.2   Laparoscopic (i.e. keyhole) surgery being performed on a
patient with hepatitis B. Disposable, nonwoven synthetic fibre
operating-room apparel and drapes being used to minimize
infection risks. (Photograph: T Browne, Killingbeck Hospital,
Leeds, UK)

Absorbent dressings

The traditional absorbent dressing is cotton gauze. A gauze dressing
allows a scab to form, but, on drying, the wound exudate absorbed by
the dressing becomes stiff and firmly attached to the wound surface. If
the dressing were left it would not be possible to remove it without further
trauma until healing was effectively complete. Dressings are not normally
left so long, as they become dirty and may be damaged or loosened.
There may be a need to examine the wound for infection and to assess
the progress of healing. Also, if moist exudate reaches the surface of

the dressing, a path is provided into the wound for infectious micro-organisms.

Changing a gauze dressing before the wound is healed results in removal of at least part of the scab and may destroy some of the regenerating tissue. The pain and danger of reinfection associated with changing such dressings has led to the development of nonadherent dressings. The simplest and best-known of these is paraffin-coated gauze or tulle gras, but this has the disadvantages that the coating tends to soak into the dry layers of the dressing and to introduce a foreign material into the wound that tends to delay healing. It has been found that dressings are less likely to adhere if the wounds can be kept dry by removing exudate as it forms. Wounds treated in this way do not develop the thick scab produced in the absence of any dressing. Some dressings make use of a perforated polymer film that acts as a barrier to excessive absorption of exudate; although a gel layer forms under such dressings, exudate dries in columns through the holes in the film, leading to nonuniform healing and further injury on removal of the dressing.

Recently, Courtaulds have developed a calcium alginate fibre for use in absorbent dressings. Calcium alginate has the advantages that it can absorb large quantities of exudate and is itself absorbed by the body. This means that it can be used in nonwoven form, as loose fibres entering the wound are not harmful.

Nonwoven materials are also used as covering layers to absorbent pads in an effort to reduce adhesion. Typically a viscose fabric is used, again with perforations, or apertures, to allow absorption of exudate. The viscose may additionally be coated with a hydrophobic material, such as aluminium or polypropylene, to reduce adhesion. Multiple-layer pads commonly use a nonwoven viscose outer layer, possibly coated to reduce adherence, though nylon has also been used. Some dressings incorporate an upper layer that acts as a partial barrier to strike-through, further extending the life of the dressings and reducing the risk of infection being transmitted to the wound through the dressing.

Small dressing pads are used in the well-known island dressings, which have an absorbent dressing pad located in the centre of an adhesive strip. Some modern island dressings have an absorbent pad made from styrene–butadiene rubber foam with a cover of nonwoven fabric coated with polyethylene, to combine high absorption with nonadherence.

Nonabsorbent dressings

Nonabsorbent, or occlusive, dressings are designed to form a barrier to liquids across the wound. This permits regenerating tissue to spread in a layer of moist exudate under the dressing and consequently scab formation does not occur. If the dressing is also permeable to oxygen, rapid healing occurs. Occlusive dressings must form a good seal with the skin around the wound to be effective, a difficulty in the presence of large amounts of exudate. The materials used for these dressings include materials designed as surgical drapes, but typically consist of polymeric films spread with an appropriate adhesive.

13.4 Absorbent medical textiles

13.4.1 Mechanisms of absorption

Medical textiles make use of a range of mechanisms to absorb liquids. Liquid may be contained and immobilized or channelled in gross structural features such as folds or pleats. The fabric structure itself offers pores and capillaries that will hold fluid, whether the 'fabric' is woven, nonwoven, or a batt of fibrous material. Where the fabric is composed of spun staple yarns, the yarns will also contain pores and capillary spaces capable of absorbing liquid. Fibres offer two potential absorption mechanisms: physical and chemical. Physical absorption takes place into the microstructure of the fibre, in particular into the amorphous, noncrystalline regions, whereas chemical absorption binds water molecules to appropriate sites on the fibre molecule. Chemical modification of the fibre is used to create a more open molecular structure capable of containing greater quantities of fluid, and physical modification is used to create the hollow, inflated and superinflated fibre forms with their macroscopic absorbent voids.

It is appropriate to mention two other materials at this point. Activated carbon fabrics are not strictly absorbent but adsorbent, immobilizing adsorbed material onto the fibre surface. Such fabrics are not used to contain large quantities of liquid, but to adsorb and control unpleasant odours and bacteria. X-ray-opaque yarns are absorbent only in the sense of absorbing clinical X-rays. They are made by incorporating a minimum of 50 per cent barium sulphate by weight into PVC or polypropylene filaments during extrusion. These yarns are used as X-ray markers in surgical swabs.

The different absorbent products utilize the different mechanisms of absorption as best suits the particular end-use. The principal end-uses of absorbent materials are swabs and dressings, baby and adult incontinence products, and feminine hygiene products. Absorbent wound dressings have been considered above in the context of their role as barriers to infection, so the following discussion will be confined to other absorbent products.

13.4.2 Surgical swabs

A swab is an absorbent textile pad used in general surgery to prepare the site of the operation, to absorb excess blood and body fluids, to pack body cavities during surgery, and to clean the incision prior to suturing. The traditional material for manufacturing swabs is cotton gauze. X-ray-opaque marker yarns are incorporated into the structure, or another X-ray opaque marker is sewn in, so that in the rare event of a swab remaining in the body of the patient after surgery, it can be located without the use of exploratory surgery.

Cotton gauze suffers from disadvantages despite its widespread use. Alternatives must retain the advantages of traditional swabs, in particular their high absorbency and nonlinting properties, to be acceptable.

13.4.3 Incontinence and hygiene products

In the developed countries over recent decades there has been a decline in the birth rate and a general increase in longevity. These changes have been accompanied by a steadily rising affluence. Incontinence is normally regarded as a problem of the very young, the very old, and the disabled or bedridden. Although there is some truth in this view, there are a significant number of incontinence sufferers of all ages, representing about 5 per cent of the adult population. The main problem is that of urinary incontinence, with stress incontinence (caused by laughing, sneezing, exertion or emotional upset) representing the largest part, followed by urge incontinence and dribble incontinence. More women than men are affected by incontinence (childbirth being blamed by many patients), though the incidence rises for both sexes in old age. Product designs need to meet the different requirements of patients with widely different levels of incontinence, different levels of activity from fully active to chair bound or bedridden, a spread of disabilities including arthritis, blindness and dementia, and the differences between male and female anatomy. Incontinence protection must not leak, or cause discomfort or skin irritation to the patient.

Rising affluence in the developed world has led to a decrease in the use of reusable products (such as the cotton terry towelling baby nappy (diaper)) and an increase in the use of disposable products, as consumers pay for increased convenience. An incontinence pad or baby's nappy has three basic functional components. These are the absorbent body itself, an outer waterproof backing and the skin contact layer or coverstock. The most common absorbent material used in disposable incontinence products is wood pulp, also referred to as fluff pulp. This consists of fibres obtained by chemical or mechanical pulping of trees and is used because it is both highly absorbent and economical. Hydrogels may be added to the fluff pulp to increase its capacity to hold urine, and thermoplastic fibres may be added to strengthen the structure of the fluff pulp and prevent it collapsing when wet. Such fibres help to maintain an open structure and, if fused, help to bind the fluff fibres and fix hydrogel particles in the structure. Melt-blown microfibre webs, normally made from polypropylene, are used as wicking layers and to fix hydrogel particles.

A wide range of fibres is used to manufacture coverstock fabrics, including viscose, polypropylene, polyester, polyethylene, polyamide and bicomponent fibres. Nonwoven fabrics are used to minimize the cost of coverstock materials. Coverstock must allow the rapid passage of urine through the material, but they must not retain liquid themselves or allow liquid to leak back to the skin ('wet-back'). It is important that the coverstock be soft to the skin and nonabrasive, especially where the patient is elderly. A wide range of nonwoven fabric manufacturing methods are used to make coverstocks, including melt-blowing, thermal and adhesive bonding and spun-lacing, using carded, air-laid or wet-laid webs. Apertured fabrics, fibrillated nets and perforated films also find application as coverstocks, with each fabric type offering a particular claimed advantage. Mechanical and chemical finishing treatments may be applied to the coverstock to improve its properties, for example embossing is used to prevent run-off, surfactants are added to improve wettability, and antimicrobial agents are added to reduce odours and risks of infection.

13.5 Prosthesis and reinforcement

13.5.1 Introduction

A prosthesis is a device that is used to overcome surgically some deficiency in the body, perhaps caused by the necessary surgical removal of diseased tissue. This implies that the prosthetic device must perform its function for the rest of the life of the patient. Reinforcement, in contrast, may be temporary or permanent, depending on need and the ability of the relevant tissue to regenerate.

The most common prostheses made from synthetic fibres are artificial tendons and ligaments and vascular prostheses. Polyester is the most common synthetic fibre used in these applications, although carbon is also used for replacement tendons and ligaments, and nylon has been used in the past for vascular replacements. PTFE finds a place in vascular replacements though not in fibrous form. Polyester fabrics also form part of other prosthetic devices such as heart valves, in the form of sewing rings. As their function is to act as a cushion, they will be reviewed later.

Similar polyester fabrics are used as patches to close gaps in the septal walls of the heart. Patches made from polyester mesh or tightly woven polyester are used to replace parts of the dura, the membrane covering the brain, following brain surgery. Meshes of polyester, polypropylene or polyamide are used to repair hernias of the anterior abdominal wall, an application that requires high strength to resist the sharp rises in pressure that accompany sneezing, coughing and other activities.

Cotton gauze is the traditional reinforcement for plaster casts, used to immobilize and support broken limbs, but glass impregnated with a water-curing resin is finding a growing role in this area. Glass-fibre-reinforced casts are lighter than plaster, develop their full strength quickly and do not deteriorate if they subsequently become wet. They are also permeable, giving greater patient comfort.

The advent of elastomeric yarns has permitted the development of acceptable support tights and stockings for patients with varicose veins. In this condition, more common in women than in men, the veins in the legs become swollen as blood accumulates in the veins rather than being returned promptly to the heart. In veins near the surface, the swelling and discoloration are unsightly and painful, but can be ameliorated by the wearing of support hoisery. This is knitted from polyamide and elastomeric filaments to give precisely controlled mechanical properties. In use, the hosiery exerts gentle pressure on the legs, decreasing from the ankles to the thighs. Modern support hosiery has acceptable appearance, little different from fashion hosiery, in contrast to earlier support hosiery based on relatively coarse rubber yarns.

13.5.2 Vascular prostheses

Vascular prostheses are used to replace important blood vessels that have become blocked or have developed an aneurism. Approval by regulatory authorities means that polyester is used almost universally. Such protheses are manufactured in woven and knitted forms, as parallel, tapered and bifurcated tubes, and are cut to length by the surgeon at the time of implantation. The grafts normally incorporate a coloured line along the length of the graft to assist the surgeon in avoiding introducing twist into the graft as it is implanted. Woven constructions (made as a double cloth tube on a narrow fabric loom to give a circular tube) offer the greatest strength and are available in low permeability forms, making them more suitable for high stress locations such as the thoracic aorta. However, woven constructions are more prone to fraying at cut ends and are more difficult to suture.

Knitted prostheses are made as single jersey structures on weft knitting machines, as raschel structures on warp knitting machines, or as velour

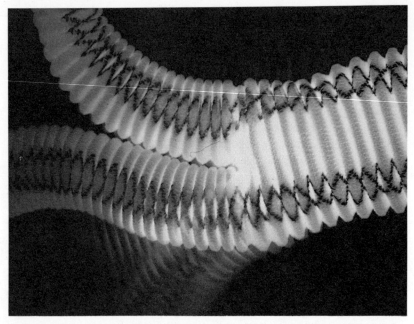

Figure 13.3   Warp knitted polyester bifurcated vascular prosthesis with internal and external velour piles of different pile heights. (Photograph: Vascutek Limited, Inchinnan, Renfrewshire)

structures. Velour structures have a rougher inner surface to encourage better adhesion of the thrombus layer (grafts are normally preclotted in the patient's own blood before implantation). The roughened surface may be obtained by raising a nap, by knitting a looped pile, or by using textured filament yarns. Knitted structures are used, for example, to replace the abdominal aorta where it branches into the iliac arteries supplying the legs with blood. A typical example is shown in Fig. 13.3.

Knitted structures are too porous to be used in the machine state, so they are 'compacted' by chemical swelling agents or thermal processing. Almost all commercial grafts are 'crimped'; this is carried out during or after compaction, either by fitting the graft on a mandrel and constraining it with a helically wound filament or by compressing the graft in a mould. This leaves the graft with a spiral or circular macroscopic crimp, whose purpose is to allow the tubes to bend easily without collapsing. It also provides a better match to the mechanical properties of the replaced blood vessel, reduces the stress at the suture line, minimizes the local fluctuations in blood pressure and reduces the chance of thrombosis. The crimp tends to be lost after implantation as tissue ingrowth proceeds. Although tissue ingrowth seals the graft against leakage, the strength of the polyester 'skeleton' remains the main protection against failure.

13.5.3 Replacement tendons and ligaments

Natural tendons and ligaments are composed mainly of collagen, a fibrous protein exhibiting crimp. Tendons perform the role of attaching muscles to bones, while ligaments hold bones together across joints to maintain the structural integrity of the skeleton. The crimp typical of collagen fibres imparts a J-shaped load–extension curve that matches the mechanical needs of tendons and ligaments. When a ligament or tendon is torn or permanently overstretched, control over the particular joint or mobility of the limb is lost and severe pain may ensue.

A successful graft must match the mechanical properties of the original ligament or tendon as closely as possible and must not induce adverse tissue reaction. Synthetic materials cannot match the fatigue resistance of natural tissue with its ability to regenerate itself, so a graft should encourage the ingrowth of new collagen to form a new tendon or ligament over time. Carbon fibre and polyester have been found to have the necessary combination of properties for such grafts, and have found application in twisted tow, braided strand and woven strip forms. When first implanted, these materials have the necessary mechanical strength and properties to perform as a replacement for the damaged ligament or tendon. After implantation the graft is invaded by fibroblasts, the cells that extrude collagen, so that the graft becomes encased in new tissue. This new tissue has a much greater volume than the graft and its development is assisted by the application of slight tension to the graft. Since the graft provides the necessary support, normal mobility is restored quickly after implantation. Normal exercise causes the new tissue that develops to become coherent and oriented in the directions of greatest stress and reduced in volume, so that in time the graft becomes almost indistinguishable from the natural tendon. A typical example, a knee ligament, is shown in Fig. 13.4.

13.5.4 Sutures and ligatures

Sutures are threads that are used to hold tissue surfaces together, while ligatures are threads tied around vessels such as arteries to close them. Both are an essential part of surgery, and are supplied in a wide variety of forms to suit different surgical needs. Sutures are normally supplied already attached to needles, which are swaged onto the suture to give a joint with the same diameter as the suture.

Sutures are used to hold tissue surfaces together while healing takes place. The suture becomes redundant when healing is complete, so, ideally, the suture should be absorbed as the strength of the healing tissue develops. Because different types of tissue heal at different rates, sutures

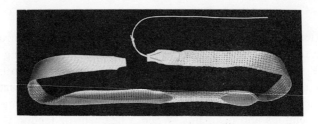

Figure 13.4   The Leeds–Keio artificial ligament. This is used to replace a
ruptured anterior cruciate ligament in the knee and is made
from woven polyester. The tubular structure enables the
ligament to be fixed by bone 'dowels', which become reinte-
grated to the host bone through the mesh of the prosthesis.
(Photograph: B B Seedhom, School of Medicine, University
of Leeds, UK)

with different absorption rates are needed. Tissue that heals slowly
requires suturing with threads that are not absorbed.

The traditional absorbable suture material is catgut, but this provokes
a strong tissue reaction and has an unpredictable absorption rate.
Polyglycolide, polylactide and polydioxanone are totally absorbed over
time at a predictable rate. Sutures that are to remain in place for the
lifetime of the patient and not be absorbed are made from both synthetic
fibres (polyester, polyamide, polypropylene and steel) and natural fibres
(silk, cotton and linen). Synthetic fibre sutures are available in monofilament, multifilament and braided forms.

Sutures require a mix of properties. The breaking strength of the suture
should be greater than the tissue it is in, though a large excess of strength
is unnecessary and undesirable, as it would mean that an oversize suture
was being used. Knot strength is of equal importance, as sutures are
secured by knotting and the knot is the weakest point. The frictional
properties of the suture affect how easily the suture can be drawn through
the tissue and how stable the knot will be. A high coefficient of friction
gives stability to the knot but increases tissue drag. To overcome this
problem, some braided sutures are coated to reduce the high coefficient
of friction typical of braided materials, using beeswax, PTFE, polybutylate
or silicone.

The reaction of the surrounding tissue to the suture is of critical
importance. A severe adverse reaction can cause serious postoperative
complications. Synthetic materials are better than natural materials in
general, and monofilaments are better than braids, in terms of minimizing
adverse tissue reactions. Minimizing the size of the suture assists in
reducing tissue reaction, with sutures as small as 30 μm in diameter being

used for microsurgical procedures. Tissue reaction and absorption rates are linked indirectly. Nonabsorbable sutures are required where the wound recovers strength at such a slow rate that an absorbable suture would have lost its strength before the wound had healed sufficiently. In other situations, the permanent presence of sutures would lead to complications arising from adverse tissue reactions and from their role in prolonging infections.

13.6 Cushioning

Cushioning fabrics are used to distribute mechanical stresses evenly in situations where stress concentrations would be harmful. The role of dressings in preventing further mechanical injury has been mentioned already. In terms of fabric usage, the prevention of decubitis ulcers (pressure sores) is probably the dominant example of the use of cushioning. Decubitis ulcers arise when tissue is subjected to external pressure for a period of time, with ulcers occurring most rapidly with increasing pressure. Such pressure sores can occur in patients recovering from surgery and are a particular problem among the physically disabled. Pressure sores can lengthen the duration of a stay in hospital significantly and hence greatly increase the cost of treatment.

A number of aids have been used in the past in the prevention and cure of decubitis ulcers, with natural sheepskin being probably the least expensive. The sheepskin was placed under the patient with the pile uppermost, to reduce the pressure on the most stressed tissue by spreading the weight of the patient more evenly. However, sheepskins are naturally limited in size and are difficult to wash and sterilize without felting (which greatly reduces effectiveness). They also retain urine, a contributory factor in causing ulcers if used with incontinent patients.

By the early 1970s, synthetic fibre pile fabrics had begun to replace natural sheepskins. Typically, the pile fabrics are sliver-knitted in polyester to give a pile height of about 20 mm. The strong backing fabric (also polyester) is treated with a polymer coating to help secure the pile, but it remains porous to prevent the retention of liquid. The fabric is made into large pads, which can be used to lift patients if necessary. Such pads can be autoclaved many times and can be used on the operating table itself, where long operations can themselves contribute to the development of decubitis ulcers.

Prosthetic heart valves, whether mechanical or biological, are mounted on a frame called a stent that is, of necessity, rigid. When fitted, the stent is surrounded by tissue that pulses with the beating of the heart. Stents are therefore fitted with a sewing ring made from woven polyester. This performs three functions: first, it provides a secure anchorage on the stent for the sutures holding the valve in position; second, it seals the valve against leaks (especially after tissue ingrowth); and third, it acts as a cushion between the rigid stent and the flexible tissue, spreading the stresses caused by the sutures more evenly.

13.7 Thermal requirements

Hypothermia is recognized as the cause of a large number of deaths each winter among the elderly. However, the chronically sick and premature babies are also at serious risk, and many occupations and leisure pursuits carry the danger of exposure to extreme cold, especially in the event of accident or injury. Synthetic fibres have found a role in the prevention and treatment of hypothermia in the form of pile fabrics, which are used as highly efficient insulators to retain body heat. The simplest form is that of an underpad for use in bed, similar to the artificial sheepskin used to treat decubitis ulcers. More complex devices take the form of sleeping-bags or cocoons. In these, the pile forms closely to the body, eliminating gaps but resisting crushing, so that a layer of insulating air is trapped in the pile. The pile also wicks away liquid, in the event of accidents involving immersion. Pads for use in bed are valuable in preventing hypothermia in the elderly and bed-ridden at home. Cocoons have been used to prevent hypothermia in patients with poor temperature regulation, and more complex cocoons have been used to treat hypothermia in accident victims in hostile environments such as arise in offshore, mountain and cave rescue.

13.8 Psychological comfort

The psychological comfort of a patient is important in promoting recovery and minimizing the stress of treatment. Modesty gowns are worn by patients in situations where medical procedures have to be carried out. The modesty gown is normally a loose-fitting garment that opens at the back and is closed by fabric ties. Single-use modesty gowns made from nonwoven fabrics are now becoming common, especially in clinics and surgeries that lack laundry facilities.

Mention has already been made of activated carbon fibre dressings in the treatment of malodorous wounds. The removal of odour has important psychological benefits; the dignity of the patient is increased and relatives and friends find visiting the patient less distressing.

Psychological comfort is also provided by a cheerful, colourful environment. Curtains and, in some areas, carpets can contribute colour and pattern. Synthetic fibres offer significant advantages over natural fibres in terms of flammability, resistance to soiling and ease of cleaning.

Further reading

It has not been possible in the space available to present more than a brief summary of the medical applications of synthetic fibres. The following publications are suggested for a fuller discussion.

Proceedings of the Conference *Medical Applications of Textiles*, University of Leeds, 6–9 July 1981. Available from the Department of Continuing Professional Education, Continuing Education Building, Springfield Mount, Leeds LS2 9NG, UK.

Proceedings of the Conference *Textiles in Health Care—An Information Exchange*, the Second Leeds Medical Applications of Textiles Conference, 17–19 September 1985. Available from the Department of Continuing Professional Education, Continuing Education Building, Springfield Mount, Leeds LS2 9NG, UK.

Proceedings of the Conference *Textiles in Medicine and Surgery*, University of Manchester Institute of Science and Technology, 11–12 July 1989. Available from UMIST, P.O. Box 88, Manchester M60 1QD, UK.

Absorbent Incontinence Products, G E Cusick and T Hopkins, *Textile Progress*, **20**(3) (1990). Published by the Textile Institute, 10 Blackfriars Street, Manchester M3 5DR, UK.

Medical Textiles, published monthly by Elsevier Science Publishers Ltd, Mayfield House, 256 Banbury Road, Oxford OX2 7DH, UK.

CHAPTER 14

Specialized uses 3: cement reinforcement

D J HANNANT

14.1 Introduction

The inclusion of synthetic fibres into cement-based products is potentially a very large worldwide market. For instance, about 90 countries produce asbestos cement for cladding, roofing or pipes and about 3.5 million tonnes of asbestos fibre are used annually in the asbestos cement and building products industries, giving about 28 million tonnes of products.[1] Most of this fibre will eventually be substituted by man-made fibres to avoid the health problems associated with asbestos fibres. Additionally, the concrete industry represents a production of about half a tonne per head of population per year in the industrialized countries and even a small fibre percentage (typically less than 1 per cent) in a small proportion of this output could represent a substantial market for a producer of polymer fibres. It is clear, therefore, that there is a potential market and since the early 1980s considerable inroads have been made by synthetic fibres into this industry.

The main objectives of the engineer in including fibres to modify the properties of cement or concrete are to improve the tensile and impact strengths, to control cracking and the mode of failure by providing postcracking ductility or to change the rheology of the material in the fresh state.[2] Major limitations of the matrix in these respects are its low tensile failure strain of less than 0.05 per cent, the high elastic modulus (~ 30 GPa) and the limited capacity for fibre incorporation, which will generally be less than 6 per cent by volume in fine grained mortar and less than 1 per cent by volume in concrete that may already contain 70 per cent of its volume as aggregate particles. Additionally, the fibre–cement bond is poor with many polymers and the high alkalinity of the paste (pH 12–13) may cause degradation of some synthetics. The properties of the cement paste will also change with time and ambient conditions, which makes an accurate theoretical prediction of composite performance difficult to achieve.

14.2 Mechanism of reinforcement

14.2.1 Fresh concrete

Extensive use has been made in the construction industry of small quantities (0.1 per cent by volume) of short (<25 mm) fibrillated or monofilament fibres to alter the properties of the fresh concrete. Fresh concrete refers to the material in the first 4 or 5 hours after mixing, either before or just after the concrete has stiffened but could not be described as hardened. 'Reinforcement' is probably not a correct term to describe the effects of the fibres and there is no theoretical treatment that adequately predicts their effects. Also, the type of polymer from which the fibre is formed is probably not very important because little use is made of the fibre stiffness or strength during this critical initial period and alkalis have had no time to cause damage. The following description of their mode of action is therefore mainly qualitative.

Badly made and cured concrete may suffer from sedimentation of the aggregate particles, with consequential bleeding of water to the surface of the cast layer. With excessive sun or wind, this water may rapidly evaporate, with an increase in suction and shrinkage on a horizontal surface that can cause substantial cracks known as plastic shrinkage cracks to form. Some tests[3] have shown that small quantities of fibres in the mix increase the cohesion and prevent sedimentation owing to their interlocking network characteristics. The result may be that in some cases plastic shrinkage cracking may be reduced, not necessarily by the reinforcing effect of the fibres but by the reduction of sedimentation and bleeding. There is an insufficient fibre volume to expect reinforcement in the hardened state.

Similar small amounts of polypropylene monofilaments have been used in highly air-entrained concretes[2] to hold a pattern rolled into the surface with a shaped roller.

Another use for polymeric fibres in the fresh cement material is as a filtration aid in replacing asbestos fibres in alternatives to asbestos cement sheeting products. In systems such as the Hatschek process where a large amount of water has to be removed from the cement slurry, it is essential to keep an open pore structure while retaining cement particles. In polymer fibre systems assistance in holding the cement grains is provided by treating the fibre surface to make it hydrophilic.

14.2.2 Hardened concrete

The principles by which hardened concretes or hardened cement mortars are reinforced by fibres depend crucially on the strength, volume, size and spacing of the fibres in the matrix, and three different failure mechanisms may occur depending on the values of these parameters.

Uniaxial tension

The simplest way to understand the mechanics of these three reinforcing mechanisms is to examine the uniaxial tensile stress–strain behaviour.

Figure 14.1 shows the three basic types of tensile stress–strain curve available to the engineer. Curves B and C are based on the assumption that the stress in the composite is increased at a constant rate.

For all three curves, the portion OX defines the elastic modulus of the uncracked composite (E_c). In curve A, after the cracks have formed at X the fibres slowly pull out and absorb energy, leading to a tough but rather weak composite typified by steel fibre concrete or short random fibre polypropylene concrete. In curve C, which is representative of asbestos cement and of some cellulose-fibre-based composites when dry, the relatively high crack propagation stress leads to a sudden large release of energy and to an almost instantaneous fibre fracture or fibre pull-out at C. However, stable microcracks may exist well before point C is reached. The mechanism of the reinforcement of cement by asbestos fibres is very complex and leads to a special case of a rather brittle composite with a high tensile strength. A full understanding of the fracture process has not yet been achieved, but it appears to require a combination of the reinforcing theory described for curve B with a knowledge of fracture mechanics.

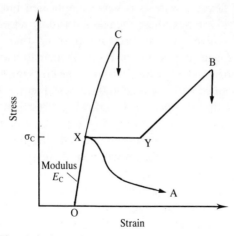

Figure 14.1 Theoretical tensile stress–strain curves for different fibre composites.

Curve B is typical of a composite in which there are sufficient fibres to maintain the load on the composite when the matrix cracks; the theoretical principles behind curve B are given in references 2 and 3. The horizontal portion X–Y is a result of multiple cracking at approximately constant stress and Y–B represents extension or pull-out of fibres up to separation of the component into two parts at point B. Curve OXYB is typified by fine grained materials such as glass-fibre-reinforced cement or some continuous fibre composites such as polypropylene-network-reinforced cement.

For the simplest case of aligned continuous fibres with frictional bond, the minimum fibre volume ($V_{f(crit)}$) required for the composite to follow curve OXY is given by Eqn 14.1.

$$V_{f(crit)} = \frac{\sigma_c}{\sigma_{fu}} \qquad [14.1]$$

where σ_c is the cracking stress of the composite and σ_{fu} is the strength of the fibre.

In the more common practical case of short, randomly oriented fibres where the fibres pull out at a crack rather than a break, the stress in the fibre causing pull-out may be substantially less than σ_{fu}, and hence $V_{f(crit)}$ for such composites may be up to 10 times that required for continuous aligned fibres. Appropriate efficiency factors for fibre length and orientation have been given in references 2 and 3.

Flexural tension

The analysis of flexural systems is very complicated but, to understand the difference between flexure and tension, it is only necessary to identify the principles involved. It has been established[2] that, because of the movement of the neutral axis and the resulting increased area of the tensile stress block in postcracked flexural systems, the cracking load in the beam can often be sustained with material in the cracked tensile zone supporting less than half of the uniaxial tensile cracking stress. This means that a material with a fibre volume of only 50–60 per cent of $V_{f(crit)}$ in tension will enable the load on a beam to be increased after the first crack point. The bending moments in the beam will then increase, resulting in further matrix cracking remote from the first crack even though, for the material in the tensile zone, $V_f < V_{f(crit)}$. Thus we may have an apparent anomaly that the material from which the beam is made may be less than $V_{f(crit)}$ while the beam, because of the loading systems, has a fibre volume apparently greater than $V_{f(crit)}$.

The result is that, in flexure, multiple cracking and hence ductility and toughness may be achieved at considerably smaller fibre volumes than in tension. Hence a composite might remain tough and ductile in flexure but could become brittle in tension.

Toughness

A major reason for including fibres in cement-based products is to increase their toughness.

Toughness is a measure of the amount of energy absorbed by the composite before a certain degree of damage is reached, and this energy depends very much on whether fracture occurs mainly by a single crack opening or by multiple cracking where energy is absorbed throughout the whole volume of the material. Equations for predicting toughness are given in reference 4.

14.3 Durability

The specific factors that affect the durability of individual fibre types in the alkaline matrix are mentioned in section 14.4. However, there are some general principles that affect the durability of the composite as a unit. It has already been stated that one of the most desirable features of fibre-reinforced cement-based materials is their toughness and ductility in relation to the unreinforced matrix. Thus a change under natural weathering exposure from ductile to brittle characteristics may be considered to be a durability problem even though the tensile strength of the composite may be little changed or even increased. The reason for such changes may be explained by reference to Eqn 14.1. The critical fibre volume $V_{f(crit)}$ for tensile strengthening is dependent not only on the fibre strength (σ_{fu}) but also on the composite cracking stress (σ_c). A characteristic of cement matrices in damp conditions is that σ_c may increase with time owing to continuing hydration of the cement grains. This effect, whether or not the fibre strengths are time stable, may be sufficient to increase $V_{f(crit)}$ in a practical composite to a value greater than the included fibre volume and thus embrittlement or lack of durability may occur after some years if this effect is not allowed for at the production stage.

14.4 Polymer fibre used in cement-based products

Typical properties of synthetic fibre used in cement-based products are shown in Table 14.1.

Table 14.1. Typical fibre properties

Material	Relative density	Diameter or thickness/μm	Length/mm	Elastic modulus/GPa	Tensile strength/MPa	Failure strain/%	Volume in composite/%
Aromatic polyamides (aramids)	1.45	10–15	5–Continuous	70–130	2900	2–4	1–5
Carbon	1.6–1.95	7–18	3–Continuous	30–390	600–2700	0.5–2.4	3–5
Polyacrylonitrile (PAN)	1.16	13–104	6	17–20	900–1000	8–11	2–10
Polyethylene pulp	0.91–0.97	1–20	1	–	–	–	3–7
HDPE filament	0.96	900	3–5	5	200	–	2–4
High modulus	0.96	20–50	Continuous	10–30	>400	>4	5–10
Polypropylene monofilament	0.91	20–100	5–20	4	–	–	0.1–0.2
Chopped film	0.91	20–100	5–50	5	300–500	10	0.1–1.0
Continuous nets	0.91–0.93	20–100	Continuous	5–15	300–500	10	5–10
Poly(vinyl alcohol) (PVA, PVOH)	1–3	3–8	2–6	12–40	700–1500	–	2–3

14.4.1 Aromatic polyamides—aramids

Chopped fibres

Aramids have a high strength and modulus of elasticity and would be expected to give excellent composite properties when used in a cement matrix. However, the price is so high that tests have mainly been restricted to laboratory trials.[5,6] In these tests the properties expected from theoretical considerations have been realized, leading to strong, tough composites with excellent creep resistance. Resistance to alkalis is not entirely clear but the 20-year performance of the uncoated fibres in a cement matrix is thought to be satisfactory.

Continuous fibres

Although not producing strictly fibre-reinforced cement, continuous aramid fibre bundles have been used both as reinforcement and as prestressing tendons in concrete structures.[7] In order to avoid damage and also to give protection against alkalis, the fibre bundles may be impregnated with epoxy resin or coated with a polymer sheath. Aramids avoid corrosion problems associated with steel tendons and show considerable potential for the future.

14.4.2 Carbon fibres

Carbon fibres of a wide variety of strengths and stiffnesses have been used in laboratory trials.[3] It is not possible to summarize the composite properties because these are so dependent on fibre length, strength and volume. However, excellent strength properties can be achieved with little change during accelerated alkali testing. Long-term durability would be expected to be good. As with aramid fibres, a major problem preventing their-large scale use in concrete products is their price. An additional problem with carbon fibres is that during mixing some of the fibres break and become shorter, thus reducing their efficiency.

14.4.3 Polyacrylonitrile fibres

Fibres of diameter 13 µm and 6 mm long have been used in cement mortars and fibres of 104 µm diameter in concretes. The fibres are kidney shaped rather than round, but equivalent diameters are stated. The main purpose of these applications has been to reduce cracking due to drying shrinkage and to increase the flexural ductility. No claims are made for improved tensile strength.

Chopped fibrillated films of polyacrylonitrile have also been used on a trial basis on a Hatschek machine as a substitute for asbestos fibres, and satisfactory flexural strengths were achieved.[8]

Although acrylic fibres are relatively stable in an alkaline environment, some tests have indicated small losses in strength and therefore there is some uncertainty about the long-term performance.[3]

14.4.4 Polyethylene fibres

Short fibre pulp

Polyethylene pulp has mainly been used as a cement retention and drainage aid as a substitute for asbestos fibres in a Hatschek type process for the manufacture of thin sheet products. Up to 12 per cent by volume has been used and at this level improvements in flexural strength and ductility have also been observed. Because the fibres do not swell in the presence of water, the durability of the products is said to be improved in comparison with similar systems using cellulose fibres.

Continuous networks of high modulus fibres

Highly oriented polyethylene fibres may be produced by gel spinning or high draw ratios and fibres have been produced with the elastic modulus of glass and the strength of steel. Commercial fibrillated tapes with initial elastic moduli of about 30 GPa have been used in thin cement sheets in a similar fashion to polypropylene nets, and improved composites in terms of stiffer postcracking performance and smaller crack widths have been observed.[9,10] However, at the time of writing there were no opened nets available with widths of about 1 m suitable for factory production of the composite.

Durability in alkalis is expected to be good but the films that have been available have suffered from high creep strain in comparison with polypropylene.

14.4.5 Polypropylene

Polypropylene has been used in many forms since 1970 to modify some properties of fresh and hardened concrete. Fibre additions have ranged from very low concentrations (~ 0.1 per cent) of 5 mm long round mono-filaments through similar volumes of 5–25 mm long split fibres to about 0.4 per cent by volume of 40 mm twisted twine and up to 5 per cent by volume of continuous opened networks. The significant features of these applications are detailed below. Polypropylene has been shown to be very durable in the alkaline environment of cement and has shown little change in strength after 10 years in natural weathering or 10 years under water.[11]

Monofilament

Small volumes (0.05–0.2 per cent) of fine (20–50 µm) round fibres have been used to modify the properties of fresh concrete. They have been used in highly air-entrained concretes to enable the surface to hold a sharp rolled-in profile and also in normal concretes to provide additional cohesion and resist bleeding plastic shrinkage cracking. There is no evidence that the tensile strength of the hardened concrete is improved but there is some controversy related to potential improvements in durability.

Chopped films

There is a large potential market in the use of about 0.1 per cent by volume of untwisted split film in lengths of 20–50 mm in the ready-mixed concrete industry. The main benefits claimed are a reduction in bleeding and reduced sedimentation of the aggregate particles, which can reduce the incidence of plastic shrinkage cracking in badly protected concrete. This is a significant problem in the industry and the fibre is thought to work by dividing into five filaments during the mixing process; these filaments produce a more cohesive mix.

Chopped twine

There are significant differences in the uses for and mode of action of chopped twine in concrete compared with chopped untwisted split film. Chopped twine is used to increase the impact resistance of precast concrete elements, typically pile shells and marine protection units. About 0.4 per cent by volume of chopped staple lengths are used in normal concrete mixers.[2] The composite is compacted into moulds by vibration and pressure and, when hardened, the concrete is provided with considerably increased resistance to impact by the chopped twine, which prevents cracks propagating and holds the finely cracked sections together.

Continuous networks of polypropylene fibrillated film

This system contains about 6 per cent by volume of networks, which is greater than the critical fibre volume in tension, the networks being arranged in layers in orthogonal directions. The material is typified by high postcracking toughness and excellent durability. Network packages are produced specifically for this application[12] and their impregnation with fine grained cement mortar is a separate specialized factory operation. The networks of polypropylene are made to be easily wettable by surface treatments and by fine powder inclusions within the highly drawn films that are typically between 50 and 80 μm thick.[13] The main market area is for flat or corrugated sheeting for roofing and cladding and products of this type based on patents of the University of Surrey have been on sale in Italy since 1987.[14] The potential world market for fibre-reinforced cement-based sheeting is in excess of 1 billion pounds sterling per annum based on the known sales of asbestos cement, which is being replaced in many industrialized countries due to the health hazards of asbestos fibres. The equivalent potential market for polypropylene as a reinforcement is therefore very substantial.

Poly(vinyl alcohol) (PVA) fibres

PVA fibres of high strength and stiffness are used widely as an asbestos replacement in asbestos cement products. However, taken by themselves in a cement slurry, they offer little retention of the cement grains and hence must be used in conjunction with cellulose pulp to keep the cement in the system as water is sucked out by vacuum.[1] The fibres are treated on the surface to enhance their compatibility with the matrix, the quantity of fibres being typically 3 per cent by volume. Flexural strengths of the sheeting are adequate to meet the requirements of the appropriate European standards. Alkali resistance has been stated to be excellent and the fibres can survive exposure to temperatures of 150°C without loss in strength.

14.5 Concluding remarks

The use of synthetic fibres in the civil engineering field and particularly in cement and concrete products is a growing market with considerable commercial potential. An immediately available market is the substitution of asbestos fibres in asbestos cement products and this will continue throughout the 1990s.

Price and product performance in relation to other materials for sheeting and pipes will determine whether market share is maintained. The bulk concrete market also has considerable potential but it will be harder to gain entry for fibres because the price of concrete is very low and the performance is already adequate in most respects. Identification of areas where low volumes of cheap fibres can produce a significant improvement in properties may be crucial to this market.

References

1. A Hodgson, *Alternatives to Asbestos and Asbestos Products*, Anjalena Publications, Crowthorne, 1987.
2. D J Hannant, *Fibre Cements and Fibre Concretes*, Wiley, Chichester, 1978.
3. A Bentur, S Mindess, *Fibre Reinforced Cementitious Composites*, Elsevier Applied Science, Barking, 1990.
4. A Hibbert, D J Hannant, *Composites*, **13**, 105 (1982).
5. A L Walton, A J Majumdar, *J. Mater. Sci.*, **18**, 2939 (1983).
6. P L Walton, A J Majumdar, *Properties of Cement Composites Reinforced with Kevlar Fibres*, Building Research Establishment Current Paper CP 57/78, 1978.
7. H J Scherhoff, A Gerritse, Aramid reinforced concrete (ARC), Paper 2.6 RILEM Symposium, Sheffield 1986, Publ. RILEM Technical Committee 49 TFR.
8. Amrotex AG, UK patent application GB 20 65 735 A (1981).
9. D C Hughes, D J Hannant, *J. Mater. Sci.*, **17**, 508 (1982).
10. P L Walton, A J Majumdar, *J. Mater. Sci. Lett.*, **3**, 718 (1984).
11. D J Hannant, The effects of changes in matrix properties with time on the measured and predicted long term properties of fibre reinforced cement. American Concrete Institute Montreal Symposium, 1991.
12. A Vittone, Industrial development of the reinforcement of cement based products with fibrillated polypropylene networks as replacement of asbestos. Paper 9-2 RILEM Symposium Sheffield 1986. Publ. RILEM Technical Committee 49 TFR.
13. D J Hannant, J J Zonsveld, *Phil. Trans. R. Soc.*, **A294**, 591 (1980).
14. University of Surrey, UK patent 1 582 945 (1981).

CHAPTER 15

Specialized uses 4: monofils

B M McINTOSH

15.1 Introduction

In relation to synthetic fibre production, monofils, or single filament yarns, are often considered to be in the 'back yard', low technology end of the business.

Such a reputation is not fully borne out by the facts, and while, at one extreme, low cost monofils of polypropylene or polyethylene can be made with minimal capital investment or technical expertise, at the other, monofils for precision filters or specialized conveyor belts are not only sophisticated in terms of the polymers used and the technology employed, but also sell at prices in excess of the most highly valued multifilament yarns.

The following section on monofil applications starts with a definition of monofils and a general consideration of the advantages and disadvantages of single filament yarns in textile and other applications. An end-use by end-use consideration will then be presented to try to elucidate the reasons behind the choice of polymers and diameters for monofils in the various industries involved.

15.2 Definition

The definition of a monofil is normally taken to be a single filament product, with a diameter in the range 100–2000 µm. Although this covers the vast majority of applications, some specialized products are produced in the diameter range 30–100 µm. These are not normally manufactured using conventional monofil spinning equipment, and this tends to exclude them from consideration as monofils.

Synthetic monofils are produced by melt extruding thermoplastic polymers. The only major exception to this is in the case of some fluoropolymers that can be produced by a powder sintering route. Natural monofils such as pig bristle or animal gut are in small and declining use in a very few specialized applications; and more than in other fibre areas, monofils can be thought of as being fundamentally the preserve of synthetics.

15.3 Features of monofils

The aspects that lead to the choice of monofils for any given application are most commonly (a) high flexural rigidity, (2) good surface release, and (3) resistance to damage in use. In addition there are processing considerations that limit the practical production or use of monofils and these will be discussed under (4) limitations in use.

15.3.1 Flexural rigidity

The most obvious difference in performance between monofil and multifil yarns is the great increase in stiffness or flexural rigidity available when using monofils.

The flexural rigidity (FR) is defined as the force required to bend a fibre about its neutral axis to unit curvature ($r = l$). See Fig. 15.1.

The equation relating this rigidity or fibre stiffness to the key fibre parameters can be expressed as

$$\text{FR} = \frac{1}{4\pi} \frac{SET^2}{\rho} \times 10^{-3} \qquad [15.1]$$

where, in practical textile units, FR is the flexural rigidity in N mm^2, ρ is the material density in g cm^{-3}, E is the specific modulus in N tex^{-1}, T is the linear fibre density in tex and S is the shape factor, a dimensionless number.

S is related to the radius of gyration (k) of the fibre cross-section about

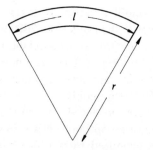

Figure 15.1 Monofil flexural rigidity, unit radius of curvature.

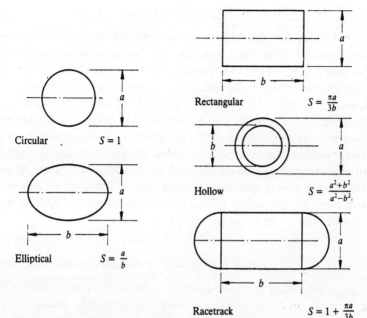

Figure 15.2 Shape factor calculations; typical cross-sections of monofils used in different applications.

the axis of bend by

$$S = \frac{4\pi k^2}{A}$$ [15.2]

where A is the area of cross-section.

Shape factors can be calculated for a number of possible monofil cross-sections, and the relative stiffness can then be compared. The axis about which bend occurs is taken for nonsymmetrical cross-sections to be the natural or easiest axis. Shape factors are shown in Fig. 15.2.

The principal influence on stiffness is seen from Eqn 15.1 to be fibre linear density, where the flexural rigidity increases as the square of the individual fibre linear density, or as the fourth power of the individual fibre diameter.

As an example, for a polymer with a density of 1.3 g cm^{-3}, a circular 40 tex monofil (diameter 0.20 mm) can be calculated to be 1600 times as stiff as a 1 tex monofil (diameter 0.0315 mm) and therefore 40 times as stiff as a 40 tex multifil with 40×1 tex filaments. In some applications multifils can be stiffened by twisting to high levels, and possibly even by resin coating to produce pseudo-monofils. These methods are expensive and cannot effectively close the stiffness gap, although the substitution of monofils by pseudo-monofils may have other advantages, especially in relation to relieving internal compressive strains on bending.

15.3.2 Surface release

A second important characteristic of monofils is that they offer very much better surface release than multifils. This is particularly important in applications that require cleaning, as in filter screens. Circular monofils present the minimum surface area for a given yarn linear density.

Taking the case of the previous example of the 40 tex monofil and multifil yarns, the monofil can be calculated to have only a sixth of the surface area of the multifil. This, combined with the typically smooth extruded surface characteristic, provides as good a release surface as is available without resorting to topical treatments. The reverse situation will of course arise if the fibre is being used to reinforce or adhere to a matrix material or coating. In this case the much greater surface area of the multifil proves to be an advantage.

15.3.3 Resistance to damage

The large differences in exposed surface areas between equivalent monofil and multifil products also dominates the product's ability to resist aggressive environments. This is particularly true for chemical or ultraviolet attack. As an example a comparison is given in Fig. 15.3 of the loss of strength suffered by two equivalent 65 tex fibre products in an exposure tester.

As with all such cases there is no simple relationship between area of

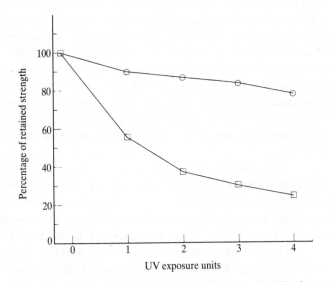

Figure 15.3   Comparison of loss of strength on exposure to UV of monofil (○, 650F1) and multifil (□, 650F90).

exposure and the rate of loss of performance. Many other factors come into play. The monofil, however, will always comfortably outperform the multifil yarn—in this actual example by a factor of three or four times.

In the situation of resistance to abrasive wear, the phenomena involved in failure are more complex still. If the mechanism of attack is primarily one of surface rubbing at a low angle of contact, monofils generally perform well, and may survive in service despite eventual local weight losses as high as 30 or 40 per cent. Multifil yarns will suffer relatively rapid filamentation and failure in this situation. However, if the mechanism of abrasion contains significant flexing and bending of the structure, the situation can be reversed—fine fibres moving to share the surface abrasion, while thick monofils, by being put under significant internal compressive loads, can suffer accelerated flex fatigue failure.

15.3.4 Limitations in use

Some of the limitations of monofils as fibres in textile structures have been suggested in the previous sections. In general, for applications that require flexibility or softness, such as in apparel or domestic textiles, monofils are unlikely to find application. Even in relation to industrial uses there are a number of further technical and trade aspects that can limit areas of application further.

Both the initial production and, where appropriate, the further textile processing of monofil products requires significantly different techniques and equipment than for finer thermoplastic fibres. Because of the high heat capacity of the individual filaments, monofil spinning machines rely mainly on liquid quenching and relatively low speed roll sets to complete the orientation and heat setting of the products. String-up, especially for higher diameters, is manual, and speeds are limited to a few hundred metres per minute. Multifil spinning machines with automatic pneumatic string-up devices and air quenching have, on the other hand, been developed to run efficiently at many thousands of metres per minute.

The economics of production and capital employed have therefore grown to be very different, and monofil spinning can no longer necessarily be taken as being the least expensive route from polymer to fibre.

A consequence of the relatively slow and uneven rates of heating and cooling through thick filaments is that optimum product strengths achievable are lower for monofils than for multifils. Compromises in process settings need to be made to accommodate the greater skin/core differences that inevitably arise in monofil production. With multifils, rapid even heat and tension treatments ensure that higher optimum properties can be achieved.

Finally, in terms of the onward textile processing of monofils, while finer fibres are fundamentally suitable for all types of weaving, knitting, spinning and texturing, specialist looms are required in the case of monofils. Otherwise, with the exception of cabling and some laying-in at knitting, monofils are basically unsuitable for normal textile processing. This clearly limits the textile trade outlets for monofil products.

15.4 Monofil applications

The major areas in which monofils find application are set out and discussed under five broad categories—(1) agriculture and fisheries, (2) paper manufacture, (3) industrial, (4) textile, (5) leisure. As far as possible the basic technical and cost reasons for the choice of polymers, specifications, and the finished structures for monofils are discussed in relation to each end-use. A summary of the polymers used in each application is given in Table 15.1.

15.4.1 Agriculture and fisheries

These areas cover the use of monofils in farming, fishing and outdoor working applications. The principal requirements here are for good UV stability, adequate robustness and low cost.

Ropes and cords, in situations where a degree of rigidity is required, are usually made up of a number of monofils, typically between 0.2 and 0.5 mm in diameter, twisted and plied together. Polyethylene (PE), normally high density, or polypropylene (PP), chemically stabilized and pigmented to resist light degradation, are mostly used. Blends of the two polymers combined with two-stage drawing can give significantly stronger products, up to 60 or 70 cN tex^{-1}, where this is required. Applications are for supporting plants, such as hops or vines; as general twine (although fibrillated tape is more common), and in stock control electric fencing, where polyethylene monofils are twisted together with copper wire.

Monofil-based marine ropes are usually polypropylene, low moisture uptake and low density (0.9 g cm^{-3}) guaranteeing flotation. Again, unyielding applications requiring little flexibility or knotting are chosen, such as for net supports, tethering or trawl ropes. A substantial rope structure of this type might, for example, contain 32 monofils of 0.30 mm diameter combined together at 50 turns per metre, 72 of these twisted ends being cabled to give a rope structure containing over 2300 monofils with a breaking strength of 8 tonnes. Nets using monofils find application in fisheries, and to some extent in general agricultural, industrial and sporting end-uses. Inexpensive pseudo-monofil nets can be produced directly in various patented integrated spinning processes, but for well specified netting products monofils are knotted or knitted together into

Table 15.1. Summary of principal applications/polymers/diameters

Applications	PVC	PE	PP	N	PET	PBTB	PPS	PEEK	PTFE	Diameter/mm
Agriculture and fisheries										
Ropes		✓	✓	✓	✓					0.2–0.5
Nets		✓	✓	✓						0.3–0.6
Strimmer lines				✓						1.0–3.0
Paper manufacture										
Forming				✓	✓					0.1–0.3
Pressing				✓						0.1–0.3
Drying					✓		✓	✓		0.4–0.6
Industrial										
Conveyors				✓	✓		✓	✓		0.4–0.8
Filters		✓	✓	✓	✓		✓	✓	✓	0.03–0.3
Brushes	·✓	✓	✓	✓	✓	✓				0.1–1.5
Rubber reinforcement				✓	✓					0.4–1.5
Screening printing				✓						0.03–0.1
Textile										
Zip fasteners				✓	✓					0.4–1.0
Sewing threads				✓						0.1–0.5
Stiffenings				✓	✓					0.03–0.1
Leisure										
Sports racket strings				✓				✓		0.1–0.7
Fishing lines				✓						0.1–0.7

net structures. Polyethylene or nylon (normally nylon 6), which are inherently much more flexible than polypropylene, are preferred. Poly-(vinyl chloride) (PVC) monofils can also be used. Typical fishing net products are manufactured by knotting 0.4 mm diameter high tenacity polyethylene or nylon monofils, which have latent shrinkage to ensure the locking of the knotted structure. Industrial netting such as debris netting, used for safety protection around building sites, is often a knitted structure based on 0.25 mm polyethylene monofils. Lower cost sports netting for tennis and soccer will also be monofil based, and invariably pigmented black.

Strimmer lines, used for cutting grass and vegetation with high speed rotating heads, are a good demonstration of the need to develop highly specified monofil products. Polymers used in this application are basically nylons, and plasticized polymers or co-polymers (nylon 6,6/6, nylon 6/9 etc.) are extensively used to provide the right balance of flexibility, abrasion resistance and temperature tolerance. In operation the flex point of the line undergoes rapid local heating, and fusing has to be avoided. Good dimensional stability and control in terms of diameter uniformity and shrinkage are also required to avoid difficulties with take-off from small diameter plastic reels. Diameters used in this application are very large, typically 1.5 mm for garden strimmer lines and between 2.0 mm and 3.0 mm for agricultural cutters. This is a performance driven rather than a price sensitive end-use.

15.4.2 Paper manufacture

Monofils woven into continuous conveyor belts are involved in the manufacture of paper. There are three separate but closely integrated stages in the continuous production of paper sheet—forming, pressing and drying—and different types of monofil conveyor belting are used for each stage.

Forming is involved with the forced drainage of the water from the cellulosic slurry, leaving the paper sheet with sufficient internal cohesion to transfer to the second stage. A forming fabric is a fine open mesh conveyor that has to be dimensionally stable over considerable machine widths (up to 10 metres) to allow for an even drainage from the slurry. It must also resist the abrasion on the underside of the fabric against vacuum extraction boxes and machine elements. These aspects are usually covered by designing a two- or three-layer woven fabric, principally constructed using polyester (PET) monofils in the diameter range 0.15–0.30 mm, leading to good dimensional stability. Nylon monofils of similar diameters are used in the underside of the conveyor to improve wear properties against the machine frame.

Pressing involves removing as much of the excess water from the paper sheet as can conveniently be done using a combination of heat and pressure. Conveyor fabrics in this stage are sponge-like in action, commonly known as felts, absorbing water from the paper sheet under compression at roll nips, and depositing this water when not in contact with the paper, before recontacting new paper sheet. Press felts are predominantly staple fibre in composition, but this fibre is needled into a resilient monofil base weave. Nylon 6,6 or nylon 6 monofils in the diameter range 0.15–0.30 mm are typically used. Nylon monofils are the best at resisting the tendency of many monofils to split longitudinally under compression. In addition, by using small groups of twisted monofils in the weave, compressive forces are further dissipated and a resilient base weave with good recovery is achieved.

Drying is the final stage of the paper manufacturing process and it typically uses a succession of woven monofil conveyors, either to hold the paper against steam-heated rolls, or to carry the paper sheet through hot-air ovens. Water is flashed off at temperatures anywhere from 100 to 200°C to leave the paper with a final moisture content below 10 per cent. Conveyors at this stage typically employ a tight mesh, double-layer woven structure, using 0.4–0.5 mm diameter monofils. Monofils may also be racetrack or rectangular in cross-section (e.g. 0.4 mm × 0.6 mm), especially for the warp yarns. Polymers used are normally hydrolytically stabilized polyesters, but for higher temperature installations (> 150°C) the use of polyphenylene sulphide (PPS) monofils, and for ultimate performance poly(ether ether ketone) (PEEK) monofils are increasingly being used. These newer polymers, although more expensive, have almost complete hydrolysis resistance, and, in the case of PEEK, also much better abrasion resistance and dimensional stability at temperature than the polyester monofils.

15.4.3 Industrial

Conveyors in industrial processes that use monofils may have similar requirements to paper conveyors, but many do not. Food preparation, nonwovens' manufacture and textile fabric finishing are typical examples of processes using monofil conveyors. The choice of polymer type and monofil diameter depends on the details of the temperature and the exposure conditions of the industrial process involved. In addition to woven structures, these types of conveyors can utilize a linked monofil coil structure. Continuous monofil coils are formed by heat setting around a mandrel; coils are then intermeshed and locked together using straight lengths of monofils inserted in the weft direction to make a belt. Predominantly PET monofils in the diameter range 0.5–0.8 mm are used. Coiling with PEEK monofils is also established.

Filters, mainly in the form of plain woven monofil fabrics, play an important role in many industries. The range of filter fabrics produced is large, going from relatively open sieve-like meshes based on 0.3 or 0.4 mm monofils to tight woven specialist 0.03 or 0.04 mm monofil fabrics. A wide variety of materials is used, with polypropylene being important in many lower temperature applications; nylon 6, polyester and PPS at higher temperature; nylon 11, poly(butylene terephthalate) (PBTP) and PEEK in the more rigorous filter press applications; and the fluoropolymers (for example PTFE) for severe chemical exposure. The convenient release of filter cake material from the fabric surface is an important reason for choosing monofils in many applications.

Brushes represent a most diverse application area for monofil products. They are produced in a complete range of diameters from less than 0.1 mm to greater than 1.0 mm, and in some cases with different cross-sections. All major polymer types are used, covering different aspects of several activities.

PVC monofils are used in fine diameters for household floor brushes. Low cost, with reasonable recovery and creep performance and the ability to colour to a full range of clear shades, are the main attractions. The abrasion performance is poor, however.

Polypropylene monofils are used in heavy diameters for mechanized street-cleaning brushes, where a stiff action with good abrasion performance and low cost is balanced against relatively poor elastic recovery and creep. Polypropylene is also one of the main monofils used in paint brushes to replace pig bristle. Profiled fibres can be produced, thinning to the tip for optimum performance.

Polyethylene fibres are low in modulus—extremely soft and flexible, with reasonable durability and cost. End-uses include automated car-wash brushes and food-cleaning brushes, where damage to vegetables or fruit is at risk.

Nylon brushes are used for their excellent abrasion resistance and superior recovery from large deflections. They make good quality industrial and road-cleaning brushes, and in the domestic situation are used in draught excluders and vacuum cleaners. In wet applications, standard nylon suffers from softening, while nylon 6.12 with lower moisture uptake has good wet performance, and is preferred for applications such as dishwasher brushes and toothbrush bristles. PBTP is used also for toothbrushes and is preferred in many agricultural machinery wet applications mainly on a cost basis. For brushes that require steam sterilization, in hygienic medical applications, nylon 11 or PET is commonly used.

One interesting aspect in the production of monofils for brushes is that, especially for the larger diameter bristles, product is often automatically cut and bundled in lengths rather than wound on reels, as is standard for most monofil production.

Rubber reinforcement is normally an area dominated by multifilament nylon, polyester or acrylic yarns. Most applications use the flex and recovery of rubber in areas such as tyres or inflatable structures, and need the reinforcement fibre to complement this flexibility. Multifils are usually twisted to improve flex fatigue and to modify stiffness, and with their high surface area are also easier to bond chemically to latex. Despite this, some applications for monofils do exist. The main areas are in hose reinforcement, where polyester monofils are used in cord form to spirally reinforce large diameter hoses—typically seven 0.45 mm monofils are twisted and heat set into a cord. In the side wall of tyres, by using elongated racetrack cross-section (e.g. 1.25 mm × 0.4 mm) high tenacity nylon monofils, greater fibre packing densities can be achieved at well controlled stiffness, with adequate flex fatigue performance.

Screen printing is an area where fine monofil wovens have effectively supplanted silk screen fabrics, because of both cost and performance factors. Polyester monofils dominate this market because of their uniformity and dimensional stability. Diameters are extremely fine by monofil standards, 0.03 to 0.1 mm, and diameter uniformity and weaving quality have to be very high to avoid blemishes or unevenness in the screen surface.

15.4.4 Textile

There are some marginal domestic textile applications for 100 per cent monofil woven fabrics in areas such as blinds, shower curtains and flexible seat fabrics. The main impact of monofils in textiles, however, is in the make-up areas, in products such as fastenings, threads and stiffenings.

Zip fasteners are fabricated directly from monofils, being remoulded to the required shape in a continuous process; diameter uniformity is particularly important for good processing. Nylon monofils at about 0.5 mm diameter are used for most apparel garment zips, with nylon 6,6 giving a significantly stiffer product than nylon 6. Nylon is chosen for its excellent abrasion resistance, and also for its ability to dye readily to a wide range of shades for garment matching. Some polyester monofils are also used, normally for the larger nongarment zips, using diameters in the range 0.6–1.0 mm. Hook-and-loop fasteners are an area expanding rapidly with the expiry of patent cover. Only the hook is a monofil, usually nylon 6 of about 0.2 mm diameter.

Sewing threads are normally multifil or spun staple in construction, but a market does exist for monofil products. Nylon monofil in a transparent form is used in diameters ranging from 0.1 to 0.3 mm, with a tenacity of about 60 cN tex^{-1}. The product has residual shrinkage in hot water, which allows the stitches to bed down, and, with the thread picking up its only colour from the surrounding fabric, stitches virtually disappear. This is especially useful in the repair of seams on existing garments. Medical threads in the form of sutures for wound closing are often monofils. Different polymer types are used, with polypropylene and nylon predominating in a diameter range of 0.3–0.5 mm.

Stiffenings or webbings in the form of narrow woven fine monofils are used in a number of apparel and domestic applications, such as in waist bands and curtain tapes. As an example, in the case of curtain tapes a substantial market exists for polyester monofil, usually undyed in a diameter range from 0.2 to 0.5 mm.

15.4.5 Leisure

Sports racket strings, at one time made with animal gut processed into monofils, are now principally constructed of nylon monofils in the diameter range 0.1–0.7 mm. These are combined together by twisting, resin coating and heat setting to give flexible, compression-resistant circular string products of about 0.8 mm diameter for badminton rackets and 1.4 mm diameter for tennis rackets. Equivalent sports strings using PEEK monofils have been shown not only to have advantages in terms of lower creep and better weathering compared to nylon, but also to have the lowest dynamic stiffness of any synthetic string product. This allows more recoverable energy to be stored, the hysteresis losses in the projectile always being much greater than the losses in the string bed, so that shots can be returned with greater pace for the same effort. Although expensive, it is likely that PEEK will become increasingly used in this application.

Fishing lines require energy absorption and strength, especially in relation to knot tenacity. The fact that this has to be combined with a soft, nonwiry characteristic for good line control has meant that nylons, and in particular nylon copolymers, have been optimized to satisfy most requirements. In terms of specifications, diameter is secondary to breaking strength, with a typical range running from 1 kg lines (0.15 mm) to 200 kg lines (0.70 mm). A special feature of the more sophisticated products is that they are tapered for minimum visibility and optimum strength. The smooth change in diameter along the length, narrowing towards the hook, is achieved at the manufacturing stage by programmed polymer metering, the pump cycling its output over the required lengths.

Further reading

J Freitag, Monofilaments—Technology and fields of application, *International Textile Bulletin—Yarn Forming*, **3**, 52 (1988).

J E Ford, Polyethylene textiles, *Textiles*, **17**(2), 30 (1988).

Monofilament lines for profile precision. *Chemiefasern* PT16, ENG E22–23 (March 1990).

W M Szomanski, R Postle, Measurement of torsion in monofilament textile structures, *Textile Res. J.* (August 1970).

Monofilament for rubber reinforcement, *High Performance Textiles*, 4 (March 1990).

D W Hadley, I M Ward, J Ward, The transverse compression of anisotropic fibre monofilaments, *Proc. R. Soc. Series A*, **285**, 275 (April 1965).

R J E Cumberbirch, J D Owen, The mechanical properties and birefringence of various monofils, *J. Textile Inst.*, **56**, 389 (July 1965).

Monofilament reinforcement for rubber, *High Performance Textiles*, 1 (May 1992).

INDEX